Lecture Notes in Mathematics

Edited by A. Dold and B. Eckmann

1165

Seminar on Deformations

Proceedings, Łódź – Warsaw 1982/84

Edited by J. Ławrynowicz

Springer-Verlag
Berlin Heidelberg New York Tokyo

Editor

Julian Ławrynowicz
Institute of Mathematics of the Polish Academy of Sciences
Łódź Branch, Narutowicza 56, 90-136 Łódź, Poland

Mathematics Subject Classification (1980): 14-xx, 30-xx, 31-xx, 32-xx, 35-xx, 49-xx, 58-xx

ISBN 3-540-16050-7 Springer-Verlag Berlin Heidelberg New York Tokyo
ISBN 0-387-16050-7 Springer-Verlag New York Heidelberg Berlin Tokyo

Printing and binding: Beltz Offsetdruck, Hemsbach / Bergstr.
2146/3140-543210

ALDO ANDREOTTI

in memoriam

Aldo Andreotti, Oregon 1974
(Courtesy of Mrs. B. Andreotti)

FOREWORD

These Proceedings contain selected papers from those submitted by a part of participants in connection with the talks at the Seminar on Deformations organized in Łódź in 1982-84 by Julian Ławrynowicz and Leszek Wojtczak. The most fruitful part of the seminar took place in 1983 in Warsaw during the International Congress of Mathematicians in the form of a Spontaneous Seminar on Complex Analysis with Physical Applications. There were four sessions of the Spontaneous Seminar: August 18 - 105 min., August 19 - 90 min., August 20 - 105 min., and August 22 - 90 min.

The participants decided to dedicate their papers to the memory of the unforgetable Mathematician and Colleague, Professor ALDO ANDREOTTI (1924-1980) who had influenced in a considerable way the development of the complex analysis, algebraic geometry, and global analysis in Poland, in particular our Seminar on Deformations, giving there talks during his visits in Poland.

The collection contains 22 papers connected with deformations of mathematical structures in the context of complex analysis with physical applications. These are research papers in final form: no version of them will be submitted for publication elsewhere. In addition, the collection contains 4 short notes devoted to the presentation of open problems.

In order to make a substitute of the organizing committee of a conference which usually recommends papers suitable for publication, the organizers asked some Colleagues to form an Advisory Committee: C. Andreian-Cazacu (Bucureşti), Z. Charzyński (Łódź), P. Dolbeault (Paris), H. Grauert (Göttingen), S. Łojasiewicz (Kraków), J. Leiterer (Berlin, GDR), W. Tutschke (Halle/Saale), and S. Walczak (Łódź).

The preparation of these Proceedings was possible thanks to the help of the Łódź Society of Sciences and Arts, especially to its President, Professor Witold Śmiech. The organizers express also their gratitude to the Springer-Verlag for kind consent of publishing the Proceedings in the series "Lecture Notes in Mathematics". Finally, the organizers thank warmly Mrs. A. Marciniak, an English philologist, for improving the language style and for typing most of the papers.

Łódź, July 1985 Julian Ławrynowicz

CONTENTS

PAPERS CONNECTED WITH TALKS DURING THE SPONTANEOUS SEMINAR NOT INCLUDED IN THIS VOLUME

Cabiria ANDREIAN-CAZACU (Bucureşti): Extremal length and definitions of quasiconformality

G. ANDRZEJCZAK (Łódź): On the Pontrjagin classes of foliations

P. CARAMAN (Iaşi): Qasiconformal mappings between infinite-dimensional manifolds in normed spaces

P. DOLBEAULT (Paris): Le théorème de Plemelj sur les hypersurfaces réelles

R. DWILEWICZ (Warszawa) and C.-H. SUNG (San Diego, CA): Extensions of CR-functions

Zerrin GÖKTÜRK (Istanbul): Some coefficient conditions on quasiconformal extendability of analytic functions

V.Ja. GUTLJANSKIĬ (Donetsk): On integration of the differential equation of Löwner-Kufarev

G.M. HENKIN (Moskva): On analytic continuation of CR-functions

K. JEZUITA and M. SKWARCZYŃSKI (Radom): Holomorphic geometry of space-time

R. LANGEVIN (Dijon): Courbure totale de feuilletages de codimension plus grande que un

LÊ DŨNG TRANG (Paris): Diagrammes de Cerf et exposants d'intégrales singulières

J. LEITERER (Berlin, GDR): On the restriction of stable vector bundles on $\mathbb{P}_n$

P. MALLIAVIN (Paris): Intégrales de volume et valeurs frontières

G.F. MANDŽAVIDZE (Tbilisi) and W. TUTSCHKE (Halle an der Saale): An open problem concerning nonlinear first order systems

I. RAMADANOV (Sofia): On the complex Monge-Ampère equation

G. ROOS (Paris): Géométrie du domaine symétrique exceptionnel de dimension 16

A. SHIELDS (Ann Arbor, MI): Multipliers on spaces of analytic functions

J.A. SIDDIQI (Québec): Uniform approximation by exponential sums on plane arcs

C.-H. SUNG (San Diego, CA): Minimal surfaces and value distribution theory

J. SZTAJNIC and S. WALCZAK (Łódź): On controllability of smooth and nonsmooth dynamical systems

L.M. TOVAR (Méxioco): Some open problems in the theory of Stein spaces

P. WALCZAK (Łódź): Local stability of holomorphic and transversely holomorphic foliations

A.B. ŽIŽČENKO (Moskva): On classification of algebraic varieties with Kodaira's dimension 0

PROGRAMME OF THE SPONTANEOUS SEMINAR ON COMPLEX ANALYSIS WITH PHYSICAL APPLICATIONS

Session I, August 18, 1983. Speakers: Opening - J. Ławrynowicz (Łódź) 3 min. - 1. M. Okada (Sendai) 10 min. - 2. P. Dolbeault (Paris) 10 min. - 3. Cabiria Andreian-Cazacu (Bucureşti) 10 min. - 4. L.M. Tovar (México) 10 min. - 5. I. Ramadanov (Sofia) 10 min. - 6. G. Roos (Paris) 13 min. - 7. K. Jezuita and M. Skwarczyński (Radom) 10 min. - 8. Z. Charzyński, J. Chądzyński, and P. Skibiński (Łódź) 12 min. - 9. Y.-T. Siu (Cambridge, MA) 15 min. - Closing - L. Wojtczak (Łódź) 2 min.

Session II, August 19. Speakers: 0.- L. Wojtczak 3 min. - 1. A.A. Gončar (Moskva) 10 min. - 2. A. Crumeyrolle (Toulouse) 15 min. - 3. G.M. Henkin (Moskva) 10 min. - 4. R.O. Wells (Houston, TX) 10 min. - 5. C.-H. Sung (San Diego, CA) 5 min. - 6. R. Dwilewicz (Warszawa) and C.-H. Sung 5 min. - 7. J. Leiterer (Berlin, GDR) 10 min. - 8. V. Tsanov (Sofia) 10 min. - 9. E.M. Čirka (Moskva) and S.I. Pinčuk (Ufa) 10 min. - C. - J. Ławrynowicz 2 min.

Session III, August 20. Speakers: 0. - J. Ławrynowicz 3 min. - 1. S. Dimiev (Sofia) 10 min. - 2. E.W. Wette (Radevormwald, FRG) 7 min. - 3. S. Axler (East Lansing, MI) 10 min. - 4. P. Malliavin (Paris) 10 min. - 5. J.A. Siddiqi (Québec) 10 min. - 6. G.F. Mandžavidze (Tbilisi) and W. Tutschke (Halle/Saale) 10 min. - 7. W. Pleśniak (Kraków) 10 min. - 8. Ewa Ligocka (Warszawa) 10 min. - 9. J. Sztajnic and S. Walczak (Łódź) 15 min. - 10. G. Andrzejczak (Łódź) 8 min. - C. - L. Wojtczak 2 min.

Session IV, August 22. Speakers: 0. - J. Ławrynowicz 3 min. - 1. L.A. Gutljanskiĭ (Donetsk) 10 min. - 2. P.M. Tamrazov (Kiev) 10 min. - 3. S. Toppila (Helsinki) 10 min. - 4. S. Agajan (Erevan) 7 min. - 5. U. Cegrell (Uppsala) 10 min. - 6. T. Tonev (Sofia) 10 min. - 7. Agnieszka Maciejkowska (Lublin) 10 min. - 8. J. Stankiewicz (Rzeszów) 10 min. - 9. P. Walczak (Łódź) 8 min. -C. - J. Ławrynowicz 2 min.

NUMERICALLY EFFECTIVE BUNDLES ON MOIŠEZON AND STRONGLY PSEUDOCONVEX MANIFOLDS

Vincenzo Ancona (Firenze)* and Alessandro Silva (Trento)*

Contents

Abstract : We prove a vanishing theorem for numerically effective invertible sheaves relative to projective morphisms between irreducible analytic spaces, by using techniques of Kawamata and by introducing intersection numbers on Moišezon spaces. We give applications to Moišezon and strongly pseudoconvex manifolds, extending results of Schneider and T. Peternell.

1. Intersection numbers on Moišezon spaces

Let X be a Moišezon space. It follows from Artin's algebraization theorem, [3], that there is an algebraic space Z, complete and of finite type over $\mathbb{C}$, such that $Z^{an} = X$. Serre's GAGA, [16], reduces the study of coherent analytic sheaves on X to the study of coherent algebraic sheaves on Z. Most of Kleiman's results established in [9] for complete schemes over an algebraically closed field easily extend to complete algebraic spaces of finite type over $\mathbb{C}$. Indeed, the essen-

* Partially supported by C.N.R.

tial tools in Kleiman's proofs, that is the "Chow lemma" and the "Dévissage lemma", hold validity also for algebraic spaces, (see [10]). We can therefore state the following results (part of which were also proved directly by Moišezon in [12]).

Threorem 1.1. Let X be a Moišezon space, $\mathcal{F}$ a coherent sheaf on X, $L_1,\ldots,L_t$ invertible sheaves on X. Then the Euler-Poincaré characteristic, $\chi(X, \mathcal{F}\otimes L_1^{n_1}\otimes\ldots\otimes L_t^{n_t})$ is a numerical polynomial in $n_1,\ldots,n_t$ of total degree $s = \dim \operatorname{supp} \mathcal{F}$. ([9], pag. 295).

Definition 1.2. Let $L_1,\ldots,L_t$ invertible sheaves on the Moišezon space X, $\mathcal{F}$ a coherent sheaf on X such that $\dim \operatorname{supp} \mathcal{F} \leq t$. The intersection number $(L_1\cdot\ldots\cdot L_t\cdot\mathcal{F}) = (L_1\cdot\ldots\cdot L_t\cdot\mathcal{F})_X$ of $L_1,\ldots,L_t$ with $\mathcal{F}$ is the coefficient of the monomial $n_1\cdot\ldots\cdot n_t$ in $\chi(X, \mathcal{F}\otimes L_1^{n_1}\otimes\ldots\otimes L_t^{n_t})$.

We have :

- $(L_1\cdot\ldots\cdot L_t\cdot\mathcal{F})$ is an integer,
- $(L_1\cdot\ldots\cdot L_t\cdot\mathcal{F})$ is a symmetric t-linear form in $L_1,\ldots,L_t$
- if $0\to\mathcal{F}'\to\mathcal{F}\to\mathcal{F}''\to 0$ is an exact sequence of coherent sheaves on X, then $(L_1\cdot\ldots\cdot L_t\cdot\mathcal{F}) = (L_1\cdot\ldots\cdot L_t\cdot\mathcal{F}') + (L_1\cdot\ldots\cdot L_t\cdot\mathcal{F}'')$,
- $(L_1\cdot\ldots\cdot L_t\cdot\mathcal{F}) = 0$ if $\dim \operatorname{supp} \mathcal{F} < t$, and $(\mathcal{F}) = h^0(\mathcal{F})$ if $\dim \operatorname{supp} \mathcal{F} = t = 0$.

Definition 1.3. Let X be a Moišezon space, L an invertible sheaf on X, Y a subspace of X of pure dimension s. We set $(L^s\cdot Y) = (L_1\cdot\ldots\cdot L_s\cdot\mathcal{O}_Y)_X$, where $L_1 = \ldots = L_s = L$. If $n = \dim X$, we set $(L^n) = (L^n\cdot X)$. (L^n) is called the highest self-intersection number of L.

Let us remark that $(L^n) = c_1(L)^n$, where $C_1(L)$ is the first Chern class of L.

Definition 1.4. Let X be a Moišezon space, L an invertible sheaf on X. L is called numerically effective if for every curve $C \subset X$, $(L \cdot C) \geq 0$.

(By a curve we mean a closed reduced irreducible subspace of X of dimension one).

Let us remark that in such a case we have $(L \cdot C) = \deg L_{|C}$.

For instance, if there exists k such that $L^{\otimes k}$ is generated by its global sections, then L is numerically effective.

Proposition 1.5. Let X be a Moišezon space, L an invertible sheaf on X.

1) If L is numerically effective, then for every closed subspace Y of X of pure dimension s, we have $(L^s \cdot Y) \geq 0$.

2) Let $f : X' \to X$ be a morphism between Moišezon spaces, and $L' = f^*L$. Then if L is numerically effective so is L'. Moreover, if f is surjective, L' numerically effective implies L numerically effective.

3) If $f : X' \to X$ is a modification and $n = \dim X = \dim X'$, then $(L'^n) = (L^n)$.

We recall, [2], that a proper morphism of complex spaces $f : X \to Y$ is called a Moišezon morphism if for every $y \in f(X)$, there is a Stein open neighborhood V_y of y such that $(X_{|V})_{red}$ has a finite number of irreducible components $X_1, \ldots, X_r$ and if $Y_j = f(X_j)$ the relative algebraic dimension $t(X_j, Y_j) = \dim X_j - \dim Y_j$.

The Definition 1.4 easily extends to the relative case as follows:

Definition 1.6. Let $f : X \to Y$ be a Moišezon morphism, L an invertible sheaf on X. We shall say that L is f-numerically effective (or L is numerically effective relatively to f) if for every $y, y \in Y$, the restriction sheaf $L_y = L_{|f^{-1}(y)}$ is numerically effective.

We leave to the care of the reader to give an explicit relative form of Proposition 1.5.

2. A relative vanishing theorem

The main object of this section is to give a proof of the following.

Theorem 2.1. Let $f : X \to Y$ be a projective morphism between irreducible analytic spaces with X smooth. Let L be an invertible sheaf on X such that:

(i) L is f-numerically effective

(ii) the highest self-intersection number of L_y is > 0 for every y in an open dense subset U of Y.

Then:

$$\mathcal{R}^q f_*(\mathbb{K}_X \otimes L) = 0$$

for $q \geq 1$.

We are going to need several lemmas.

Lemma 2.2. Let X be a Moišezon space of dimension n and L be an invertible sheaf on X.

Then $(L^n) > 0$ if and only if the L-dimension of X $\kappa(X,L) = n$.

Proof. With no loss of generality we can suppose X reduced, irreducible and normal. Let then $f : X' \to X$ be a projective desingularization of X and $L' = f^*L$. It follows from 1.5,3) that $(L'^n) = (L^n)$, and, furthermore, $\kappa(X',L') = \kappa(X,L)$.

Lemma 2.2 follows then from ([8], lemma 3).

q.e.d.

Let $f : X \to Y$ be a morphism between analytic spaces and L be an invertible sheaf on X.

We set:

$N(X|Y,L) = \{i \in \mathbb{Z}, i > 0$: there exists $x \in X$ such that $(f^*f_*L^{\otimes i})_x \to (L^{\otimes i})_x$ is surjective$\}$. On the other hand, for every $i \in N(X|Y,L)$ there is a meromorphic map $f_i : X \to \mathbb{P}(f_*L^{\otimes i})$ such that the diagram

$$\begin{array}{ccc} X & \xrightarrow{f_i} & \mathbb{P}(f_*L^{i}) \\ & {\scriptstyle f}\searrow \quad \swarrow & \\ & Y & \end{array}$$

commutes.

Let $I(f_i)$ be the indeterminary locus of f_i and put:

$$\kappa(X/Y,L) = \begin{cases} \sup\limits_{i \in \mathbb{N}(X/Y,L) \neq \emptyset} \dim \overline{f_i(X-I(f_i))} - \dim Y \\ -\infty \quad \text{if } N(X/Y,L) = \emptyset \end{cases}$$

Then $\kappa(X/Y,L)$ is called the relative L-dimension of X (cfr. [11]).

We have

Lemma 2.3. Let $f : X \to Y$ be a proper morphism between irreducible analytic spaces, L an invertible sheaf on X. Let $g : Y' \to Y$ be a modification and

$$\begin{array}{ccc} X' & \xrightarrow{g'} & X \\ {\scriptstyle f'}\downarrow & & \downarrow{\scriptstyle f} \\ Y' & \xrightarrow{g} & Y \end{array}$$

be the corresponding cartesian diagram. Then if $L' = g'^*L$,

$$\kappa(X/Y,L) = \kappa(X'/Y',L').$$

Proof. By taking normalizations if needed, we can suppose X normal. For every integer m, we have then $g'_*g'^*L^{\otimes m} = L^{\otimes m}$, whence $\Gamma(X,L^{\otimes m}) = \Gamma(X',L'^{\otimes m})$. The contention then easily follows.

q.e.d.

Lemma 2.4. Let $f : X \to Y$ be a Moišezon morphism between irreducible analytic spaces and L be an invertible sheaf on X. If assumption (ii) in Theorem 2.1 holds, then $\kappa(X/Y,L) = \dim X - \dim Y$.

Proof. By Hironaka's flattening theorem, [7], there exists a cartesian diagram:

$$\begin{array}{ccc} X' & \xrightarrow{g'} & X \\ {\scriptstyle f'}\downarrow & & \downarrow{\scriptstyle f} \\ Y' & \xrightarrow[g]{} & Y \end{array}$$

where g is a sequence of monoidal transformations and f' is a flat morphism. If we put $L' = g'^*L$, we have $\kappa(X/Y,L) = \kappa(X'/Y',L')$ by 2.3. We remark also that $\dim X' = \dim X$, $\dim Y' = \dim Y$ and $V' = g^{-1}(V)$ is an open dense subset of Y'. If $y' \in V'$, $f'^{-1}(y') = f^{-1}(y)$, where $y = g(y')$. It follows that it is enough to show our contention for the flat morphism f' and the sheaf L'. By theorem 3.4 in [11], there exists a subset $W \subset Y$, such that $\complement_Y W = \bigcup_{n\in\mathbb{N}} Z_n$, with Z_n a proper analytic subspace of Y, and for every $y \in W$, $\kappa(X/Y,L) = \kappa(X_y,L_y)$, while if $y \notin W$, $\kappa(X/Y,L) < \kappa(X_y,L_y)$. Moreover, since W is dense in Y, there exists $y, y \in V \cap W$, such that $X_y = f^{-1}(y)$ is reduced and irreducible with $\dim X_y = \dim X - \dim Y$. It follows from Lemma 2.2 that $\kappa(X_y,L_y) = \dim X_y$, hence the conclusion.

q.e.d.

We are going to need also a relative version of a theorem of Norimatsu ([13]):

Theorem 2.5. Let f be a proper, projective morphism between irreducible analytic spaces with X smooth. Let D be a divisor with simple normal crossings, an H and f-ample sheaf on X. Then:

$$\mathcal{R}^q f_*(\mathbb{K}_X \otimes H \otimes [D]) = 0 \quad \text{for} \quad q \geq 1.$$

Proof. (cfr. [13]). Let $n = \dim X$, $D = \Sigma D_j$ be the decomposition of D into its irreducible components.
We put:

$$D^k = \{D_J,\ J \subset \{1,\dots,r\},\ |J| = k\},$$

where $D_J = D_{i_1} \cap \dots \cap D_{i_k}$ if $J = (i_1,\dots,i_k)$.

Following [4], we define a filtration of $\Omega^n_X(D)$ if we put $W^n_k = \Omega^k_X(D) \wedge \Omega^{n-k}_X$. We obtain then in particular: $W^n_o = \Omega^n_X = \mathbb{K}_X$ and $W^n_n = \Omega^n_X[D] = \mathbb{K}_X \otimes [D]$. The exact sequence:

$$0 \to W^n_{k-1} \to W^n_k \to \Omega^{n-k}_{D^k} \to 0$$

tensored with H gives the exact sequence

$$(2.6) \qquad 0 \to W^n_{k-1} \otimes H \to W^n_k \otimes H \to \Omega^{n-k}_{D^k} \otimes H \to 0.$$

If we apply the relative Kodaira vanishing theorem in ([18], lemma II-A), we obtain:

$$\mathcal{R}^q f_*(D^k, \Omega^{n-k}_{D^k} \otimes H) = 0, \quad \text{for } q \geq 1.$$

On the other hand, applying the same result for $k = 0$, we obtain:

$$\mathcal{R}^q f_*(X, \Omega^n \otimes H) = 0, \quad \text{for } q \geq 1.$$

If we proceed by induction on k, we obtain the conclusion from the exact sequence 2.6.

q.e.d.

Lemma 2.7. Let $f : X \to Y$ be a proper, projective morphism between irreducible analytic spaces with X smooth. Let L be a f-numerically effective sheaf on X and H an invertible f-ample sheaf on X. Then, for every positive integer m, $L^{\otimes m} \otimes H$ is f-ample.

Proof. To show: for every $y, y \in Y$, $(L^{\otimes m} \otimes H)_y$ is ample. If we apply Seshadri's criterion to the fibre X_y, we can see that we can suppose H very ample relatively to f. We can suppose then that X is embedded into $Y \times \mathbb{P}_N$ and $H = \mathcal{O}_{Y \times \mathbb{P}_N}(1)\big|_X$. X_y is then embedded into $\{y\} \times \mathbb{P}_N \simeq \mathbb{P}_N$. Let now C be a curve in X_y; then, $(H_y \cdot C) = \deg H_y\big|_C = \deg_{\mathbb{P}_N} C = m(C)$, with $m(C)$ = multiplicity of C.

For every positive integer m we have then:

$$((L_y^{\otimes m} \otimes H_y) \cdot C) = m(L \cdot C) + H \cdot C \geq H \cdot C \geq m(C),$$

so that $L_y^{\otimes m} \otimes H_y$ is ample by Seshadri's criterion.

q.e.d.

Lemma 2.8. Let X be a quasi-projective smooth algebraic variety, D a divisor on X with normal crossings. Let m_i be positive integers each corresponding with an irreducible component D_i of D. There exist then a quasi-projective non singular algebraic variety X' and a flat finite morphism $f : X' \to X$ such that:

(i) $f^*D_i = m_i D_i'$, with D_i' effective and reduced,

(ii) $D' = \Sigma D_i'$ is a divisor with normal crossings in X'.

Proof. Let Z be a projective compactification of X that we may suppose smooth. Let $\bar{D}_i$ be the closure of D_i in Z and $\bar{D} = \Sigma \bar{D}_i$. We are reduced to the case where $\bar{D}$ is still with normal crossings by a finite sequence of monoidal transformations with centers lying outside X. We apply then ([8], lemma 5) to the pair $(Z,\bar{D})$ to obtain a finite flat morphism $g : Z' \to Z$ such that $g^*\bar{D}_i = m_i \tilde{D}_i$ ($\tilde{D}_i$ effective and reduced) and $\tilde{D}' = \Sigma \tilde{D}_i$ is a divisor with normal crossings in Z'.

If we set $X' = g^{-1}(X)$ and $f = g_{|X}$, we obtain our assertion,

q.e.d.

We are now ready to prove theorem 2.1.

The theorem being local in nature, we may suppose that there exists an invertible sheaf H on X very ample relatively to f. Let D be the divisor associated with L. By lemma 2.4 there exists a positive integer m such that $mD - H$ is linearly equivalent to an effective divisor E. Now, there exists a finite sequence of monoidal transformations with smooth centers $f' : X' \to X$ with X' a quasi-projective smooth algebraic variety, and $\mathrm{red}(f'^*(H+E))$ has only normal crossings. Then $\kappa(X/Y,D) = \kappa(X'/X,f^*D)$, by lemma 2.3.

There exists then an invertible sheaf H' on X', very ample relatively to f'. By lemma 2.4, there exists a positive integer m' such that $m'f^*D - H'$ is linearly equivalent to an effective divisor E'. Therefore we have the equalities of analytic sets $\mathrm{red}(f'^*(E+H)) =$ $= \mathrm{red}(m'f^*D) = \mathrm{red}(H'+E')$, hence $\mathrm{red}(H'+E')$ has only normal crossings.

Let $E' = \Sigma n_i E_i'$, n a common multiple of the n_i's, $m_i = nm'/n_i$. By lemma 2.8, there exists a finite covering map $g : X'' \to X'$ such that $g^*E_i' = m_i E_i''$, $g^*H' = m'H''$ with H'' very ample relatively to $g \circ f' \circ f$ and $E'' = \Sigma E_i''$ with normal crossings. Therefore $m'g^*f^*D$ is linearly equivalent to $g^*E' + g^*H'$ that, in its turn, is equal to $m'H'' + \Sigma n_i g^*E_i' = m'H'' + nm'\Sigma E_i''$ and nmg^*f^*D is linearly equivalent to $nm'H'' + n^2m'E''$, so that $D'' = g^*f^*D$ is linearly equivalent to $H'' + nE''$. By lemma 2.7 some multiple of $((n-1)/n)D'' = (1-\frac{n-1}{n})H'' =$ $= H'' + (n-1)E''$ is $(g \circ f' \circ f)$-ample. Then g^*f^*D is linearly equivalent to $H^* + E''$ with H^* relatively ample. The theorem then follows from lemma 2.5.

q.e.d.

3. Applications to Moišezon manifolds

Theorem 3.1. Let X be a Moišezon manifold of dimension n, L be a numerically effective invertible sheaf on X such that $(L^n) > 0$. Then $H^q(X, \mathbb{K}_X \otimes L) = 0$ for $q \geq 1$.

Proof. Let $f : X' \to X$ be a projective desingularization of X and $L' = f^*L$; then L' is numerically effective (prop.1.5,2)) and $(L'^n) =$ $= (L^n)$ (prop. 1.5,3)). Moreover $f_*(\mathbb{K}_{X'} \otimes L') = \mathbb{K}_X \otimes L$ satisfies the assumptions of theorem 2.1 (assumption (ii) is automatically satisfied f being a modification). Theorem 3.1 then follows by standard arguments, q.e.d.

If we proceed as in [17], using theorem 3.1 and Chow's lemma for Moišezon morphisms [1] , we have the following relative version of 3.1 :

Theorem 3.2. Let $f : X \to Y$ be a Moišezon morphism with X smooth. Let L be an invertible sheaf satisfying assumptions (i) and (ii) in theorem 2.1. Then:

$$\mathcal{R}^q f_*(\mathbb{K}_X \otimes L) = 0 \quad \text{for } q \geq 1.$$

Let now X be an analytic space and E a locally free sheaf on X. We denote by $\mathbb{P}(E)$ the associated projective bundle and by $\mathcal{O}_{\mathbb{P}(E)}(1)$ the corresponding hyperplane bundle. We can then extend to Moišezon spaces two theorems of [15], with analogous proofs.

Theorem 3.3. Let X be a Moišezon space of dimension n, E,F locally free sheaves of rank $r,1$ respectively. We suppose that $\mathcal{O}_{\mathbb{P}(E)}(1)$ and $L = (\det E)^{-1} \otimes \mathbb{K}_X^{-1} \otimes F$ are numerically effective. We assume moreover that either $c_1(L)^n > 0$ or $\tilde{c}_n(E) > 0$. Then :

$$H^q(X,S^k(E) \otimes F) = 0$$

for $q \geq 1$ and $k \geq 0$.

(We denote here by $\tilde{c}_n(E)$ the n-th inverse Chern class of E).

Theorem 3.4. In the same notation as in theorem 3.3, we assume that there exist $k_o \geq 1$ such that either $S^{k_o}(E)$ is generated by its global sections of that $L = F \otimes \mathbb{K}_X^{-1} \otimes (\det E)^{-1}$ is numerically effective and $c_1(L)^n > 0$ (or $\tilde{c}_n(E) > 0$).

Then:

$$H^q(X,S^k(E) \otimes F) = 0 \quad \text{for } q \geq 1 \text{ and } k \geq 0.$$

4. Applications to strongly pseudoconvex manifolds

Theorem 4.1. Let X be a connected strongly pseudoconvex manifold, $T \subset X$ its exceptional subspace. Let L be an invertible sheaf on X such that $L|_T$ is numerically effective, then:

$$H^q(X, \mathbb{K}_X \otimes L) = 0 \quad \text{for} \quad q \geq 1.$$

proof. Let $f : X \longrightarrow S$ be the Remmert reduction of X. Since f is a modification, S is a Stein space and there exists $s_o \in S$ such that $f^{-1}(s_o) = T$ and $f|_{X-T}$ induces an isomorphism with $S-\{s_o\}$, the conclusion then follows at once from theorem 2.1.

q.e.d.

Theorem 4.1. applies, for instance, to the case where the invertible sheaf L is such that some tensor power $L|_T^{\otimes k}$ is generated by its global sections.

Theorem 4.2. Let X be a connected strongly pseudoconvex manifold, $T \subset X$ its exceptional subspace, E, F locally free sheaves on X of rank r and 1 respectively. We assume that the restrictions to T of $\mathcal{O}_{\mathbb{P}(E)}(1)$ and of $L = (\det E)^{-1} \otimes \mathbb{K}_X^{-1} \otimes F$ are numerically effective.

Then:

$$H^q(X, S^k(E) \otimes F) = 0 \quad \text{for} \quad q \geq 1$$

and $k \geq 0$.

Proof. Let $P = \mathbb{P}(E)$, $p : P \to X$ the projection, $f : X \to S$ the Remmert reduction of X (with $T = f^{-1}(s_o)$, $s_o \in S$) and $\pi : P \to S$, $\pi = f \circ p$. π is a Moišezon morphism. For every integer $k \geq 0$, we put: $M_k = \mathcal{O}_{\mathbb{P}(E)}(r+k) \otimes p^*L$. Then: $M_k|_{p^{-1}(T)} = \mathcal{O}_{\mathbb{P}(E|_T)}(r+k) \otimes p^*(L|_T)$ is numerically effective. Since $p^{-1}(T) = \pi^{-1}(s_o)$, $M_k|_{\pi^{-1}(s_o)}$ is then numerically effective. If $s \in S$, $s \neq s_o$, and $x \in X$ is the unique point such that $f(x) = s$, we have $M_k|_{\pi^{-1}(s)} = \mathcal{O}_{\mathbb{P}(E_x)}(r+k) \simeq \mathcal{O}_{\mathbb{P}^{r-1}}(r+k)$, thus $M_k|_{\pi^{-1}(s)}$ is ample and thus numerically effective and its highest intersection number is positive.

From theorem 3.2, we have :

$$\mathcal{R}^q \pi_*(\mathbb{K}_P \otimes M_k) = 0 \quad \text{for} \quad q \geq 1.$$

But, $\mathbb{K}_P = \mathcal{O}_{\mathbb{P}(E)}(-r) \otimes p^*(\mathbb{K}_X \otimes \det E)$ (see [6]), whence we obtain

$$\mathcal{R}^q \pi_*(\mathcal{O}_{\mathbb{P}(E)}(k) \otimes p^*F) = 0 \quad \text{for} \quad q \geq 1$$

and S being Stein it follows:

$$H^q(\mathbb{P}(E), \mathcal{O}_{\mathbb{P}(E)}(k) \otimes p^*F) = 0 \qquad \text{for } q \geq 1.$$

This last group is isomorphic to:

$$H^q(X, S^k(E) \otimes F) \text{ (by [4])},$$

hence the conclusion.

q.e.d.

In particular we have obtained:

Theorem 4.3. Let X be a connected strongly pseudoconvex manifold, $T \subset X$ its exceptional subspace, E, F locally free sheaves on X of rank $r, 1$, respectively. We assume that for some $k \geq 1$, $S^k(E_{|T})$ is generated by its global sections and that the restriction to T of $L = (\det E)^{-1} \otimes \mathbb{K}_X^{-1} \otimes F$ is numerically effective.

Then

$$H^q(X, S^k(E) \otimes F) = 0 \qquad \text{for } q \geq 1.$$

A particular case of Theorem 4.3 (X embedded in $\mathbb{C}^n \times \mathbb{D}^m$, E generated by its global sections on X, $L_{|T}$ ample) has been shown in [14].

Added in proof: Some results of the paper are consequences of the Kawamata-Kielweg results. This remark does not apply to the results connected with pseudoconvex manifolds.

Bibliography

[1] Ancona, V. Espaces de Moisezon relatifs et algebrisation des modifications analytiques. Math. Ann. 246, 155-165 (1980).

[2] Ancona, V. e Tomassini, G. Modifications analytiques. Lecture Notes on Math. n° 943. Berlin, Heidelberg, New York: Springer 1982.

[3] Artin, M. Algebraization of formal moduli, II. Existence of modifications. Ann. of Math., 91, 88-135. (1970).

[4] Deligne, P. Theorie de Hodge II. Publ. Math. IHES 40, 5-58 (1971).

[5] Hartshorne, R. Ample vector bundles. Publ. IHES 29, 63-94 (1966).

[6] Hartshorne, R. Ample subvarieties of algebraic varieties . Lecture Notes in Math. 156. Berlin, Heidelberg, New York: Springer 1970.

[7] Hironaka, H. Flattening theorem in complex analytic geometry. Am. J. of Math. 97, 503-547 (1975).

[8] Kawamata, Y. A generalization of Kodaira - Ramanujam's vanishing theorem. Math. Ann. 261, 43-46 (1982).

[9] Kleiman, S.L. Toward a numerical theory of ampleness. Annals of Math. 84, 293-344 (1966)

[10] Knutson, D. Algebraic spaces. Lecture Notes in Math. n°203. Berlin, Heidelberg, New York Springer 1971.

[11] Lieberman, D. - Sernesi, E. Semicontinuity of L-dimension. Math. Ann. 225, 77-88 (1977).

[12] Moišezon, B.G. On n-dimensional compact complex varieties with n algebraically independent meromorphic functions. Amer. Math. Soc. Transl. 63, 51-177 (1967)

[13] Norimatsu, Y. Kodaira vanishing theorem and Chern classes for ∂-manifolds. Proc. Japan Acad. 54, Ser.A, 107-109 (1978).

[14] Peternell, T. On strongly pseudoconvex Kähler manifolds. Inv. Math. 70, 157-168 (1982).

[15] Schneider, M. Some remarks on vanishing theorems for holomorphic vector bundles. Preprint (1983).

[16] Serre, J.P. Géometrie algébrique et géometrie analytique. Ann. Inst. Fourier.

[17] Silva, A. Relative vanishing theorems I: application to ample divisors. Comment. Math. Helvetici 52, 483-489 (1977).

[18] Sommese, A.J. On manifolds that cannot be ample divisors. Math. Ann. 221.

Dipartimento di Matematica
Università di Firenze, Viale Morgagni 67/A
I-50134 Firenze, Italy

REMARQUES SUR LES IDÉAUX DE POLYNÔMES

Bruno Bigolin (Brescia)

Résumé. L'objet de cet article est d'exposer en detail la construction par blocs de la variété représentative des idéaux de polynômes.

1. J'expose ici avec les preuves completes une partie de mon article [1] (à paraître dans les "Rendiconti Sem. Mat. Brescia"), précisément la construction (par blocs) de la variété représentative des idéaux de polynômes. Dans cet article on démontre en outre que, parmi tous les idéaux, ceux qui admettent pour ensemble des zéros une intersection complète, forment un ouvert de Zariski (non vide) et que le corps des invariants projectifs est un corps unirationnel. Bien que ce soit ce dernier le résultat important de l'article, je pense toutefois qu'il y a de l'intérêt à reproduire ici la construction de la variété $J(l_1,\dots,l_q;n)$, avec des démonstrations détaillées, quoique élémentaires.

Pour d'autres remarques sur le corps des invariants projectifs, on peut voir Spinelli [2].

2. Envisageons: un corps K algébriquement clos (de caractéristique quelconque); un ensemble $(x)=(x_o,\dots,x_n)$ d'indéterminées sur K; une suite

$$l_1=\dots=l_{m_1}<l_{m_1+1}=\dots=l_{m_2}<\dots<l_{m_p+1}=\dots=l_q$$

de nombres entiers positifs.

Soit $N_\alpha=\binom{n+l_\alpha}{n}$ et identifions les polynômes homogènes

$$f(a;x)=\sum_{i_1\le\dots\le i_{l_\alpha}=0}^{n} a_{i_1\dots i_{l_\alpha}}x_{i_1}\dots x_{i_{l_\alpha}}$$

de degré l_α dans les (x) avec les points $(a)=(a_{i_1\dots i_{l_\alpha}})$ de l'espace affine $A^{N_\alpha}(K)$; la famille des systèmes

$$(f) = \begin{pmatrix} f_1^{(l_1)} \\ \vdots \\ f_q^{(l_q)} \end{pmatrix},$$

où $f_1^{(l_1)},\dots,f_q^{(l_q)}$ sont des polynômes homogènes des degrés

$$l_1 = \dots = l_{m_1} < l_{m_1+1} = \dots = l_{m_2} < \dots < l_{m_p+1} = \dots = l_q,$$

est en correspondance biunivoque avec le produit

$$A = A^{N_{m_1}\cdot m_1}(K) \times A^{N_{m_2}\cdot(m_2-m_1)}(K) \times \dots \times A^{N_q\cdot(q-m_p)}(K).$$

PROPOSITION 1. L'ensemble des

$$(f) = \begin{pmatrix} f_1^{(l_1)} \\ \vdots \\ f_q^{(l_q)} \end{pmatrix}$$

de A tels que, pour tout β

(i) ni $f_{m_\beta+1}^{(l_{m_\beta+1})},\dots,f_{m_{\beta+1}}^{(l_{m_{\beta+1}})}$ ni aucune leur combinaison linéaire $\lambda f_{m_\beta+1} + \dots + \mu f_{m_{\beta+1}}$ $(\lambda,\dots,\mu \in K)$ ne peuvent s'exprimer comme des combinaisons linéaires, à coefficients formes homogènes de degré convenable, des $f_{m_\alpha+1}^{(l_{m_\alpha+1})},\dots,f_{m_{\alpha+1}}^{(l_{m_{\alpha+1}})}$, avec $\alpha < \beta$;

(ii) $f_{m_\beta+1}^{(l_{m_\beta+1})},\dots,f_{m_{\beta+1}}^{(l_{m_{\beta+1}})}$ sont K-linéairement indépendantes parmi les formes de degré $l_{m_\beta+1} = \dots = l_{m_{\beta+1}}$,

est un ouvert de Zariski (éventuellement vide) $\Omega \subset A$.

Preuve. Ordonnons les degrés $l_1,\dots,l_q$:

$$l_1 = \dots = l_{m_1} < l_{m_1+1} = \dots = l_{m_2} < \dots < l_{m_p+1} = \dots = l_q \quad (m_0 = 0).$$

Si $p=0$, c'est à dire si les polynômes $f_1(a^{(1)};x),\dots,f_q(a^{(q)};x)$ ont le même degré l, il suffit d'exprimer la condition de dépendance linéaire

$$\sum_{j=1}^{q} \lambda_j f_j(a^{(j)};x) \equiv 0$$

entre $f_1,\dots,f_q$. Soient $(\Lambda)=(\Lambda_1,\dots,\Lambda_q)$, $(A^{(j)})=(A^{(j)}_{i_1\dots i_l})$, $j=1,\dots,q$, des ensembles d'indéterminées sur K.

Le système des $\binom{n+l}{n}$ équations

$$\begin{cases} \Lambda_1 A^{(1)}_{i_1\dots i_l} + \Lambda_2 A^{(2)}_{i_1\dots i_l} + \dots + \Lambda_q A^{(q)}_{i_1\dots i_l} = 0, \\ i_1 \le \dots \le i_l, \\ i_1,\dots,i_l = 0,\dots,n, \end{cases}$$

est résoluble par des valeurs non toutes nulles des Λ si et seulement si tous les mineurs d'ordre q de la matrice (de type $\binom{n+l}{n}\times q$) $\| A^{(1)}_{i_1\dots i_l}\ A^{(2)}_{i_1\dots i_l}\ \dots\ A^{(q)}_{i_1\dots i_l} \|$ sont nuls:

$$(*_1) \qquad \begin{cases} R_\lambda(A^j)=0, \\ \lambda = 1,2,\dots\ . \end{cases}$$

En ce cas il suffit de prendre comme Ω le complémentaire du fermé défini dans

$$A = A^{\binom{n+l}{n}\cdot q}(K)$$

par les équations $(*_1)$. Si $q \le \binom{n+l}{n}$, cet ensemble Ω n'est pas vide. On procède maintenant par induction sur p, en supposant l'existence d'un ouvert

$$\Omega' \subset A' = A^{\binom{n+l_{m_1}}{n}\cdot m_1}(K) \times \dots \times A^{\binom{n+l_{m_p}}{n}\cdot (m_p - m_{p-1})}(K)$$

tel que (i) et (ii) sont satisfaites pour les systèmes de m_p formes, appartenant à Ω'. Les conditions algébriques pour qu'une forme $f^{(l_q)}$ de degré l_q soit du type

$$(*_2)\qquad f^{(l_q)} \equiv \zeta_1^{(l_q-l_1)} f_1^{(l_1)} + \ldots + \zeta_{m_p}^{(l_q-l_{m_p})} f_{m_p}^{(l_{m_p})}$$

s'obtiennent immédiatement par identification.

Notons comme précédemment

$$(A) = (A_{i_1\ldots i_{l_q}}),\quad (A^{(\alpha)}) = (A^{(\alpha)}_{i_1\ldots i_{l_\alpha}}),\quad (B^{(\alpha)}) = (B^{(\alpha)}_{i_1\ldots i_{(l_q-l_\alpha)}}),$$

$\alpha = 1,\ldots,m_p$, des indéterminées sur K; $(X) = (X_o,\ldots,X_n)$ des indéterminées sur le corps $K((A),(A^\alpha),(B^{(\alpha)}))$; et écrivons l'égalité en $K((A),(A^\alpha),(B^\alpha))[X_o,\ldots,X_n]$:

$$\Sigma A_{i_1\ldots i_{l_q}} X_{i_1}\ldots X_{i_{l_q}} =$$

$$= \Sigma B^{(1)}_{i_1\ldots i_{(l_q-l_1)}} X_{i_1}\ldots X_{i_{(l_q-l_1)}} \cdot \Sigma A^{(1)}_{i_1\ldots i_{l_1}} X_{i_1}\ldots X_{i_{l_1}} + \ldots$$

$$+ \Sigma B^{(m_p)}_{i_1\ldots i_{(l_q-l_{m_p})}} X_{i_1}\ldots X_{i_{(l_q-l_{m_p})}} \cdot \Sigma A^{(m_p)}_{i_1\ldots i_{l_{m_p}}} X_{i_1}\ldots X_{i_{l_{m_p}}} .$$

En égalant les coefficients des monômes dans les deux membres, on obtient un système d'équations

$$(*_3)\qquad \begin{cases} S_\lambda((A),(A^{(\alpha)}),(B^{(\alpha)})) = 0, \\ \lambda = 1,2,\ldots, \end{cases}$$

qui sont linéaires séparément dans les $(A^{(\alpha)})$ et dans les $(B^{(\alpha)})$.

Spécialisons les $A^{(\alpha)}$ dans les points $a^{(\alpha)}$ d'un convenable ouvert de Zariski (non vide): alors la matrice $\|a^{(\alpha)}\|$ est de rang constant; imposant à la matrice complète $\|(A),(a^{(\alpha)})\|$ du système

$$\begin{cases} S_\lambda((A),(a^{(\alpha)}),(B^{(\alpha)})) = 0, \\ \lambda = 1,2\ldots \end{cases}$$

d'avoir la même caractéristique que la matrice $\|a^{(\alpha)}\|$, on obtient des conditions

$$(*_4)\qquad \begin{cases} T_\lambda(A,(a^{(\alpha)})) = 0, \\ \lambda = 1,2,\ldots \end{cases}$$

nécessaires et suffisantes pour que, lorsqu'une spécialisation (a) de (A) vérifie $(*_4)$, le système (linéaire dans les $B^{(\alpha)}$)

$$\begin{cases} S_\lambda((a),(a^{(\alpha)}),(B^{(\alpha)})) = 0, \\ \lambda = 1,2,\ldots \end{cases}$$

soit résoluble. Il n'est pas restrictif de supposer que

$$(f_1^{(l_1)},\ldots,f_{m_p}^{(l_{m_p})})$$

individué par les coefficients $(a^{(\alpha)})$ soit dans l'ouvert Ω' de notre hypothèse inductive (autrement il suffira de prendre l'intersection de Ω' avec l'ouvert où le rang de $\|a^{(\alpha)}\|$ se maintient constant). Soit

$$\Omega_1^*(f_1,\ldots,f_{m_p})$$

le complémentaire du fermé défini dans $A^{\binom{n+1}{n}q}(K)$ par les équations $(*_4)$: si $f^{(l_q)}$ appartient à $\Omega_1^*(f_1,\ldots,f_{m_p})$, $f^{(l_q)}$ n'est du type $(*_2)$ pour aucun choix des formes $\zeta_1,\ldots,\zeta_{m_p}$. Exprimons maintenant qu'une combinaison linéaire de $q-m_p$ éléments

$$f_{m_p+1}^{(l_q)},\ldots,f_q^{(l_q)} \quad \text{de} \quad \Omega_1^*(f_1,\ldots,f_{m_p})$$

est du type $(*_2)$:

$$\begin{cases} T_\lambda(\Lambda_{m_p+1}\, A^{(m_p+1)}_{i_1\ldots i_{l_q}} + \ldots + \Lambda_q\, A^{(q)}_{i_1\ldots i_{l_q}}, (a^{(\alpha)})) = 0, \\ \lambda = 1,2,\ldots; \end{cases}$$

ces équations sont linéaires et homogènes dans les (Λ); l'élimination des (Λ) donne les conditions sur

$$f_{m_p+1}^{(l_q)},\ldots,f_q^{(l_q)}$$

pour qu'une combinaison linéaire entre elles soit du type $(*_2)$.

En appelant $\Omega_1(f_1,\ldots,f_{m_p})$ l'ouvert du produit

$$A^{\binom{n+1}{n}q\,(q-m_p)},$$

complémentaire du fermé défini par les équations ci-dessus, on a alors: si

$$\begin{pmatrix} f_{m_p+1} \\ \vdots \\ f_q \end{pmatrix} \in \Omega_1(f_1,\dots,f_{m_p}),$$

aucune combinaison linéaire des $f_{m_p+1},\dots,f_q$ n'est du type $(*_2)$, quel que soit le choix des $\zeta_1,\dots,\zeta_{m_p}$. Or je dis: il existe un voisinage $\Omega'(f_1,\dots,f_{m_p})$ de

$$\begin{pmatrix} f_1 \\ \vdots \\ f_{m_p} \end{pmatrix}$$

tel que, pour tout

$$\begin{pmatrix} \bar{f}_1 \\ \vdots \\ \bar{f}_{m_p} \end{pmatrix} \in \Omega'(f_1,\dots,f_{m_p})$$

et pour tout

$$\begin{pmatrix} f_{m_p+1} \\ \vdots \\ f_q \end{pmatrix} \in \Omega_1(f_1,\dots,f_{m_p}),$$

aucune combinaison linéaire des $f_{m_p+1},\dots,f_q$ n'est du type

$$\bar{\zeta}_1\bar{f}_1 + \dots + \bar{\zeta}_{m_p}\bar{f}_{m_p}.$$

Il s'agit de démontrer l'observation suivante: si on a, par exemple,

$$f_{m_p+1} \neq \zeta_1 f_1 + \dots + \zeta_{m_p} f_{m_p},$$

alors, à cause de la continuité (topologie produit) de l'application

$$((\bar{f}_{m_p+1}),(\bar{\zeta}_1,\dots,\bar{\zeta}_{m_p}),(\bar{f}_1,\dots,\bar{f}_{m_p})) \longmapsto$$

$$\longmapsto \bar{f}_{m_p+1} - (\bar{\zeta}_1\bar{f}_1 + \dots + \bar{\zeta}_{m_p}\bar{f}_{m_p}),$$

on aura aussi

$$\overline{f}_{m_p+1} \neq \overline{\zeta}_1 \overline{f}_1 + \ldots + \overline{\zeta}_{m_p} \overline{f}_{m_p} .$$

pour tout $(\overline{f}_1,\ldots,\overline{f}_{m_p})$ appartenant à un voisinage $\Omega'(f_1,\ldots,f_{m_p})$ de

$$\begin{bmatrix} f_1 \\ \vdots \\ f_{m_p} \end{bmatrix}$$

dans A', et pour tout $\overline{f}_{m_p+1}$ appartenant à (un voisinage de f_{m_p+1} dans

$$A^{\binom{n+1}{n}q}(K)$$

qu'il n'est pas restrictif d'identifier avec) $\Omega_1^*(f_1,\ldots,f_{m_p})$.

Enfin il existe, comme avant,

$$\Omega_2 \subset A^{\binom{n+1}{n}q \cdot (q-m_p)}(K)$$

avec la propriété que, pour tout

$$\begin{bmatrix} f_{m_p+1} \\ \vdots \\ f_q \end{bmatrix} \in \Omega_2,$$

$f_{m_p+1},\ldots,f_q$ sont linéairement indépendantes. L'ouvert cherché est alors

$$\Omega = \bigcup_{\begin{bmatrix} f_1 \\ \vdots \\ f_{m_p} \end{bmatrix} \in \Omega'} (\Omega' \cap \Omega'(f_1,\ldots,f_{m_p})) \times (\Omega_1(f_1,\ldots,f_{m_p}) \cap \Omega_2) .$$

3. Dans l'anneau gradué $A = K[x_o,\ldots,x_n]$, $A = \bigoplus_{\alpha=0}^{\infty} A_\alpha$, soit

$$I = \bigoplus_{\alpha=0}^{\infty} I_\alpha \quad (I_\alpha = I \cap A_\alpha)$$

un idéal homogène; posons

$$v_I(l) = \dim_K(I_l / I_{l-1} A \cap A_l).$$

(Remarquons que, par le théorème de la base de Hilbert, $v_I(l) = 0$ si $l >> 0$).

Indiquons avec $J(l_1,\ldots,l_q; n)$ la famille des idéaux homogènes I de A pour lesquels

$$v_I(l) = \begin{cases} 0 \quad , & \text{si } 0 \leq l \leq l_1, \\ m_1 \quad , & \text{si } l = l_1 = \ldots = l_{m_1}, \\ 0 \quad , & \text{si } l_{m_1} < l < l_{m_1+1}, \\ m_2 - m_1, & \text{si } l = l_{m_1+1} = \ldots = l_{m_2}, \\ 0 \quad , & \text{si } l_{m_2} < l < l_{m_2+1}, \\ m_3 - m_2, & \text{si } l = l_{m_2+1} = \ldots = l_{m_3}, \\ \ldots\ldots\ldots\ldots\ldots\ldots , \\ \ldots\ldots\ldots\ldots\ldots\ldots , \\ 0 \quad , & \text{si } l_{m_p} < l < l_{m_p+1}, \\ q - m_p \, , & \text{si } l = l_{m_p+1} = \ldots = l_q, \\ 0 \quad , & \text{si } l > l_q. \end{cases}$$

Si

$$f_{m_\alpha+1}^{(l_{m_\alpha+1})}, \ldots, f_{m_{\alpha+1}}^{(l_{m_{\alpha+1}})} \qquad (\alpha = 0,1,2,\ldots)$$

sont telles que les images correspondantes

$$\dot{f}_{m_\alpha+1}^{(l_{m_\alpha+1})}, \ldots, \dot{f}_{m_{\alpha+1}}^{(l_{m_{\alpha+1}})}$$

forment une base dans l'espace vectoriel

$$I_{l_{m_\alpha+1}} \Big/ I_{l_{m_\alpha+1}-1} A \cap A_{l_{m_\alpha+1}},$$

alors $(f_1^{(l_1)},, \ldots, f_q^{(l_q)})$ est une base de longueur minima pour l'idéal I de $J(l_1,\ldots,l_q; n)$, et réciproquement; donc l'ouvert $\Omega \subset A$ de la Proposition 1 peut être regardé comme la totalité des bases de longueur minima des idéaux de $J(l_1,\ldots,l_q; n)$.

Deux systèmes de polynômes

$$(f) = \begin{pmatrix} f_1^{(l_1)} \\ \vdots \\ f_q^{(l_q)} \end{pmatrix} \quad \text{et} \quad (g) = \begin{pmatrix} g_1^{(l_1)} \\ \vdots \\ g_q^{(l_q)} \end{pmatrix}, \quad (f) \text{ et } (g) \in \Omega,$$

individuent le même élément de $J(l_1,\ldots,l_q;n)$ si et seulement si $(g) = \Phi \cdot (f)$, où

$$\Phi = \|\zeta_{ij}^{(h)}\|_{1<i,\, j\leq q}$$

est une matrice de type $q \times q$, dont les éléments $\zeta_{ij}^{(h)}$ sont (i) formes homogènes en $(x_o,\ldots,x_n)$, de degré h (éventuellement $h = 0$) donné par le tableau ci-dessous (Tab. 1), pour les valeurs i et j

$$h = \begin{cases} 0, \ \text{si} \ 1 \leq i, j \leq m_1 \\[2ex] l_{m_\beta} - l_{m_1} \ (2 \leq \beta \leq p), \ \text{si} \begin{cases} m_{\beta-1} + 1 \leq i \leq m_\beta \\ 1 \leq j \leq m_1 \end{cases} \\[4ex] l_{m_\beta} - l_{m_\alpha} \ (2 \leq \alpha \leq \beta \leq p), \ \text{si} \begin{cases} m_{\beta-1} + 1 \leq i \leq m_\beta \\ m_{\alpha-1} + 1 \leq j \leq m_\alpha \end{cases} \\[4ex] l_q - l_{m_1}, \ \text{si} \begin{cases} m_p + 1 \leq i \leq q \\ 1 \leq j \leq m_1 \end{cases} \\[4ex] l_q - l_{m_2}, \ \text{si} \begin{cases} m_p + 1 \leq i \leq q \\ m_1 + 1 \leq j \leq m_2 \end{cases} \\[2ex] \ldots\ldots\ldots\ldots\ldots \\[2ex] 0, \ \text{si} \ m_p + 1 \leq i, j \leq q \end{cases}$$

Tab. 1

indiquées dans la tableau, et vérifiant en outre les conditions

$$\begin{cases} \det \|\zeta_{ij}^{(o)}\|_{1 \le i, j \le m_1} \neq 0, \\ \det \|\zeta_{ij}^{(o)}\|_{m_1+1 \le i, j \le m_2} \neq 0, \\ \ldots\ldots\ldots\ldots, \\ \ldots\ldots\ldots\ldots, \\ \det \|\zeta_{ij}^{(o)}\|_{m_p+1 \le i, j \le q} \neq 0; \end{cases}$$

(ii) $\zeta_{ij}^{(h)} = 0$, pour toutes les i, j différentes de celles du tableau.

Les matrices Φ de ce type forment un groupe G; l'application

$$(\Phi, (f)) \longrightarrow (g) = \Phi \cdot (f)$$

définit une opération de G dans le produit

$$A = A^{N_{m_1} \cdot m_1}(K) \times A^{N_{m_2} \cdot (m_2 - m_1)}(K) \times \ldots \times A^{N_q \cdot (q - m_p)}(K);$$

l'ensemble quotient Ω/G est en correspondance biunivoque avec $J(l_1, \ldots, l_q; n)$.

PROPOSITION 2. $J(l_1, \ldots, l_q; n)$ *est en correspondance biunivoque avec une variété algébrique rationnelle, et la projection canonique* $p : \Omega \longrightarrow J(l_1, \ldots, l_q; n)$ *est une application ouverte*.

Preuve. Ordonnons les degrés $l_1, \ldots, l_q$:

$$l_1 = \ldots = l_{m_1} < l_{m_1+1} = \ldots = l_{m_2} < \ldots < l_{m_p+1} = \ldots = l_q \quad (m_o = 0).$$

(α) Si $p = 0$, les polynômes $f_1, \ldots, f_q$ ont tous le même degré l. L'idéal $(f_1, \ldots, f_q)$ s'identifiera avec

$$\{\Phi \cdot (f)\}_{\Phi \in PGL_K(q-1)} \qquad \left((f) = \begin{pmatrix} f_1 \\ \vdots \\ f_q \end{pmatrix} \right),$$

c'est à dire avec la variété linéaire de dimension $q - 1$ engendrée par $f_1, \ldots, f_q$ dans l'espace projectif

$$\mathbb{P}_{\binom{n+1}{n}-1}(K),$$

dont les points représentent les formes de degré 1, non identiquement nulles et à une constante multiplicative $\neq 0$ près, dans les $(x_o,\dots,x_n)$.

Ainsi $J(l_1,\dots,l_q;n)$ est en correspondance biunivoque avec la grassmannienne

$$G(q-1;\binom{n+1}{n}-1)$$

des sous-espaces de dimension $q-1$ de $\mathbb{P}_{\binom{n+1}{n}-1}(K)$, qui est une variété rationnelle.

L'ouvert Ω de $A^{\binom{n+1}{n}\cdot q}(K)$ est formé des systèmes de formes $(f_1,\dots,f_q)$ dont les composantes sont K-linéairement indépendantes; il s'agit de démontrer que la projection naturelle

$$p:\Omega \longrightarrow G(q-1;\binom{n+1}{n}-1)$$

est ouverte.

Si $q=1$, l'application p est l'identité, donc la thèse est vraie. Supposons - la prouvée pour $q-2$. Etant donnés un ouvert $A\subset\Omega$ et un de ses éléments $(f_1,\dots,f_{q-1},f_q)$, soient U' et U'' (Prop. 1) deux voisinages ouverts de $(f_1,\dots,f_{q-1})$ et (f_q) respectivement dans

$$A^{\binom{n+1}{n}\cdot(q-1)}(K) \text{ et } A^{\binom{n+1}{n}}(K),$$

tels que

$$\forall(\bar f_1,\dots,\bar f_{q-1})\in U',$$

$$\forall(\bar f_q)\in U'';$$

$(\bar f_1,\dots,\bar f_{q-1},\bar f_q)$ soient linéairement indépendants. Désignons ensuite

$$p':U'\longrightarrow G(q-2;\binom{n+1}{n}-1),$$

$$p'':U''\longrightarrow G(0;\binom{n+1}{n}-1)$$

les projections correspondantes.

Alors $p''(U'')$ et $p'(U')$ sont tous les deux ouverts, par l'hypothèse inductive; donc leur produit aussi, $p'(U') \times p''(U'')$, est ouvert dans

$$G(q-2;\binom{n+1}{n}-1) \times G(0;\binom{n+1}{n}-1).$$

Or remarquons que les $p'(U') \times p''(U'')$ sont parmi les ouverts du type

$$V = V' \times V'',$$

$$V' \in G(q-2;\binom{n+1}{n}-1),$$

$$V'' \in G(0,\binom{n+1}{n}-1),$$

V' et V'' étant caractérisés par la propriété suivante:

$$\forall L' \in V',$$

$$\forall L'' \in V''$$

on a $L'' \not\subset L'$, donc (L',L'') détermine un élément de $G(q-1;\binom{n+1}{n}-1)$; remarquons en outre que $G(q-1;\binom{n+1}{n}-1)$, en tant qu'espace topologique, est un quotient $\bigcup V/R$; et que l'on a $p(U' \times U'') = p'(U') \times p''(U'')/R$. Un voisinage de $p(f_1,\ldots,f_q)$, qui soit entièrement formé de points provenants de A, est donné alors par

$$p(U' \times U'') \setminus p((U' \times U'') \setminus A).$$

(β) Procédons par induction sur p, en supposant le theorème vrai pour les idéaux définis par des systèmes de formes avec $p-1$ degrés distincts. Soit

$$\pi : J(l_1,\ldots,l_q;n) \longrightarrow J(l_1,\ldots,l_{m_p};n)$$

la projection canonique:

$$I = (f_1^{(l_1)},\ldots,f_{m_p}^{(l_{m_p})},f_{m_p+1}^{(l_{m_p+1})},\ldots,f_q^{(l_q)}) \longrightarrow$$

$$\longrightarrow I' = (f_1^{(l_1)},\ldots,f_{m_p}^{(l_{m_p})}).$$

Pour $I' = (f_1^{(l_1)},\ldots,f_{m_p}^{(l_{m_p})})$ fixé dans $J(l_1,\ldots,l_{m_p};n)$, on a évidemment

$$\pi^{-1}(I') = \{(f_1^{(l_1)}, \ldots, f_{m_p}^{(l_{m_p})}, f_{m_p+1}, \ldots, f_q)\}, \quad \text{où} \quad (f_{m_p+1}, \ldots, f_q)$$

sont des formes de degré l_q linéairement indépendantes et telles que aucune de leurs combinaisons linéaires ne soit dans I'.

Soit $L_{(f_{m_p+1},\ldots,f_q)}$ le sous-espace linéaire engendré par $f_{m_p+1}, \ldots, f_q$ dans le

$$\mathbb{P}_{\binom{n+l_q}{n}-1}(K)$$

des formes de degré l_q; nous pouvons remarquer deux choses: que I est univoquement déterminé par $I' = \pi(I)$ et

$$L_{(f_{m_p+1},\ldots,f_q)},$$

et qu'on obtient de cette façon une application injective de $\pi^{-1}(I')$ dans la grassmannienne

$$G(q - m_p - 1; \binom{n+l_q}{n} - 1).$$

Donc on peut identifier $\pi^{-1}(I')$ avec un sous-ensemble de

$$G(q - m_p - 1; \binom{n+l_q}{n} - 1);$$

on remarquera qu'il est ouvert dans la grassmannienne. Nous savons en effet que, dans

$$A^{\binom{n+l_q}{n}\cdot(q-m_p)}(K),$$

les points $(f_{m_p+1},\ldots,f_q)$ pour lesquels $f_{m_p+1},\ldots,f_q$ sont linéairement indépendantes (parmi les formes de degré l_q qui n'appartiennent pas a I')telles qu'aucune combinaison linéaire n'est dans I', constituent (dem. de la Prop. 1) un ouvert $\Omega''_{I'}$; mais pour ce qu'on a vu en (α), l'application

$$\begin{cases} \Omega \longrightarrow G(q - m_p - 1; \binom{n+l_q}{n} - 1), \\ (f_{m_p+1},\ldots,f_q) \longmapsto L_{(f_{m_p+1},\ldots,f_q)} \end{cases}$$

est ouverte; comme $\pi^{-1}(I')$ est l'image de $\Omega''_{I'}$ dans

$$G(q - m_p - 1; \binom{n+l_q}{n} - 1)$$

par cette application, s'ensuit la thèse.

Nous noterons encore $\Omega''_{I'}$ l'image de $\Omega''_{I'}$ dans la grassmannienne.

De la démonstration de la Prop. 1 il s'ensuit en outre l'existence d'un voisinage $\Omega'_{I'}$ de $(f_1,\ldots,f_{m_p})$ dans

$$A' = A^{\binom{n+l_{m_1}}{n}\cdot m_1}(K) \times \ldots \times A^{\binom{n+l_{m_p}}{n}\cdot(m_p - m_{p-1})}(K)$$

et d'un voisinage Ω''' de $(f_{m_p+1},\ldots,f_q)$ dans

$$A^{\binom{n+l_q}{n}(q-m_p)}(K),$$

tels que

$(*_1)$ $\quad \forall(\bar{f}_1,\ldots,\bar{f}_{m_p}) \in \Omega'_{I'}$,

$(*_2)$ $\quad \forall(\bar{f}_{m_p+1},\ldots,\bar{f}_q) \in \Omega'''$

aucune combinaison linéaire des $\bar{f}_{m_p+1},\ldots,\bar{f}_q$ n'est dans $\overline{I'} = (\bar{f}_1,\ldots,\bar{f}_{m_p})$.

Or nous pouvons remplacer $(*_1)$ (en utilisant l'hypothèse inductive que la projection sur $J(l_1,\ldots,l_{m_p}; n)$ est ouverte) et $(*_2)$ (en utilisant la propriété démontrée en (α)) par

$(*'_1)$ $\quad \forall\, \overline{I'} \in \Omega'_{I'}$,

$(*'_2)$ $\quad \forall\, L_{(\bar{f}_{m_p+1},\ldots,\bar{f}_q)} \in \Omega'''$,

respectivement, où maintenant $\Omega'_{I'}$ est un voisinage ouvert de I' dans $J(l_1,\ldots,l_{m_p}; n)$ et Ω''' est un ouvert de

$$G(q - m_p - 1, \binom{n+l_q}{n} - 1),$$

qu'il n'est pas restrictif d'identifier avec $\Omega''_{I'}$.

Ayant posé

$$\Omega_{I'} = \Omega'_{I'} \times \Omega''_{I'},$$

$$\Omega_{J'} = \Omega'_{J'} \times \Omega''_{J'},$$

les identités définies presque partout

$$\begin{cases} \Omega'_{I'} \cdots \xrightarrow{i} \Omega'_{J'}, \\ \Omega''_{I'} \cdots \xrightarrow{i} \Omega''_{J'} \end{cases}$$

déterminent par produit des applications birationelles

$$\tau_{I',J'} : \Omega_{I'} \cdots \longrightarrow \Omega_{J'}$$

vérifiant la condition de transitivité; la variété cherchée, en correspondance biunivoque avec l'ensemble $J(l_1,\ldots,l_q;n)$, est donnée alors (Weil [3], pag. 2-5) par

$$\{\Omega_{I'},\tau_{I',J'}\}_{I',J'\in J(l_1,\ldots,l_{m_p};n)}.$$

De la rationalité de tout $\Omega'_{I'}$, par l'hypothèse d'induction, et des fibres $\Omega''_{I'}$, par la propriété analogue des variétés grassmanniennes, il s'ensuit (Weil [3], pag. 10) la rationalité de $J(l_1,\ldots,l_q;n)$.

(γ) Il nous reste à démontrer: si $\tilde{\Phi}$ est un ouvert de Zariski (non vide) de Ω, alors $p(\tilde{\Phi})$ est un ouvert de Zariski (non vide) de $J(l_1,\ldots,l_q;n)$.

Raisonnement standard. Le cas $p=0$, c'est à dire $f_1,\ldots,f_q$ du même degré, a été vu en (α). Etant donnés un ouvert $\subset\Omega$ et l'un de ses éléments

$$(f_1^{(l_1)},\ldots,f_{m_p}^{(l_{m_p})},f_{m_p+1}^{(l_{m_p+1})},\ldots,f_q^{(l_q)}),$$

soient, comme avant, U' et U'' deux voisinages ouverts de $(f_1,\ldots,f_{m_p})$ dans A' et de $(f_{m_p+1},\ldots,f_q)$ dans

$$A^{\binom{n+l_q}{n}\cdot(q-m_p)}(K)$$

respectivement, tels que

$$\forall(\bar{f}_1,\ldots,\bar{f}_{m_p})\in U',$$

$$\forall(\bar{f}_{m_p+1},\ldots,\bar{f}_q)\in U''$$

aucune combinaison linéaire des $\overline{f}_{m_p+1},\ldots,\overline{f}_q$ ne soit dans $\overline{I}'=(\overline{f}_1,\ldots,\overline{f}_{m_p})$, et que les $\overline{f}_{m_p+1},\ldots,\overline{f}_q$ soient linéairement indépendantes.

Ayant noté

$$p' : U' \longrightarrow J(l_1,\ldots,l_{m_p};n),$$

$$p'' : U'' \longrightarrow G(q-m_p-1;\binom{n+l_q}{n}-1)$$

les projections canoniques, aussi bien $p'(U')$ (par l'induction) que $p''(U'')$ (pour ce qu'on a démontré en (α)) sont ouverts; donc $p'(U')\times p''(U'')$ est aussi ouvert dans $J(l_1,\ldots,l_{m_p};n)\times G(q-m_p-1;\binom{n+l_q}{n}-1)$.

On conclut précisément comme au point (α), en rappelant comment, au point (β), on avait défini la topologie de $J(l_1,\ldots,l_q;n)$.

Bibliographie

[1] BIGOLIN, B., Osservazioni sugli invarianti proiettivi delle intersezioni complete, à paraître dans Rend. Sem. Mat. Brescia.

[2] SPINELLI, A., Due osservazioni di teoria dei corpi, Rend. Sem. Mat. Brescia, Vol. 6.

[3] WEIL, A., Fibre spaces in algebraic geometry, University of Chicago, Chicago, Ill. 1953.

Università Cattolica del Sacro Cuore
Sede di Brescia
Via Trieste, 17
I-25100 Brescia, Italy

CONNECTIONS ON FOLIATED MANIFOLDS

Robert A. Blumenthal (Saint Louis, MO)

Summary. The results reported here, whose detailed versions will appear in [1, 2], concern connections in bundles over a foliated manifold and the interplay between the geometry of the connection and the topology of the foliation. Here we only give the proofs of Theorems 1 and 3.

1. Let M be a connected smooth manifold and let F be a smooth codimension q foliation of M. Let Q be the normal bundle of F and let ∇ be a connection on Q. Let T and R be the torsion and curvature respectively of ∇. Then ∇ is basic (in the sense of [6]) if $T(X,Y) = R(X,Y)Z = 0$ whenever X is tangent to F. We say that F has a locally reductive normal bundle if Q admits a basic connection ∇ satisfying $\nabla T = \nabla R = 0$.

THEOREM 1. *Let (M,F,∇) be a foliated manifold with a complete (in the sense of [9]) locally reductive normal bundle. Let $p : \widetilde{M} \to M$ be the universal cover of M. Then there is a simply connected reductive homogeneous space G/H and a locally trivial fiber bundle $f : \widetilde{M} \to G/H$ whose fibers are the leaves of $p^{-1}(F)$. Moreover, the lift of ∇ to $\widetilde{M}$ agrees with the basic connection obtained by pulling back via f the canonical connection of the second kind on G/H.*

When F is zero-dimensional we obtain from Theorem 1 the theorem of Kobayashi [8] which states that a simply connected manifold with a complete linear connection with parallel torsion and curvature is isomorphic to a reductive homogeneous space with the canonical connection of the second kind.

It is elementary that a codimension one foliation of a compact manifold defined by a nonsingular closed one-form has a complete locally reductive normal bundle. As a corollary of Theorem 1 we obtain Reeb's structure theorem for such codimension one foliations.

COROLLARY (Reeb [10]). *Let M be a compact manifold and let F be a codimension one foliation of M defined by a nonsingular closed one-form. Then the universal cover* $\tilde{M}$ *is diffeomorphic to a product* $\tilde{L} \times R$ *and the lift of F to* $\tilde{M}$ *is the foliation by leaves of the form* $\tilde{L} \times \{t\}$, $t \in R$.

2. When F is a Riemannian foliation, Q admits a smooth metric g invariant under the natural parallelism along the leaves and there is a unique torsion-free metric-preserving basic connection ∇ on Q. We say g is complete if ∇ is complete and we say F is Riemannian locally symmetric if $\nabla R = 0$. For each $x \in M$ and each two-dimensional subspace p of Q_x, the (transverse) sectional curvature of p is defined by $K(p) = -g(R(\tilde{X}_1, \tilde{X}_2)X_1, X_2)$ where $\{X_1, X_2\}$ is an orthonormal basis of p and $\tilde{X}_1, \tilde{X}_2 \in T_x(M)$ project to X_1, X_2 respectively.

THEOREM 2. *Let F be a complete Riemannian locally symmetric foliation of a manifold M. If* $K > 0$, *then M/F is compact. If in addition F has a compact leaf with finite fundamental group, then M is compact with finite fundamental group.*

3. The study of connections on the normal bundle of a foliation is a special case of the more general study of Cartan connections in foliated bundles. We define a Cartan connection in a foliated bundle in a way which generalizes the notion of a Cartan connection in an ordinary principal bundle and provides a unified setting for the study of Riemannian, conformal, and projective foliations as well as other types of geometric structures for foliations.

Let M be a connected smooth manifold and let F be a smooth codimension q foliation of M. Let T(M) be the tangent bundle of M and let $E \subset T(M)$ be the tangent bundle of F. Let G be a Lie group and let $H \subset G$ be a closed subgroup with $\dim(G/H) = q$. Let $\pi : P \to M$ be a foliated principal H-bundle in the sense of [6]. That is, there is a foliation $\tilde{F}$ of P satisfying

(i) $\tilde{F}$ is H-invariant,

(ii) $\tilde{E}_u \cap V_u = \{0\}$ for all $u \in P$,

(iii) $\pi_{*_u}(\tilde{E}_u) = E_{\pi(u)}$ for all $u \in P$,

where $\tilde{E} \subset T(P)$ is the tangent bundle of $\tilde{F}$ and $V \subset T(P)$ is the bundle of vertical vectors. Let $\mathfrak{g}$ be the Lie algebra of G and let $\mathfrak{h}$ be the Lie algebra of H. For $A \in \mathfrak{h}$, let A^* be the corresponding fundamental vector field on P.

Definition. A Cartan connection in the foliated bundle $\pi : P \to M$ is a $\mathfrak{g}$-valued one-form ω on P satisfying

(i) $\omega(A^*) = A$ for all $A \in \mathfrak{h}$,

(ii) $(R_a^*)\omega = \mathrm{ad}(a^{-1})\omega$ for all $a \in H$, where R_a denotes the right translation by a acting on P and $\mathrm{ad}(a^{-1})$ is the adjoint action of a^{-1} on $\mathfrak{g}$,

(iii) for each $u \in P$, $\omega_u : T_u(P) \to \mathfrak{g}$ is onto and $\omega_u(\tilde{E}_u) = 0$,

(iv) $L_X\omega = 0$ for all $X \in \Gamma(\tilde{E})$, where $\Gamma(\tilde{E})$ denotes the smooth sections of $\tilde{E}$ and L_X is the Lie derivative.

Note that if $\dim F = 0$, then ω is a Cartan connection in the bundle $\pi : P \to M$ in the sense of Ehresmann [4].

Following Molino [9] we say that a section $\tilde{Y}$ of the normal bundle $\tilde{Q}$ of $\tilde{F}$ is complete if there exists a complete vector field Y on P which projects to $\tilde{Y}$ under natural projection $T(P) \to \tilde{Q}$.

Definition. We say ω is complete if each section $\tilde{Y}$ of $\tilde{Q}$ such that $\omega(\tilde{Y})$ is constant is complete.

Note that if $\dim F = 0$, this reduces to the notion of completeness given in [7].

THEOREM 3. *Let ω be a complete Cartan connection in P. Then all the leaves of F have the same universal cover. In particular, if F has a compact leaf with finite fundamental group, then all the leaves of F are compact with finite fundamental group.*

As a corollary of Theorem 3, we obtain the Reinhart Stability Theorem.

COROLLARY (Reinhart [11]). *Let F be a Riemannian foliation of a compact connected manifold M. Then all the leaves of F have the same universal covering space.*

The curvature of ω is the g-valued 2-form Ω on P defined by $\Omega = d\omega + 1/2[\omega,\omega]$. We say ω is flat if $\Omega = 0$.

THEOREM 4. *Let ω be a complete flat Cartan connection in P. Let $p : \tilde{M} \to M$ be the universal cover of M and let $\widetilde{(G/H)}$ be the universal cover of G/H. There is a locally trivial fiber bundle $\tilde{M} \to \widetilde{(G/H)}$ whose fibers are the leaves of $p^{-1}(F)$.*

As a corollary of Theorem 4 we obtain Fédida's structure theorem for Lie-foliations.

COROLLARY (Fédida [5]). *Let* $\mathfrak{g}$ *be a finite-dimensional real Lie algebra and let* F *be a Lie-*$\mathfrak{g}$ *foliation of a compact manifold* M. *Then the leaves of the lift of* F *to the universal cover* $\widetilde{M}$ *of* M *are the fibers of a locally trivial fiber bundle* $\widetilde{M} \to G$, *where* G *is the simply connected Lie group whose Lie algebra is* $\mathfrak{g}$.

We construct a class of examples as follows. Let G be a Lie group, let $H \subset G$ be a closed subgroup and let Γ be a discrete subgroup of G. The foliation of G whose leaves are the left cosets of H induces a foliation F on $M = \Gamma \backslash G$. We construct a foliated H-bundle $(P, \widetilde{F})$ over (M, F) and a complete flat Cartan connection in P. The Roussarie example [3] is the case where $G = SL(2, \mathbb{R})$, $G/H \cong S^1$, Γ is a uniform discrete subgroup of $SL(2, \mathbb{R})$, and M is the unit tangent bundle of a compact connected Riemann surface of genus $\geqq 2$. The generalized Roussarie example is the case where $G = SL(q+1, \mathbb{R})$, Γ is a uniform discrete subgroup of G, and $G/H \cong \mathbb{R}P^q$.

4. We consider projective and conformal foliations from the point of view of Cartan connections in foliated bundles by considering the normal projective (respectively, conformal) connection in the projective (respectively, conformal) normal bundle of a projective (respectively, conformal) foliation.

THEOREM 5. *Let* F *be a complete projective or conformal foliation of codimension* q ($q \geqq 2$ *in the projective case*, $q \geqq 3$ *in the conformal case*). *If* F *has a compact leaf with finite holonomy group, then all the leaves of* F *are compact with finite holonomy group*.

THEOREM 6. *Let* F *be a complete flat projective or conformal foliation of codimension* q ($q \geqq 2$ *in the projective case*, $q \geqq 3$ *in the conformal case*) *of a connected manifold* M. *Then the universal cover of* M *fibers over the sphere* S^q, *the fibers being the leaves of the lifted foliation*.

5. We prove Theorem 1. Let U be an open set in M and let $f: U \to V$ be a submersion defining $F|U$, where V is an open set in $\mathbb{R}^q$. A straightforward computation shows that ∇ induces a linear connection $\overline{\nabla}$ on V such that $f^{-1}(\overline{\nabla}) = \nabla$. Moreover, $\overline{\nabla}\overline{T} = \overline{\nabla}\overline{R} = 0$, where $\overline{T}$ and $\overline{R}$ are the torsion and curvature of $\overline{\nabla}$ and hence we may assume, by shrinking U if necessary, that V is an open set in a simply connected reductive homogeneous space G/H and that $\overline{\nabla}$ is the canonical connection of the second kind. Thus we can find an open cover $\{U_\alpha\}_{\alpha \in A}$

of M such that for each $\alpha \in A$ the leaves of $F|U_\alpha$ are the level sets of a submersion $f_\alpha : U_\alpha \to V_\alpha$, where V_α is an open set in a simply connected reductive homogeneous space $(G/H)_\alpha$ and $f_\alpha^{-1}(\overline{\nabla}_\alpha) = \nabla$, where $\overline{\nabla}_\alpha$ is the canonical connection of the second kind on $(G/H)_\alpha$. If $U_\alpha \cap U_\beta \neq \emptyset$, there is a diffeomorphism $g_{\alpha\beta} : f_\beta(U_\alpha \cap U_\beta) \to f_\alpha(U_\alpha \cap U_\beta)$ satisfying $f_\alpha = g_{\alpha\beta} \circ f_\beta$ on $U_\alpha \cap U_\beta$ and an elementary argument shows that $g_{\alpha\beta}$ is an affine transformation. Since $(G/H)_\alpha$, $(G/H)_\beta$ are simply connected analytic manifolds and $\overline{\nabla}_\alpha$, $\overline{\nabla}_\beta$ are complete analytic linear connections, $g_{\alpha\beta}$ can be uniquely extended to an affine isomorphism from $(G/H)_\beta$ to $(G/H)_\alpha$. Since M is arcwise connected it is easy to see that $(G/H)_\alpha$ and $(G/H)_\beta$ are affinely isomorphic for all α, $\beta \in A$. Hence there exists a simply connected reductive homogeneous space G/H such that F is defined by a (G/H)-cocycle $\{(U_\alpha, f_\alpha, g_{\alpha\beta})\}_{\alpha,\beta \in A}$ such that $f_\alpha^{-1}(\overline{\nabla}) = \nabla$, where $\overline{\nabla}$ is the canonical connection of the second kind on G/H and each $g_{\alpha\beta}$ is the restriction of an affine isomorphism of G/H. A monodromy argument now yields a submersion $f : \tilde{M} \to G/H$ which is constant along the leaves of $p^{-1}(F)$ and satisfies $f^{-1}(\overline{\nabla}) = p^{-1}(\nabla)$. Using the completeness of $p^{-1}(\nabla)$, one can show that f is a locally trivial bundle map.

6. We prove Theorem 3. Let Y be a vector field on P such that $\omega(Y)$ is constant. Let $X \in \Gamma(\tilde{E})$. Then $-\frac{1}{2}\omega[X,Y] = \frac{1}{2}(X\omega(Y) - Y\omega(X) - \omega[X,Y])$ $= d\omega(X,Y) = (L_X\omega)(Y) - (d(\omega(X)))(Y) = 0$ and so $[X,Y] \in \Gamma(\tilde{E})$. Let $X_1,\dots,X_r$ be a basis of $\mathfrak{g}$ and let $Y_1,\dots,Y_r$ be complete vector fields on P such that $\omega(Y_i) = X_i$ for $i = 1,\dots,r$. For $i = 1,\dots,r$ let ϕ_t^i, $t \in \mathbb{R}$, be the flow generated by Y_i. Since $[\Gamma(\tilde{E}),Y_i] \subset \Gamma(\tilde{E})$ it follows that ϕ_t^i maps leaves of $\tilde{F}$ to leaves of $\tilde{F}$. The group generated by the diffeomorphisms ϕ_t^i acts transitively on the set of leaves of each connected component of P. Since we can get from one component of P to another by a suitable element of H, it follows that all the leaves of $\tilde{F}$ are diffeomorphic. Let L be a leaf of F. Then $\tilde{E}$ defines a flat connection in the principal H-bundle $\pi^{-1}(L) \to L$. Thus if $\tilde{L}$ is any leaf of $\tilde{F}$ satisfying $\pi(\tilde{L}) = L$, then $\pi : \tilde{L} \to L$ is a principal bundle whose structure group is the holonomy group of the flat connection $\tilde{E}$ in $\pi^{-1}(L) \to L$. Hence $\pi : \tilde{L} \to L$ is a regular covering and so all the leaves of F have the same universal cover.

References

[1] BLUMENTHAL, R.A., Foliations with locally reductive normal bundle, Illinois J. of Math., to appear.

[2] ——, Cartan connections in foliated bundles, Michigan Math. J., to appear.

[3] BOTT, R., Lectures on characteristic classes and foliations, Lectures Notes in Math. 279, Springer-Verlag, New York 1972, 1-80.

[4] EHRESMANN, C., Les connexions infinitésimales dans un espace fibré différentiable, Colloq. Topologie, Bruxelles 1950, 29-55.

[5] FÉDIDA, E., Sur la théorie des feuilletages associée au repère mobile: cas des feuilletages de Lie, Lectures Notes in Math. 652, Springer-Verlag, Berlin 1976, 183-195.

[6] KAMBER, F.W. and P. TONDEUR, Foliated bundles and characteristic classes, Lecture Notes in Math. 493, Springer-Verlag, Berlin 1975.

[7] KOBAYASHI, S., Espaces à connexion de Cartan complets, Proc. Japan Acad. Sci. 30 (1954), 709-710.

[8] ——, Espaces à connexions affines et riemanniens symmétriques, Nagoya Math. J. 9 (1955), 25-37.

[9] MOLINO, P., Propriétés cohomologiques et propriétés topologiques des feuilletages à connexion transverse projetable, Topology 12 (1973), 317-325.

[10] REEB, G., Sur certaines propriétés topologiques des variétés feuilletées, Actualités Sci. Indust. 1183, Hermann, Paris 1952.

[11] REINHART, B., Foliated manifolds with bundle-like metrics, Annals of Math. 69 (1959), 119-132.

Department of Mathematics
Saint Louis University
Saint Louis, MO 63103

A CONTRIBUTION TO KELLER'S JACOBIAN CONJECTURE

Zygmunt Charzyński, Jacek Chądzyński
and Przemysław Skibiński (Łódź)

Contents

Summary. In the paper there have been investigated polynomial mappings $(P,Q) : \mathbb{C}^2 \to \mathbb{C}^2$ in the aspect of the connection between the structure of the jacobian and the coordinates of the mapping. There have been obtained some informations on coefficients of an expansion of the Newton-Puiseux type of one of the coordinates with respect to the other one. Starting from these informations, a theorem is proved stating that if d denotes the degree of the jacobian of the mapping (P,Q), then each of the coordinates P and Q has at most $d + 2$ zeros at infinity. There have been obtained some equations connecting the homogeneous components of the polynomials P and Q.

Introduction

In the circle of the well-known Keller's jacobian conjecture (see [2], [3]) there are investigated polynomial mappings with a constant jacobian.

A Contribution to Keller's Jacobian Conjecture

In this paper there are considered more general polynomial mappings $(P,Q) : \mathbb{C}^2 \to \mathbb{C}^2$ for which the jacobian may be a polynomial of an arbitrary degree. The influence of the jacobian on the structure of the mapping is studied here.

Starting from an expansion of the Newton-Puiseux type of one of the coordinates with respect to the other one, there are obtained some equations connecting the homogeneous components of both the coordinates P and Q (Theorem 2.1). Hence some informations about the form of the coefficients of the expansion are derived (Theorem 2.2).

Next, it is stated, through differentiation and use of the properties of the jacobian, that some of these coefficients are constants. Moreover, it is observed that some of the coefficients simply vanish (Theorems 2.3 and 3.1), particularly, those with the indices relatively prime with the degree of the correspondent coordinate (Corollary 3.2).

The critical coefficient from the border line between the constant and nonconstant coefficients is examined in detail (Theorem 4.1). Hence, as a consequence, the following result is obtained: if d denotes the degree of the jacobian of the mapping (P,Q), then either of the coordinates P and Q has at most $d+2$ zeros at infinity; what is more, the zeros are identical in both the coordinates (Theorem 5.1). The particular case of the above result for mappings with a constant jacobian was obtained earlier in some other way (the remark following Theorem 5.1).

In a similar direction researches of mappings with the constant jacobian were carried out in [4] by A. Magnus who obtained the results crossing those of this paper.

1. Auxiliary results

Let a holomorphic mapping (U,V) of the variables (x,y) and t ranging over an open set in $\mathbb{C}^2$ and over an open set in $\mathbb{C}$, respectively, be given. Besides, let C be a given closed regular curve lying in the domain of variability of t. Let further M be a holomorphic function of the variable (ξ,η) ranging over an open set in $\mathbb{C}^2$, containing the image of the domain of definition of the mapping (U,V).

LEMMA 1.1. *Under the above assumptions, the following equality holds:*

$$(1.1)\qquad \frac{\partial}{\partial x}\int_C M(U,V)\,\frac{\partial U}{\partial t}\,dt = \int_C \frac{\partial M}{\partial V}\,\frac{D(V,U)}{D(x,t)}\,dt \qquad \textit{where}$$

$$\frac{D(V,U)}{D(x,t)} = \frac{\partial V}{\partial x}\frac{\partial U}{\partial t} - \frac{\partial V}{\partial t}\frac{\partial U}{\partial x} .$$

P r o o f. The difference between the left- and right-hand sides of (1.1) is equal to

$$\int_C \left(\frac{\partial M}{\partial U}\frac{\partial U}{\partial t}\frac{\partial U}{\partial x} + \frac{\partial M}{\partial V}\frac{\partial V}{\partial t}\frac{\partial U}{\partial x} + M \frac{\partial^2 U}{\partial x \partial t} \right) dt.$$

Now, taking into account the equality of the mixed partial derivatives, we can write the above integral in the form

$$\int_C \frac{\partial}{\partial t}\left(M \frac{\partial U}{\partial x} \right) dt.$$

Since the curve C is closed, the last integral vanishes. This completes the proof.

Let us notice that formula (1.1) is still valid if we replace in it the partial derivative with respect to x by the partial derivative with respect to y; the same thing takes place when we replace the derivative $\partial U/\partial t$ on the left-hand side by the derivative $\partial V/\partial t$ with the simultaneous replacement of the derivative $\partial M/\partial V$ by $-\partial M/\partial U$ on the right-hand side.

2. Polynomial mappings. Expansion of the Newton-Puiseux type

We shall show here some relations between the coordinates of polynomial mappings, connected with the form of the corresponding jacobian.

Let $(P,Q) : \mathbb{C}^2 \to \mathbb{C}^2$ be a polynomial mapping and $J : \mathbb{C}^2 \to \mathbb{C}$ the jacobian of this mapping.

Let further

(2.1) $P(\xi,\eta) = P_m(\xi,\eta) + \ldots + P_1(\xi,\eta),$

(2.2) $Q(\xi,\eta) = Q_n(\xi,\eta) + \ldots + Q_1(\xi,\eta),$

(2.3) $J(\xi,\eta) = E_d(\xi,\eta) + \ldots + E_0(\xi,\eta),$

where the right-hand sides are expansions in the sums of homogeneous polynomials of the successive degrees when the components of the highest degrees P_m, Q_n do not vanish identically.

Let us consider, for instance, the polynomial P. Let us further take a point (x_0,y_0) such that $P_m(x_0,y_0) \neq 0$ and a neighbourhood Ω

of this point so small that the lower bound of $|P_m|$ in Ω is positive and there exist in Ω a branch of the m-th root $P_m^{1/m}$ and corresponding branches of the powers $P_m^{h/m}$ where h are integers.

Let us form the functions

$$(2.4)\qquad \tilde{P}(x,y,t) = P(xt,yt) = t^m P_m(x,y) + \ldots + tP_1(x,y),$$

$$(2.5)\qquad \tilde{Q}(x,y,t) = Q(xt,yt) = t^n Q_n(x,y) + \ldots + tQ_1(x,y),$$

$$(2.6)\qquad \tilde{J}(x,y,t) = J(xt,yt) = t^d E_d(x,y) + \ldots + E_0(x,y)$$

for arguments (x,y) ranging over Ω and for $t \in \mathbb{C}$ with a sufficiently great modulus. Then, as one can see, in the domain of variability of the arguments there exists a holomorphic branch of the m-th root of (2.4)

$$(2.7)\qquad s = \tilde{p}(x,y,t) = tP_m^{1/m}(x,y)\left[1 + \frac{P_{m-1}(x,y)}{P_m(x,y)}\frac{1}{t} + \ldots + \frac{P_1(x,y)}{P_m(x,y)}\frac{1}{t^{m-1}}\right]^{\frac{1}{m}}$$

with the natural choice of the branch of the m-th root of the expression in square brackets.

Moreover, with arbitrary but fixed (x,y) and with variable t, we can determine the inverse function of (2.7)

$$(2.8)\qquad t = \tilde{p}^{-1}(x,y,s)$$

defined and holomorphic in the corresponding domain of variability of the arguments, in particular, for (x,y) ranging over Ω and for s with a sufficiently great modulus. In this situation, we have the expansion

$$(2.9)\qquad \tilde{Q}(x,y,p^{-1}(x,y,s)) = u_n(x,y)s^n + u_{n-1}(x,y)s^{n-1} + \ldots + u_0(x,y) + \\ + u_{-1}(x,y)s^{-1} + \ldots$$

for (x,y) ranging over Ω and s with a sufficiently great modulus, where

$$(2.10)\qquad u_k(x,y) = \frac{1}{2\pi i}\int_D \frac{Q(x,y,p^{-1}(x,y,s))}{s^{k+1}}\,ds, \qquad k = n, n-1, \ldots$$

and D denotes an arbitrary Jordan curve surrounding the point ∞, lying in the domain of variability of s.

Taking account of (2.7), we obtain, by (2.9),

(2.11) $$\tilde{Q}(x,y,t) = u_n(x,y)(\tilde{p}(x,y,t))^n + \ldots + u_0(x,y) + u_{-1}(x,y)(\tilde{p}(x,y,t))^{-1} + \ldots$$

for $(x,y) \in \Omega$ and t with a sufficiently great modulus.

THEOREM 2.1 (comp. also [4]. Between the homogeneous components of the coordinates of the mapping (P,Q) from (2.1), (2.2) and the coefficients u_k from (2.10) of the expansion of the Newton-Puiseux type (2.9) the following effective relations hold

(2.12) $$\frac{Q_\ell(x,y)}{P_m^{\ell/m}(x,y)} = \sum_{k,\lambda_{m-1},\ldots,\lambda_1} u_k(x,y) \binom{\frac{k}{m}}{\lambda_{m-1}+\ldots+\lambda_1} \frac{(\lambda_{m-1}+\ldots+\lambda_1)!}{\lambda_{m-1}!\cdot\ldots\cdot\lambda_1!} \times \left(\frac{P_{m-1}(x,y)}{P_m^{(m-1)/m}(x,y)}\right)^{\lambda_{m-1}} \cdot\ldots\cdot \left(\frac{P_1(x,y)}{P_m^{1/m}(x,y)}\right)^{\lambda_1},$$

$\ell = n, n-1, \ldots,$

where we assume $Q_\ell = 0$ for $\ell < 0$ and where the sum is taken over all integers satisfying the following conditions

(2.13) $$\ell \leq k \leq n, \quad \lambda_{m-1} \geq 0, \ldots, \lambda_1 \geq 0; \quad \lambda_{m-1} + 2\lambda_{m-2} + \ldots + (m-1)\lambda_1 = k - \ell.$$

P r o o f. By virtue of (2.7), relation (2.11) can be written in the form

(2.14) $$\tilde{Q}(x,y,t) = \sum_{k=-\infty}^{n} u_k(x,y) t^k P_m^{k/m}(x,y) \left[1 + \frac{P_{m-1}(x,y)}{P_m(x,y)} \frac{1}{t} + \ldots + \frac{P\ (x,y)}{P_m(x,y)} \frac{1}{t^{m-1}}\right]^{k/m}$$

Enlarging, if necessary, the modulus of t and using twice the Newton formula, we can at first write the right-hand side of (2.14) in the form

$$\sum_{k=-\infty}^{n} u_k(x,y)t^k P_m^{k/m}(x,y) \sum_{j=0}^{\infty} \binom{\frac{k}{m}}{j} \left[\frac{P_{m-1}(x,y)}{P_m(x,y)} \frac{1}{t} + \ldots + \frac{P_1(x,y)}{P_m(x,y)} \frac{1}{t^{m-1}}\right]^j$$

and next, in the form

$$(2.15) \qquad \sum_{k=-\infty}^{n} u_k(x,y)t^k P_m^{k/m}(x,y) \sum_{j=0}^{\infty} \binom{\frac{k}{m}}{j} \sum_{\lambda_{m-1},\ldots,\lambda_1} \frac{j!}{\lambda_{m-1}! \cdot \ldots \cdot \lambda_1!} \times$$

$$\times \left(\frac{P_{m-1}(x,y)}{P_m(x,y)} \frac{1}{t}\right)^{\lambda_{m-1}} \cdot \ldots \cdot \left(\frac{P_1(x,y)}{P_m(x,y)} \frac{1}{t^{m-1}}\right)^{\lambda_1}$$

where the last sum is taken over all integers fulfilling the following conditions

$$\lambda_{m-1} \geqq 0, \ldots, \lambda_1 \geqq 0; \quad \lambda_{m-1} + \ldots + \lambda_1 = j.$$

Further, we can rewrite this in the form

$$\sum_{k=-\infty}^{n} \sum_{\lambda_{m-1},\ldots,\lambda_1} u_k(x,y) t^{k-\lambda_{m-1}-2\lambda_{m-2}-\ldots-(m-1)\lambda_1} P_m^{k/m}(x,y) \times$$

$$\times \binom{\frac{k}{m}}{\lambda_{m-1}+\ldots+\lambda_1} \frac{(\lambda_{m-1}+\ldots+\lambda_1)!}{\lambda_{m-1}! \cdot \ldots \cdot \lambda_1!} \left(\frac{P_{m-1}(x,y)}{P_m(x,y)}\right)^{\lambda_{m-1}} \cdot \ldots \cdot \left(\frac{P_1(x,y)}{P_m(x,y)}\right)^{\lambda_1}$$

where the second sum is taken over all non-negative integers $\lambda_{m-1},\ldots,\lambda_1$. Finally, ordering with respect to ℓ, we can write the right-hand side of (2.14) in the form

$$\sum_{\ell=-\infty}^{n} t^{\ell} \sum_{k=\ell}^{n} \sum_{\lambda_{m-1},\ldots,\lambda_1} u_k(x,y) P_m^{k/m}(x,y) \binom{\frac{k}{m}}{\lambda_{m-1}+\ldots+\lambda_1} \frac{(\lambda_{m-1}+\ldots+\lambda_1)!}{\lambda_{m-1}! \cdot \ldots \cdot \lambda_1!} \times$$

$$\times \left(\frac{P_{m-1}(x,y)}{P_m(x,y)}\right)^{\lambda_{m-1}} \cdot \ldots \cdot \left(\frac{P_1(x,y)}{P_m(x,y)}\right)^{\lambda_1}$$

where the third sum is taken over all integers satisfying conditions (2.13).

Simultaneously, the left-hand side of (2.14) can be written according to (2.5), in the form

$$\sum_{\ell=1}^{n} t^{\ell} Q_{\ell}(x,y).$$

Comparing the coefficients at t on both sides, we obtain

$$(2.16)\quad Q_{\ell}(x,y) = \sum_{k,\lambda_{m-1},\ldots,\lambda_1} u_k(x,y)\, P_m^{k/m}(x,y) \binom{\frac{k}{m}}{\lambda_{m-1}+\ldots+\lambda_1} \times$$

$$\times \frac{(\lambda_{m-1}+\ldots+\lambda_1)!}{\lambda_{m-1}!\cdot\ldots\cdot\lambda_1!} \left(\frac{P_{m-1}(x,y)}{P_m(x,y)}\right)^{\lambda_{m-1}} \cdot\ldots\cdot \left(\frac{P_1(x,y)}{P_m(x,y)}\right)^{\lambda_1}$$

From (2.16) we immediately arrive at (2.12).

Now, let us consider the factorization into prime factors of the homogeneous component of the highest degree P_m of (2.1)

$$(2.17)\quad P_m(x,y) = \prod_{j=1}^{r} (\varkappa_j x - \omega_j y)^{\gamma_j}$$

where γ_j are positive integers.

THEOREM 2.2. *For every* k, *either the coefficient* u_k *vanishes identically or the following equality holds*

$$(2.18)\quad u_k(x,y) P_m^{k/m}(x,y) = \frac{A_k(x,y)}{B_k(x,y)}$$

where A_k *and* B_k *are relatively prime homogeneous polynomials of two variables non-vanishing identically*, *Moreover, the denominator* B_k *does not contain other prime factors than those from* (2.17), *i.e.*

$$(2.19)\quad B_k(x,y) = b_k \prod_{j=1}^{r} (\varkappa_j x - \omega_j y)^{\beta_j^{(k)}}$$

where $b_k \neq 0$, $\beta_j^{(k)} \geq 0$.

What is more, the difference of degrees of A_k *and* B_k *is equal to* k.

Proof. Putting in (2.12) $\ell = n$, we obtain

$$(2.20)\quad Q_n(x,y) = u_n(x,y)P_m^{n/m}(x,y).$$

Hence, it is easily seen that, for $k = n$, it suffices to put $A_n(x,y) = Q_n(x,y)$ and $B_n(x,y) = 1$.

Assume that (2.18) holds for $\ell + 1 \leq k \leq n$. Because of (2.16), we have

$$(2.21)\quad Q_\ell(x,y) = u_\ell(x,y)P_m^{\ell/m}(x,y) + \sum_{k=\ell+1}^{n} \sum_{\lambda_{m-1},\ldots,\lambda_1} u_k(x,y)P_m^{k/m}(x,y) \times$$
$$\times \binom{\frac{k}{m}}{\lambda_{m-1},\ldots,\lambda_1} \frac{(\lambda_{m-1}+\ldots+\lambda_1)!}{\lambda_{m-1}!\cdot\ldots\cdot\lambda_1!} \left(\frac{P_{m-1}(x,y)}{P_m(x,y)}\right)^{\lambda_{m-1}} \cdot\ldots\cdot \left(\frac{P_1(x,y)}{P_m(x,y)}\right)^{\lambda_1}.$$

We notice that the expressions $(P_{m-j}/P_m)^{\lambda_{m-1}}$, $j = 1,\ldots,m-1$, either vanish identically or are some quotients of homogeneous polynomials non-vanishing identically with the difference of degrees equal to $-j\lambda_{m-j}$; simultaneously, according to the assumption, the expressions $u_k(x,y)P_m^{k/m}(x,y)$, analogously, either vanish identically or are some quotients of homogeneous polynomials non-vanishing identically with the difference of degrees equal to k. Therefore, after rejecting the summands vanishing identically on the right-hand side of (2.21), under the summation sign there possibly remain, because of (2.13), expressions being some quotients of homogeneous polynomials non-vanishing identically with the difference of degrees equal to $k + (\ell-k) = \ell$. So, the above-mentioned sum is either equal identically to zero or is a quotient of homogeneous polynomials non-vanishing identically with the difference of degrees equal to ℓ. Hence, according to (2.21), we obtain that the expression $u_\ell(x,y)P_m^{\ell/m}(x,y)$ either vanishes identically or, after some reductions, is a quotient of relatively prime homogeneous polynomials non-vanishing identically with the difference of degrees equal to ℓ. At the same time, the polynomial from the denominator does not contain any other factors than these of (2.17). This follows immediately from the inductional assumption and from the shape of the terms under the summation sign on the right-hand side of (2.21) and the form of the term on the left-hand side of (2.21). Thus (2.18) holds for $k = \ell$ and, as a consequence, for $\ell \leq k \leq n$.

Induction ends the proof.

In the sequel, we shall assume that the degree d of the jacobian J of the mapping (P,Q) satisfies the inequality $d \leq m + n - 3$.

THEOREM 2.3. *If* $-(m-3) + d \leq k \leq n$, *then the coefficients* $u_n(x,y)$ *from* (2.10) *are constant for* $(x,y) \in \Omega$; *moreover,* $u_n(x,y) \neq 0$.

Proof. From (2.9), by substitution (2.7) and a suitable choice of the curve D, we obtain

$$(2.22)\quad u_k(x,y) = \frac{1}{2\pi i}\int_C \tilde{Q}(x,y,t)(\tilde{p}(x,y,t))^{-k-1}\frac{\partial}{\partial t}\tilde{p}(x,y,t)dt, \qquad k=n,n-1,\ldots,$$

where C denotes an arbitrary circle with centre at the origin, lying in the domain of variability of t.

Now, making use of Lemma 1.1, we shall evaluate the partial derivative $(\partial/\partial x)u_k(x,y)$. Namely, putting in (1.1)

$$U(x,y,t) = \tilde{p}(x,y,t), \qquad V(x,y,t) = \tilde{Q}(x,y,t), \qquad M(U,V) = U^{-k-1}V$$

and taking into account that, by (2.4), (2.5) and (2.6),

$$\frac{D(V,U)}{D(x,t)} = -\frac{y}{m}t(\tilde{p}(x,y,t))^{1-m}\tilde{J}(x,y,t),$$

we obtain, by virtue of (2.22),

$$(2.23)\quad \frac{\partial u_k(x,y)}{\partial x} = -\frac{1}{2\pi i}\frac{y}{m}\int_C t(\tilde{p}(x,y,t))^{-k-m}\tilde{J}(x,y,t)dt.$$

On the ground of (2.5) and (2.6), the right-hand side of (2.23) can be rewritten in the form

$$(2.24)\quad -\frac{1}{2\pi}\frac{y}{m}\int_C t^{-k-m+1}P_m^{(-k-m)/m}(x,y)\left[1+\frac{P_{m-1}(x,y)}{P_m(x,y)}\frac{1}{t}+\ldots+\right.$$

$$\left.+\frac{P_1(x,y)}{P_m(x,y)}\frac{1}{t^{m-1}}\right]^{\frac{-k-m}{m}}\left[E_d(x,y)t^d+\ldots+E_o(x,y)\right]dt.$$

Since, for k fulfilling the assumptions of the theorem, we have in the integrand only the terms with negative powers of t smaller than -1, therefore the integral must vanish for these k. Thus

$$(2.25)\qquad \frac{\partial u_k(x,y)}{\partial x} = 0, \qquad -(m-3)+d \leq k \leq n.$$

In a similar way we show that

$$(2.26)\qquad \frac{\partial u_k(x,y)}{\partial y} = 0, \qquad -(m-3)+d \leq k \leq n.$$

From (2.25) and (2.26) it follows that u_k is constant.

The last part of our assertion results directly from (2.20) and from the assumption that Q_n does not vanish identically.

Let us now notice that if we confine ourselves in Theorem 2.1 to ℓ satisfying the inequality $\ell \geq -(m-3)+d$, we may treat the coefficients u_k as constants in the effective relations, according to Theorem 2.2.

Let us further notice that by confining ourselves additionally to negative ℓ, we obtain the effective relations for homogeneous components of the polynomial P only.

3. Study of the constant coefficient

We shall give here some theorems on the vanishing of some constant coefficients from (2.10) for $-(m-3)+d \leq k \leq n$. For this purpose, we shall consider factorization (2.17). Let δ denote the greatest common factor of $\gamma_1,\dots,\gamma_r$ and let $\gamma_j = \sigma_j\delta$, $j = 1,\dots,r$; then $\sigma_1,\dots,\sigma_r$ are relatively prime numbers. By virtue of the condition $\gamma_1 + \dots + \gamma_r = m$, we obtain

$$(3.1)\qquad m = \delta(\sigma_1 + \dots + \sigma_r).$$

<u>THEOREM 3.1</u>. <u>If</u> k <u>is not divisible by</u> $\sigma_1 + \dots + \sigma_r$, <u>then</u> $u_k = 0$.

<u>Proof</u>. From Theorems 2.2 and 2.3 it follows that either u_k vanishes identically or

$$(3.2)\qquad (u_k(x,y))^m(P_m(x,y))^k(B_k(x,y))^m = (A_k(x,y))^m$$

where A_k and B_k do not vanish identically and are relatively prime, while u_k is a constant different from zero. Suppose that equality (3.2) holds. Then B_k^m is a divisor of A_k^m, but this is possible only in the case when B_k is a constant. At the same time, the degree of A_k is equal to k. From the assumptions of the theorem it follows that $k \neq 0$. So, admitting that the factorization of A_k has the form

$$(3.3)\qquad A_k(x,y) = a_k \prod_{i=1}^{s^{(k)}} (\nu_i^{(k)}x - \tau_i^{(k)}y)^{\alpha_i^{(k)}}, \quad \alpha_i^{(k)} > 0, \quad a_k \neq 0, \quad i=1,\dots,s^{(k)},$$

we can write (3.2), taking into account (2.17) and (2.19), in the form

$$(3.4)\qquad (u_k(x,y))^m b_k^m \prod_{j=1}^{r} (\varkappa_j x - \omega_j y)^{k\gamma_j} = a_k^m \prod_{i=1}^{s^{(k)}} (\nu_i^{(k)}x - \tau_i^{(k)}y)^{\alpha_i^{(k)}m} .$$

From the uniqueness of the factorization we conclude that the terms on both sides of (3.4) have the same prime factors with the same power exponents.

This, because of (3.1), gives after dividing by δ:

$$k\sigma_j = \alpha_j^{(k)}(\sigma_1 + \dots + \sigma_r), \qquad j = 1,\dots,r.$$

Hence it follows that each power of a prime number dividing the sum $\sigma_1 + \dots + \sigma_r$ is also a diviser of k, since, in the opposite case, this power of the prime number would be the divisor of all numbers σ_j, $j = 1,\dots,r$. and they would not be relatively prime. Thus $\sigma_1 + \dots \sigma_r$ is the divisor of k, but this contradicts the assumption of the theorem. So our supposition was wrong. This implies that u_k vanishes identically.

COROLLARY 3.1. *If* $\sigma_1 + \dots + \sigma_r > 1$, *then, for* k *relatively prime with* m, *there is* $u_k = 0$.

Indeed, from (3.1) and the definition of σ_j, $j = 1,\dots,r$, it follows that $\sigma_1 + \dots + \sigma_r$ is a divisor of m. Thus, using the assumptions, we conclude that k is not divisible by $\sigma_1 + \dots + \sigma_r$.

This, according to Theorem 3.2, gives our corollary.

From the above we easily obtain

COROLLARY 3.2. *If in factorization (2.17) there are at least two factors, i.e.* $r \geq 2$, *then, for* k *relatively prime with* m, *there is* $u_k = 0$.

This, by the inequality $u_n \neq 0$ (Comp. Theorem 2.3), immediately gives

COROLLARY 3.3. *If in factorization (2.17) there are at least two factors, i.e.* $r \geq 2$, *then the degrees* m *and* n *of coordinates (2.1) and (2.2) cannot be relatively prime.*

COROLLARY 3.4. *If in factorization (2.17) there are at least two factors, i.e.* $r \geq 2$, *and the degree* m *of coordinate (2.1) is a prime number, then the degree* n *of coordinate (2.2) must be divisible by* m.

4. Critical coefficient

Here we shall study the first coefficient from (2.9), not being a constant, namely $u_{-(m-2)+d}$, which will be called the critical coefficient and denoted by u^*.

Beginning from this section, we assume in the sequel that the homogeneous component of the highest degree E_d of the jacobian J (comp. (2.3)) does not vanish identically.

At first, let us notice that

$$(4.1') \quad \frac{\partial u^*(x,y)}{\partial x} = -\frac{y}{m} \frac{E_d(x,y)}{P_m^{(2+d)/m}(x,y)} ,$$

$$(4.1'') \quad \frac{\partial u^*(x,y)}{\partial y} = \frac{x}{m} \frac{E_d(x,y)}{P_m^{(2+d)/m}(x,y)} .$$

Indeed, because of (2.23) and (2.7), we have

$$(4.2) \quad \frac{\partial u^*(x,y)}{\partial x} = -\frac{1}{2\pi i} \frac{y}{m} \int_C t^{-1-d} P_m^{(-2-d)/m}(x,y) \times$$

$$\times \left[1 + \frac{P_{m-1}(x,y)}{P_m(x,y)} \frac{1}{t} + \ldots + \frac{P_1(x,y)}{P_m(x,y)} \frac{1}{t^{m-1}}\right]^{(-2-d)/m} \times$$

$$\times \; t^d \left[E_d(x,y) + E_{d-1}(x,y) \frac{1}{t} + \ldots + E_o(x,y) \frac{1}{t^d}\right] dt,$$

which immediately gives (4.1′). Similarly we obtain (4.1″).

Next, let us notice that, according to (4.1′) and (4.1″), $u^*(x,y)$ does not vanish identically since $E_d(x,y)$ does not vanish identically.

With reference to (2.19), let us put

$$(4.3) \qquad A^*(x,y) = A_{-(m-2)+d}(x,y), \qquad B^*(x,y) = B_{-(m-2)+d}(x,y),$$

and, by referring further to (3.3) and (2.17), let

$$(4.4) \qquad \begin{aligned} A^*(x,y) &= a^* \prod_{i=1}^{s^*} (\nu_i x - \tau_i y)^{\alpha_i^*}, \\ B^*(x,y) &= b^* \prod_{j=1}^{r} (\varkappa_j x - \omega_j y)^{\beta_j^*} \end{aligned}$$

denote factorizations (3.3) and (2.19), respectively.

According to (2.18), we have

$$(4.5) \qquad u^*(x,y)\, P_m^{(2+d)/m}(x,y) = \frac{A^*(x,y)}{B^*(x,y)}.$$

THEOREM 4.1. *In equality (4.5) the term on the right-hand side is a polynomial of degree* $d + 2$. *Moreover, this polynomial is divisible by the product* $\prod_{j=1}^{r} (\varkappa_j x - \omega_j y)$ *of all prime factors from factorization (2.17) of the homogeneous component* P_m.

P r o o f. Differentiating both sides of (4.5) with respect to x and taking account of (4.1′), we get

$$(4.6)\qquad \frac{\partial(A^*(x,y))/\partial x}{A^*(x,y)} - \frac{\partial(B^*(x,y))/\partial x}{B^*(x,y)} = -\frac{y}{m}\,\frac{E_d(x,y)B^*(x,y)}{A^*(x,y)} + \frac{2+d}{m}\,\frac{\partial(P_m(x,y))/\partial x}{P_m(x,y)} .$$

Taking (4.4) into account, we easily check that the member on the left-hand side of (4.6) is equal to

$$(4.7)\qquad \sum_{i=1}^{s^*} \frac{\alpha_i^* \nu_i^*}{\nu_i^* x - \tau_i^* y} - \sum_{j=1}^{r} \frac{\beta_j^* \varkappa_j}{\varkappa_j x - \omega_j y} .$$

At the same time, according to (2.17), there is

$$(4.8)\qquad \frac{\partial(P_m(x,y))/\partial x}{P_m(x,y)} = \sum_{j=1}^{r} \frac{\gamma_j \varkappa_j}{\varkappa_j x - \omega_j y} .$$

Substituting (4.7) and (4.8) in (4.6), after an easy transposition of the suitable terms, we obtain

$$(4.9')\qquad \frac{yE_d(x,y)}{m}\,\frac{b^* \prod_{j=}^{r} (\varkappa_j x - \omega_j y)^{\beta_j^*}}{a^* \prod_{i=1}^{s^*} (\nu_i^* x - \tau_i^* y)^{\alpha_i^*}} + \sum_{i=1}^{s^*} \frac{\alpha_i^* \nu_i^*}{\nu_i^* x - \tau_i^* y} = \sum_{j=1}^{r} \frac{\varkappa_j \left(\frac{2+d}{m} \gamma_j + \beta_j^* \right)}{\varkappa_j x - \omega_j y} .$$

Analogously, differentiating both sides of (4.5) with respect to y and taking account of (4.1″), we get

$$(4.9'')\qquad \frac{xE_d(x,y)}{m}\,\frac{b^* \prod_{j=1}^{r} (\varkappa_j x - \omega_j y)^{\beta_j^*}}{a^* \prod_{i=1}^{s^*} (\nu_i^* x - \tau_i^* y)^{\alpha_i^*}} + \sum_{i=1}^{s^*} \frac{\alpha_i^* \tau_i^*}{\nu_i^* x - \tau_i^* y} =$$

$$= \sum_{j=1}^{r} \frac{\omega_j \left(\frac{2+d}{m}\gamma_j + \beta_j^*\right)}{\varkappa_j x - \omega_j y} .$$

At the same time, one can readily see that equalities (4.9′) and (4.9″) holding in the neighbourhood Ω of the point (x_0, y_0) can be extended to the whole space $\mathbb{C}^2$, excluding the set of zeros of the polynomial P_m. Consider an arbitrary factor $(\varkappa_j x - \omega_j y)$ and the point $(\omega_j, \varkappa_j)$ being a zero of this factor. Let us investigate either equality (4.9′) or (4.9″) according to whether $\varkappa_j$ or ω_j is different from zero. We see that the right-hand side of the equality is there unbounded; this implies that also the left-hand side of this equality is unbounded. But this is only possible when one of the prime factors from the first of relations (4.4) coincides with the factor $\varkappa_j x - \omega_j y$. Because of the arbitrariness of the latter, we see that among the prime factors from the first of relations (4.4) there must be all those, from (2.17), in other words, the polynomial $A^*(x,y)$ must be divisible by the product $\Pi_{j=1}^{r} (\varkappa_j x - \omega_j y)$. From this we further conclude that the function B^* reduces to a constant since, in the opposite case, the polynomials A^* and B^* would have a common factor, which contradicts the assertion of Theorem 2.2. Hence, according to Theorem 2.2, the degree of the polynomial A^* is equal to $d + 2$. Collecting all these informations, we obtain the assertion of our theorem.

Finally, let us notice that the results analogous to those from sections 2,3 and 4 can be obtained by changing the roles of the coordinates P and Q in the expansion of the Newton-Puiseux type.

5. Structure of the homogeneous components of the highest degree in the coordinates of the mapping

We shall give here a relation between the degree d of jacobian (2.3) and the terms P_m and Q_n from (2.1) and (2.2).

THEOREM 5.1. *In the factorizations of the components of the highest degree* P_m *and* Q_n *of coordinates* (2.1) *and* (2.2) *the number of different factors in either factorization is at most* $d + 2$. *Moreover, in both factorizations there are the same factors whose power exponents* are proportional to m and n.

P r o o f. From Theorem 4.1 it follows that in factorization (2.17) of P_m there are at most $d+2$ prime factors. Changing the roles of the coordinates, we obtain the analogous result for Q_n. This gives the first part of the assertion of the theorem.

From equality (2.20) we now have

$$(Q_n(x,y))^m = (u_n(x,y))^m (P_m(x,y))^n,$$

which, because of the fact that the coefficient $u_n(x,y)$ is constant and different from zero, immediately gives the second part of the assertion of the theorem.

Let us notice that the theorem stated above can be written in the form: if the polynomial mapping (P,Q) has the jacobian of degree d, then the coordinates P and Q have at most $d+2$ zeros at infinity. Moreover, both coordinates have the same zeros with orders proportional to the degrees of the corresponding coordinates.

In the particular case $d = 0$, from Theorem 5.1 we immediately obtain the earlier result for mappings with the constant jacobian which was derived in some other way (comp. [1], p.139).

R e f e r e n c e s

[1] S.S. ABHYANKAR, Expansion techniques in algebraic geometry, Tata Institute of Fundamental Research, Bombay 1977.

[2] H. BASS, E.H. CONNEL, and D.WRIGHT, The jacobian conjecture: reduction of degree and formal expansion of the inverse, Bull. of Amer.Mat.Soc. Vol.7, No 2, (1982), pp.287-330.

[3] O.H. KELLER, Ganze Cremona-Transformationen, Monats.Math.Physik 47 (1939), pp.299-306.

[4] A. MAGNUS, Volume-preserving transformations in several complex variables, Proc. Amer. Math. Soc. Vol.5 (1954), pp.256-266.

Institute of Mathematics
Łódź University
Banacha 22,
PL-90-238 Łódź, Poland

CONSTRUCTIONS D'ALGÈBRES DE LIE GRADUÉES ORTHOSYMPLECTIQUES ET CONFORMOSYMPLECTIQUES MINKOWSKIENNES

Albert Crumeyrolle (Toulouse)

Table des matières

RÉSUMÉ

Les algèbres de Clifford orthosymplectiques et leurs représentations spinorielles, ou mieux les fibrations correspondantes au dessus de quelque espace-temps, nous permettent de justifier l'introduction des algèbres de Lie graduées en physique mathématique et d'obtenir un cadre réunissant à la fois bosons et fermions. Nous expliciterons ici des principes de construction de telles algèbres de Lie graduées orthosymplectiques et conformosymplectiques minkowskiennes par des méthodes dérivées du principe de trialité de E. Cartan. Les résultats obtenus peuvent en particulier être utiles pour l'étude des déformations d'espace-temps de Minkowski réels ou complexifiés.

INTRODUCTION

Dans un précédent article [3e] nous avons donné les principes généraux de construction d'algèbres de Lie graduées orthosymplectiques en utilisant un principe d'interaction dérivé du "principe de trialité" de E. Cartan [2]. Cet article se propose d'expliciter plus concrètement diverses constructions qui avaient été simplement suggérées ou indiquées dans leur lignes générales et d'en tirer quelques conséquences.

Le but poursuivi est de donner un fondement logique à l'introduction des algèbres de Lie graduées en physique mathématique et d'obtenir ainsi un cadre réunissant tout à la fois bosons et fermions; ce cadre peut être fourni par les algèbres de Clifford orthosymplectiques et leurs représentations spinorielles (spineurs orthogonaux et symplectiques) ou mieux par les fibrations correspondantes au-dessus de quelque espace-temps. Nous pensons qu'une des clefs possibles d'accès à cette conception unitaire se trouve dans le principe de trialité convenablement élargi [3d], laissant ainsi au second plan l'analyse graduée.

Cet article présente donc en détails les cas "conforme-symplectique" et "minkowskien-symplectique" qui se rattachent à une littérature extrêmement abondante remontant aux années 1970, ou dominent les conceptions heuristiques (Volkov et Akulov [6], Wess et Zumino [7] et autres) et qui a conduit à l'introduction des variétés graduées [4], [5] . Après avoir explicité nos tables avec détails nous justifions certain formalisme largement répandu. Nous comparons la notion de variété graduée, dans la version particulière de Kostant, avec celle de préfaisceaux des sections d'un fibré spinoriel au sens large, selon l'optique de nos travaux antérieurs [3], et expliquons à l'aide de notre approche la raison profonde de l'introduction de l'analyse et des dérivations graduées qui correspondent naturellement, dans tous les cas, à des représentations spinorielles, orthogonales ou symplectiques. Ces dernières régissent la quantification des bosons, et les premières celles des fermions faisant apparaître ainsi une remarquable unité des lois naturelles de la physique.

Nous saisissons l'opportunité de la publication de cet article pour signaler quelques erratas ou négligences de rédaction du précédent article [3e]; le lecteur voudra bien noter p. 188, ligne 3, $(-1^{|a|\,|c|})$, p. 189, ligne 11, si $\beta(x) + x = 0$, p. 191, ligne 4, $y_2 = \frac{1}{2}(-ie_2 + e_3)$ et ligne 23 "23 paramètres complexes". On notera aussi dans ce qui suit que l'identification de $\{uf,vf\}$ à $-(uf \circ vf)$ demande quelques précautions, ce qui n'avait pas été signalé antérieurement.

I. RAPPELS SUR LE PRINCIPE DE TRIALITÉ AFFAIBLI ET SUR CERTAINES ALGÈBRES DE LIE GRADUÉES

1. Notations et définitions générales. Le cadre géométrique

Nous utilisons systématiquement l'approche spinorielle que nous avons développée dans des articles antérieurs, [3] et certains résultats classiques sur les algèbres de Clifford et les spineurs [2, 3a].

E est un space vectoriel réel de dimension $n = 2r$, $E_{\mathbb{C}}$ ou E' désigne son complexifié, B est une forme bilinéaire symétrique non dégénerée de signature quelconque, Q la forme quadratique associée: $B(x,x) = Q(x)$, $x \in E$. $C(Q)$ et $C(Q')$ désignent les algèbres de Clifford respectives de (E,Q) et (E',Q'), Q' complexifiée de Q. On appellera f un r-vecteur isotrope déterminant un espace spinoriel $S = S^+ \oplus S^-$, $S = C(Q')f$, $S^{\pm} = S \cap C^{\pm}(Q')$, uf étant l'élément général de S, on écrira $uf = u^+f + u^-f$, $u^{\pm} \in C^{\pm}(Q')$.

Il existe sur $S \times S$ une forme bilinéaire $\mathcal{B}$, telle que si β est l'anti-automorphisme principal de $C(Q')$,

(1) $\mathcal{B}(uf,vf)f = \beta(uf) \cdot vf$, $\mathcal{B}$ est non dégénérée.

On peut aussi introduire, α étant l'automorphisme principal de $C(Q')$, $\tilde{\beta} = \beta \circ \alpha$ et

(2) $\tilde{\mathcal{B}}(uf,vf) = \beta(uf) \cdot vf$; on voit que:

$$\mathcal{B}(uf,vf) = (-1)^{\frac{r(r-1)}{2}} \mathcal{B}(vf,uf), \quad \tilde{\mathcal{B}}(uf,vf) = (-1)^{\frac{r(r+1)}{2}} \tilde{\mathcal{B}}(vf,uf).$$

F sera un sous espace vectoriel de $C(Q')$ tel qu'il soit possible de faire opérer sur lui le groupe $O(Q')$, F sera muni de quelque métrique étendant naturellement Q, par exemple si F est l'algèbre de Lie d'un groupe spinoriel on pourra choisir la forme de Killing.

On écrira aussi $F^{\pm} = F \cap C^{\pm}(Q')$.

2. Principe de trialité affaibli: $n = 2r$, $r = 1,2,3 \pmod 4$

Si $r = 2,3 \pmod 4$ β est antisymétrique, elle est nulle sur $S^+ \times S^-$ et $S^- \times S^+$ si r est pair, nulle sur $S^+ \times S^+$ et $S^- \times S^-$ si r est impair.

$\tilde{\beta}$ donne lieu aux mêmes propriétés mais pour $r = 1,2 \pmod 4$.

Dans ce qui suit β pourra se lire parfois $\tilde{\beta}$ indifféremment.

Sur $A = F \oplus S$ introduisons la forme bilinéaire Λ:

(3) $\Lambda(x + uf, x' + u'f) = B(x,x') + \beta(uf,u'f)$

($\tilde{\Lambda}$ sera définie de manière analogue avec $\tilde{\beta}$).

Posons:

(4) $F_0(x + u^+f + u'^-f) = \beta(xu^+f, u'^-f)$,

$x \in F^+$ si r est impair,

$x \in F^-$ si r est pair.

(On remarquera que si $\beta(x) + x = 0$, il est indifférent d'écrire au deuxième membre de (4), $\beta(xu^+f, u'^-f)$ ou $\beta(xu'^-f, u^+f)$; si on utilise $\tilde{\beta}$ c'est $\tilde{\beta}(x) + x = 0$ qui intervient.)

On définit pour ξ, η, $\zeta \in A$:

(5) $\Phi_0(\xi,\eta,\zeta) = F_0(\xi + \eta + \zeta) + F_0(\xi) + F_0(\eta) + F_0(\zeta) - F_0(\xi + \eta) - F_0(\eta + \zeta) - F_0(\xi + \zeta)$ et l'algèbre d'interaction:

$$(\xi,\eta) \longmapsto \xi \circ \eta$$

par

(6) $\Phi_0(\xi,\eta,\zeta) = \Lambda(\xi \circ \eta, \zeta)$, $\forall \zeta \in A$.

Il est alors aisé de voir que:

$$(7) \quad \begin{cases} F^- \circ S^+ \subseteq S^-, \quad S^+ \circ S^- \subseteq F^-, \quad S^- \circ F^- \subseteq S^+ & (r \text{ pair}), \\ F^+ \circ S^+ \subseteq S^+, \quad S^+ \circ S^- \subseteq F^+, \quad S^- \circ F^+ \subseteq S^- & (r \text{ impair}) \end{cases}$$

et que les autres compositions sont nulles.

On peut aussi bien utiliser $\tilde{\Lambda}$ au lieu de Λ.

On sait, que si $r = 4$, il existe un isomorphisme d'ordre 3 qui échange E', S^+, S^-, cette propriété est perdu si $r \neq 4$.

3. Les algèbres de Lie graduées orthosymplectiques

V est un espace vectoriel gradué ($\mathbb{Z}_2$-gradué):

$$V = V_0 \oplus V_1 = \oplus V_i, \quad i \in \mathbb{Z}_2.$$

Soit h une forme bilinéaire sur V, non dégénérée, telle que:

$h/_{V_0} = \Phi$ soit antisymétrique,

$h/_{V_1} = g$ soit symétrique,

et $h(V_0, V_1) = 0$.

End V est une algèbre de Lie $\mathbb{Z}_2$-graduée avec $[a,b] = ab - (-1)^{|a|\,|b|} ba$, $|a|$ désigne le degré de a. Posons:

(8) $\mathcal{G}_0 = \{a \in \mathrm{End}_0 V,\ h(ax,y) + h(x,ay) = 0,\ \forall x \in V_i,\ \forall y \in V_i\}$,

(9) $\mathcal{G}_1 = \{a \in \mathrm{End}_1 V,\ h(ax,y) = h(x,ay),\ \forall x \in V_1,\ \forall y \in V_0\}$,

conditions qui se résument pour tout a ayant un degré en:

(10) $h(x,ay) + (-1)^{|a|\,|x|} h(ax,y) = 0$.

$\mathcal{G}_0 \oplus \mathcal{G}_1 = \mathrm{Osp}(V_0, V_1)$ est une algèbre de Lie graduée orthosymplectique.

4. Le principe général d'une construction de certaines algèbres graduées orthosymplectiques

Si $r = 1,2,3 \pmod 4$, en utilisant soit β, soit $\tilde{\beta}$ on peut toujours considérer sur l'espace S des spineurs une forme bilinéaire antisymétrique, non dégénérée, invariante par l'action des groupes G_0^+ et $G_0'^+$. Dans ce qui suit $\mathcal{B}$ sera lue soit β soit $\tilde{\beta}$ selon la valeur de r, elle définira sur S une structure symplectique.

Prenons $V_0 = S$, $V_1 = F^{\pm}$, Λ du (2.) jouant le rôle de h du (3.). Tenant compte de la remarque du (2.) on se bornera à considérer des espaces $F^{\pm}$ tels que pour leurs éléments x, $\beta(x) + x = 0$ (ou $\tilde{\beta}(x) + x = 0$ si on utilise $\tilde{\mathcal{B}}$). $G_0'^+$ opère sur $F^{\pm}$ par extension isométrique de l'action sur E' et par produit à gauche sur S comme une transformation symplectique. Si nous posons par définition; uf et vf étant de parités différentes:

$$(11)\quad \begin{cases} uf(vf) = uf \circ vf, \\ uf(x) = xuf, \quad \text{pour} \quad x \in F^{\pm}, \quad uf, \quad vf \in S, \end{cases}$$

nous vérifions que:

$$\Lambda(uf(x), u'f) = \Lambda(x, uf(u'f)),$$

car ceci équivaut à:

$$\mathcal{B}(x\ uf, u'f) = B(x, uf \circ u'f) = F_o(x + uf + u'f).$$

Si uf et vf sont de même parité, leur composition est nulle. Ainsi nous obtenons une algèbre de Lie graduée orthosymplectique dont les éléments de degré 1 sont ceux de S et les éléments de degré 0 les endomorphismes obtenus par la "double" représentation de l'algèbre de Lie $G_o'^{+} = \text{Spin}\, Q'$ de $G_o'^{+}$, dans les espaces $F^{\pm}$ et S.

Il est loisible et intéressant de vérifier directement que $\{uf, vf\}_{|Y_o}$ est bien dans l'algèbre de Lie du groupe symplectique introduit car:

$$(12)\quad \{uf, vf\}(wf) = uf \circ (vf \circ wf) + vf \circ (uf \circ wf)$$

et, d'une part:

$$\mathcal{B}((vf \circ wf)uf + (uf \circ wf)vf, w'f) + \mathcal{B}(wf, (vf \circ w'f)uf + (uf \circ w'f)vf) = 0,$$

car cela s'écrit:

$$B(vf \circ wf, uf \circ w'f) + B(wf \circ uf, vf \circ w'f) - B(wf \circ vf, uf \circ w'f)$$
$$- B(w'f \circ vf, uf \circ wf)$$

selon la définition même de $\circ$ et de ses propriétés élémentaires données dans [3e]. D'autre part:

$$(13)\quad \{uf, vf\}(a) = auf \circ vf + avf \circ uf, \quad a \in F^{\pm},$$

$$B(auf \circ vf + avf \circ uf, b) + B(a, buf \circ vf + bvf \circ uf) = \mathcal{B}(bvf, auf)$$
$$+ \mathcal{B}(buf, avf) + \mathcal{B}(avf, buf) + \mathcal{B}(auf, bvf) = 0,$$

$\{uf, vf\}_{|Y_1}$ est dans l'algèbre de Lie du groupe orthogonal.

II. CONSTRUCTIONS EFFECTIVES D'ALGÈBRES DE LIE GRADUÉES

1. Généralités

Afin de préciser nos idées nous allons supposer que F^{+} est contenu dans un espace linéairement isomorphe à l'algèbre de Lie de Spin Q',

$r = 1$ ou $3 \pmod 4$).

Alors d'après (13), en prenant a tangent à un groupe à un paramètre de $G_0'^+$ et tenant compte de la propriété d'équivariance donnée dans [3e]:

$$guf \circ gvf = g(uf \circ vf)g^{-1}, \quad g \in G_0'^+,$$

on obtient: $\{uf,vf\}(a) = a(uf \circ vf) - (uf \circ vf)a$ et $\{uf,vf\}_{|V_1}$ s'identifie à $-ad(uf \circ vf)$.

Il est donc intéressant de chercher à identifier $\{uf,vf\}$ et $-(uf \circ vf)$ opérant par produit à gauche sur l'espace des spineurs. $\{uf,vf\}_{|V_0}$ est a priori un élément cliffordien z, pair (puisque dans l'espace des endomorpismes engendrés par la représentation spinorielle de $\underline{G}_0'^+$), et tel que $\beta(z) + z = 0$, selon (12).

Dans le cas étudié au (III, 2.) ci-dessous z appartiendra à $\underline{\text{Spin}}\, Q' \oplus \mathbb{C} e_N$, $e_N = e_1 e_2 \ldots e_n$ étant le produit des éléments d'une base de E, plus généralement le produit par e_N est une transformation symplectique si et seulement si $r = 1,3 \pmod 4$).

L'identification de $\{uf,vf\}$ et de $-(uf \circ vf)$ qui pourra être triviale se fera en général au moyen d'une modification de la loi $\circ$. Ainsi plus bas (formule (15)), on ajoutera des éléments contenant e_N sur la diagonale secondaire, mais rien ne sera modifié par ailleurs.

Notons que $\{guf,gvf\} = g\{uf,vf\}g^{-1}$, $g \in G_0'^+$ et que $[uf,a] = -auf$, $\forall a \in \underline{\text{Spin}}\, Q'$, venant de l'équivariance.

2. Une algèbre de Lie graduée conformosymplectique minkowskienne complexe ($r = 3$)

L'espace de Minkowski (réel) (E,B) est muni d'un repère e_1, e_2, e_3, e_4 avec $(e_1)^2 = 1$, $(e_2)^2 = (e_3)^2 = (e_4)^2 = -1$. On lui adjoint un espace $E_1 = (E_1, B_1)$ de dimension 2: (e_0, e_5), avec $(e_0)^2 = 1$, $(e_5)^2 = -1$. Après complexification apparaît un espace de dimension 6: E_1' que nous munissons de la base de Witt spéciale:

$$x_0 = \frac{e_0 + e_5}{2}, \quad x_1 = \frac{e_1 + e_4}{2}, \quad x_2 = \frac{ie_2 + e_3}{2}$$

$$y_0 = \frac{e_0 - e_5}{2}, \quad y_1 = \frac{e_1 - e_4}{2}, \quad y_2 = \frac{ie_2 - e_3}{2}$$

$$x_\alpha y_\alpha + y_\alpha x_\alpha = 1, \quad B_1(x_\alpha, y_\alpha) = \tfrac{1}{2}.$$

On observera que:

$$xy - yx = -e_o e_5,$$

$$x_1 y_1 - y_1 x_1 = -e_1 e_4,$$

$$x_2 y_2 - y_2 x_2 = -ie_2 e_3.$$

Une base de l'algèbre de Lie $\underline{O}(2,4)$ isomorphe à celle du groupe conforme $C_4(1,3)$ est:

$e_i e_j$, $1 \leq i < j \leq 4$ (correspond à $\underline{O}(1,3)$);

$e_o e_5$ (correspond aux homothéties infinitésimales);

$(e_o + e_5)e_k = xe_k$, $1 \leq k \leq 4$ (id^o translations);

$(e_o - e_5)e_k = ye_k$, $1 \leq k \leq 4$ (id^o transformations conformes spéciales).

On prendra $S = C(Q_1')fy$, avec $f = y_1 y_2$, pour espace spinoriel et

$$C(Q_1')fy = x\, C'_{1,3}\, fy \oplus C'_{1,3}\, fy, \quad (y = y_o), \quad (x = x_o).$$

L'extension $\hat{B}_1$ de B_1 à $\Lambda^2(E_1')$ linéairement isomorphe à $F^+ = \underline{O}'(2,4)$ est à un coefficient numérique près la forme de Killing: si on calcule $\mathrm{ad}(e_r e_s)\ \mathrm{ad}(e_i e_j)(e_k e_\ell)$ on trouve en dimension n pour la matrice de Killing:

$$4(n-2)(g_{is}g_{jr} - g_{ir}g_{js}), \quad g_{ij} = B(e_i, e_j);$$

nous prendrons:

$$\hat{B}_1(e_r e_s, e_i e_j) = 2(g_{rj}g_{is} - g_{sj}g_{ir}) = -4(\Lambda^2 B_1)(e_r \wedge e_s, e_i \wedge e_j),$$
$$(r < s,\ i < j).$$

Il est indispensable de calculer les $ufy \circ vfy$, au moyen d'une table que nous allons former pour les éléments d'une base spinorielle.

Nous expliciterons un seul calcul, le lecteur voundra bien vérifier les autres s'il le souhaite.

Pour calculer $fy \circ x_1 fy$, on considère:

(14) $\beta(zfy, x_1 fy) = \hat{B}_1(fy \circ x_1 fy, z), \quad \forall z \in F^+,$

$\beta(zfy)x_1 fy = yfz\ x_1 fy$, car $\beta(f) = -f$ et $\beta(z) = -z$, et on utilise la base de $\Lambda^2(E_1')$: $xx_1, xx_2, x_1x_2, xy_1, xy_2, x_1y_2, yy_1, yy_2, y_1y_2, yx_1, yx_2, y_1x_2$, e_0e_5, e_2e_3, e_1e_4, qui fait apparaître une décomposition de Witt.

On voit qu'il faut prendre z colinéaire à xx_2 pour obtenir dans (14) un résultat non nul et que dés lors $fy \circ x_1fy$ est colinéaire à yy_2.

Nous obtenons tous calculs faits le tableau (15) ci-dessous, où nous avons omis les compositions nulles et où le coefficient de e_N a été choisi de manière que l'on puisse identifier $-(ufy \circ vfy)$ et $\{ufy, vfy\}$, on sait que $e_N fy = (-i)fy$ dans le cas étudié.

(15)

$\circ$	xfy	x_1fy	x_2fy	xx_1x_2fy
fy	$2y_1y_2$	$-2yy_2$	$2yy_1$	$-1/2(J+J_1+J_2)$ / $-3/2ie_N$
xx_1fy	$2xy_2$	$2x_1y_2$	$1/2(-J-J_1+J_2)$ / $+3/2ie_N$	$2xx_1$
xx_2fy	$-2xy_1$	$1/2(J-J_1+J_2)$ / $-3/2ie_N$	$-2x_2y_1$	$2xx_2$
x_1x_2fy	$1/2(J-J_1-J_2)$ / $+3/2ie_N$	$2x_1y$	$2x_2y$	$2x_1x_2$

avec

$$J = xy - yx,$$

$$J_1 = x_1y_1 - y_1x_1,$$

$$J_2 = x_2y_2 - y_2x_y.$$

<u>Remarques</u>. a) On aurait pu introduire $F^+ = \underline{O}'(2,4) \oplus \mathbb{C}\, e_N$ et munir cet espace d'un produit scalaire prolongeant celui que l'on a utilisé pour $\underline{O}'(2,4)$. Un coefficient arbitraire apparaît pour $\hat{B}(e_N, e_N)$, on le choisit de manière à assurer l'identification de $\{ufy, vfy\}$ et de $-(ufy, vfy)$, d'où le tableau (15).

b) Un calcul de routine montre que:

$$(v^-fy \circ u^+fy)\, w^+fy + (v^-fy \circ w^+fy)\, u^+fy$$

$$= (v^-fy \circ u^+fy)\, w^-fy + (v^-fy \circ w^+fy)\, v^-fy = 0,$$

ce qui assure que

$$\sum_{P.C} (ufy \circ vfy) \circ wfy = 0.$$

c) Si on prenait $B(x_i, y_i) = \frac{k}{2}$, $x_i y_i + y_i x_i = k$, rien d'essentiel ne serait modifié, il viendrait:

$$fy \circ xfy = 2ky_1y_2,$$

$$x_1fy \circ x_1x_2fy = 2k^2x_1y,$$

$$xx_2fy \circ x_1fy = \frac{k^2}{2}(J - J_1 + J_2) - \frac{3i}{2}k^3 e_N, \quad \text{etc.}$$

Finalement, si on veut explicitier l'algèbre de Lie graduée orthosymplectique on a à calculer pour $a \in \underline{O}'(2,4)$, σ_a tel que

$$\begin{cases} \sigma_a(x') = ax' - x'a, & x' \in F^+, \\ \sigma_a(vf) = a\,vf; \end{cases}$$

on en déduit tenant compte de l'équivariance:

$$[uf, \sigma_a](vf) = -(auf) \circ (vf)$$

et banalement:

$$[uf, \sigma_a](x') = -(auf) \circ x';$$

donc:

$$[uf, \sigma_a] = -auf,$$

ce qui permet d'écrire:

$$(16) \quad \begin{cases} [xufy,\ a] = -axufy, \quad a \in \underline{O}'(1,3), \\ [ufy,\ a] = -aufy, \\ [xufy,\ xe_k] = [ufy,\ ye_k] = 0, \\ [xufy,\ ye_k] = e_k ufy, \quad [ufy,\ xe_k] = e_k xufy, \\ [xufy,\ yx - xy] = xufy, \\ [ufy,\ yx - xy] = -ufy. \end{cases}$$

L'action de e_N se traduit aisément à la façon de a. On a donc une algèbre graduée dont l'élément général dépend de 23 paramètres contenue dans une algèbre de dimension complexe 24.

3. Algèbre de Lie graduée conformosymplectique minkowskienne à composantes réelles (r = 3)

a) Les conditions de type Majorana

On sait définir sur l'espace des spineurs une forme sesquilinéaire hermitienne $\mathcal{H}$ par:

$$\tilde{\beta}(\overline{ufy})vfy = \varepsilon\varepsilon' \mathcal{H}(ufy, vfy)\gamma fy,$$

où γ est un spineur pur tel que $\gamma fy\gamma^{-1} = \overline{fy}$,

$$\bar{\gamma}\gamma fy = \varepsilon' fy, \quad \varepsilon = (-1)^{\frac{r(r-1)}{2}}, \quad \varepsilon' = \pm 1.$$

Dans le cas examiné au (2.): $\varepsilon = \varepsilon' = -1$ et on peut choisir $\gamma fy = -ix_2fy = e_2fy$.

On définit la conjugaison de charge ζ, comme une application semi-linéaire de S dans S telle que

$$\zeta(ufy) = \bar{u}\gamma fy,$$

et ici $\zeta^2 = -\mathrm{Id}$. Notons:

$$(17)\quad \begin{cases} \zeta(fy) = -ix_2fy, & \zeta(x_2fy) = ify, \\ \zeta(x_1x_2fy) = ix_1fy, & \zeta(x_1fy) = -ix_1x_2fy, \\ \zeta(xx_1fy) = -ixx_1x_2fy, & \zeta(xx_1x_2fy) = ixx_1fy, \\ \zeta(xx_2fy) = ixfy, & \zeta(xfy) = -ixx_2fy. \end{cases}$$

Dans le cas envisagé ζ change la parité.

Définitions. Nous dirons qu'un spineur uf est de "type Majorana" si:

$$\zeta(u^+f) = (u^-f)$$

(ou de manière équivalente: $\zeta(u^-f) = -(u^+f)$). Un repère spinoriel

$$\{x_{i_1 \dots i_k} f\}$$

sera dit "de Majorana" si les conjugués des éléments pairs du repère en sont les éléments impairs.

Dans un tel repère un spineur de composantes paires (a,b,c,d) et non paires respectives (a´,b´,c´,d´) sera de Majorana si $a' = \bar{a}, b' = \bar{b}, c' = \bar{c}, d' = \bar{d}$. Ces définitions s'appliquent en toute dimension $n = 2r$ pourvu que la conjugaison de charge change la parité, c'est-à-dire si $|r - p|$ (où p est le nombre de carrés positifs dans la signature de Q) est impair. Elles s'adaptent aisément à tous les cas pour toute dimension et signature comme le lecteur pourrait s'en assurer *. Notons que si uf est de Majorana, également auf si a est pair et réel, de plus si $f' = gfg^{-1}$, $g \in \mathrm{Pin}\, Q$, $gufg^{-1}$ est de Majorana si uf l'est aussi. Ainsi: La propriété pour un spineur d'être "de Majorana" se conserve par les transformations réelles naturelles.

Revenant au cas particulier considéré au (2.), à partir d'un repère de Majorana on en déduit un second, qu'on pourra appeler "réel", relativement auquel les spineurs de Majorana ont des composantes réelles; on utilisera plus loin:

$$(18)\quad \begin{cases} \dfrac{fy - ix_2fy}{\sqrt{2}}, & \dfrac{xx_1fy - ixx_1x_2fy}{\sqrt{2}}, \\[2ex] \dfrac{x_1x_2fy + ix_1fy}{\sqrt{2}}, & \dfrac{xx_2fy + ixfy}{\sqrt{2}}, \\[2ex] \dfrac{-x_2fy + ify}{\sqrt{2}}, & \dfrac{-xx_1x_2fy + ixx_1fy}{\sqrt{2}}, \\[2ex] \dfrac{x_1fy + ix_1x_2fy}{\sqrt{2}}, & \dfrac{xfy + ixx_2fy}{\sqrt{2}}. \end{cases}$$

b) Structure symplectique sur l'espace des spineurs de Majorana

L'opposé de la partie imaginaire de $\mathcal{H}$ fournit une forme symplectique σ sur un espace réel de dimension 8. Il est facile d'obtenir:

$$(19)\quad \begin{cases} \sigma(fy, xx_1fy) = 1, & \sigma(x_1fy, xfy) = 1, \\ \sigma(x_2fy, xx_1x_2fy) = 1, & \sigma(x_1x_2fy, xx_2fy) = 1, \end{cases}$$

* Avec peut-être de légères modifications. Nous ne cherchons pas à retrouver systématiquement ici les définitions utilisées par les physiciens pour les spineurs de Majorana.

et le repère (18) constitue pour σ un repère "réel" symplectique.

On remarque que σ est nulle sur $C'^{+}_{2,4} fy \times C'^{-}_{2,4} fy$, situation qu'on n'avait pas avec β ou $\tilde{\beta}$ ce qui rend nécessaire la vérification d'une condition de cohérence

$$\sigma(aufy,vfy) = \sigma(avfy,ufy)$$

quand ufy et vfy sont de même parité et a pair. Mais $\sigma(aufy,vfy) = \sigma(ufy,\tilde{\beta}(\bar{a})vfy) = \sigma(avfy,ufy)$; elle est satisfaite si a est réel avec $\tilde{\beta}(a) = \tilde{\beta}(\bar{a}) = -a$, donc pour $a \in \underline{O}(2,4)$ et également si $a \in \mathbb{R}i$, de sorte que l'on peut introduire pour la suite, l'espace $\underline{O}(2,4) \circ i\,\mathbb{R}$, qui est l'algèbre de Lie d'un groupe spinoriel élargi.

On rappelle que $\mathcal{H}$, donc σ, est invariante par l'action de G_o^+ et même par le groupe des éléments obtenus en multipliant ceux de G_o^+ par quelque facteur $e^{i\theta}$, θ réel, donc si on construit à partir de σ un principe affaibli de trialité pour lequel les définitions seront:

$$\sigma(z\ uf, vf) = \hat{B}_1(z, uf \circ vf),$$

$$z \circ uf = zuf,$$

$uf, vf \in S$ de même parité, $z \in F^+$, et les autres compositions $S^+ \circ S^- = 0$, $F^+ \circ F^+ = 0$, on pourra dresser le tableau suivant pour les éléments d'une base spinorielle (en gardant les conventions faites au (2.)):

(20) Table 1, voir p. 73.

Il faut maintenant former la table concernant les éléments réels du repère symplectique, ce qui est immédiat, il suffit d'écrire les termes au-dessous de la diagonale principale:

(21) Table 2, voir p. 74.

Le lecteur remarquera que dans le tableau (21) les colonnes et lignes de rang pair (respectivement de rang impair) permettent de définir une sous-algèbre de Lie graduée dont la partie paire s'identifie à l'espace de Minkowski réel. De même si on barre les cases dont la somme des rangs de ligne et de colonne est paire (tableau en damier) on obtient en négligeant e_N, après identification de x u f y et u f y une table d'algèbre de Lie graduée où intervient la conjugaison de charge (par exemple $x(fy + ix_2\, fy)$ s'identifie à $fy - ix_2\, fy = \mathcal{C}(fy + ix_2\, fy)$.

On comparera aux tables données heuristiquement par Wess-Zumino dans [7] et dans "Recent developments in gravitation" - Cargèse, 1978

(20)

Table 1

$\circ$	fy	xx_1fy	xx_2fy	x_1x_2fy	xfy	x_1fy	x_2fy	xx_1x_2fy	$\circ$
fy	$2yy_1$	$-\frac{1}{2}(J+J_1)$	$-(x_2-y_2)y_1$	$(x_2-y_2)y$	0	$\frac{1}{2}iJ_2-ki$	$-i(x_2+y_2)y_1$	$-i(x_2+y_2)x$	$ixfy$
xx_1fy	$-\frac{1}{2}(J+J_1)$	$2xx_1$	$-(x_2-y_2)x$	$-(x_2-y_2)x_1$	$-\frac{1}{2}J_2+ki$	0	$i(x_2+y_2)y$	$-i(x_2+y_2)x_1$	ix_1fy
xx_2fy	$-(x_2-y_2)y_1$	$-(x_2-y_2)x$	$2xy_1$	$\frac{1}{2}(-J+J_1)$	$i(x_2+y_2)y_1$	$-i(x_2+y_2)y$	0	$\frac{1}{2}J_2+ki$	ix_2fy
x_1x_2fy	$(x_2-y_2)y$	$-(x_2-y_2)x_1$	$\frac{1}{2}(-J+J_1)$	$-2x_1y$	$i(x_2+y_2)x$	$i(x_2+y_2)x_1$	$-\frac{1}{2}J_2-ki$	0	ixx_1x_2fy
fy	0	$\frac{1}{2}J_2-ki$	$-i(x_2+y_2)y_1$	$i(x_2+y_2)y$	$2xy_1$	$-\frac{1}{2}(J-J_1)$	$(x_2-y_2)y_1$	$(x_2-y_2)x$	xfy
xx_1fy	$-\frac{1}{2}J_2+ki$	0	$-i(x_2+y_2)x$	$-i(x_2+y_2)x_1$	$-\frac{1}{2}(J-J_1)$	$-2x_1y$	$-(x_2-y_2)y$	$(x_2-y_2)x_1$	x_1fy
xx_2fy	$i(x_2+y_2)y_1$	$i(x_2+y_2)x$	0	$\frac{1}{2}J_2+ki$	$(x_2-y_2)y_1$	$-(x_2-y_2)y$	$2yy_1$	$-\frac{1}{2}(J+J_1)$	x_2fy
x_1x_2fy	$-i(x_2+y_2)y$	$i(x_2+y_2)x_1$	$-\frac{1}{2}J_2-ki$	0	$(x_2-y_2)x$	$(x_2-y_2)x_1$	$-\frac{1}{2}(J+J_1)$	$2xx_1$	xx_1x_2fy
$\circ$	ify	ixx_1fy	ixx_2fy	ix_1x_2fy	xfy	x_1fy	x_2fy	xx_1x_2fy	$\circ$

(21)

Table 2

$\circ$	$fy-ix_2fy$	$xx_1fy -ixx_1x_2fy$	$x_1x_2fy +ix_1fy$	$xx_2fy+ixfy$	$-x_2fy+ify$	$-xx_1x_2fy +ixx_1fy$	$x_1fy +ix_1x_2fy$	$xfy+ixx_2fy$
$fy-ix_2fy$	$4yy_1$							
$xx_1fy -ixx_1x_2fy$	$-(J+J_1)$	$4xx_1$						
$x_1x_2fy +ix_1fy$	$2(x_2-y_2)y$	$-2(x_2-y_2)x_1$	$-4x_1y$					
$xx_2fy+ixfy$	$-2(x_2-y_2)y_1$	$-2(x_2-y_2)x$	$-J+J_1$	$4xy_1$				
$-x_2fy+ify$	0	$-iJ_2$ / $3e_N$	$-2i(x_2+y_2)y$	$2i(x_2+y_2)y_1$	$4yy_1$			
$-xx_1x_2fy +ixx_1fy$	iJ_2 / $-3e_N$	0	$2i(x_2+y_2)x_1$	$2i(x_2+y_2)x$	$-(J+J_1)$	$4xx_1$		
$x_1fy +ix_1x_2fy$	$2i(x_2+y_2)y$	$-2i(x_2+y_2)x_1$	0	iJ_2 / $3e_N$	$2(x_2-y_2)y$	$-2(x_2-y_2)x_1$	$-4x_1y$	
$xfy+ixx_2fy$	$-2i(x_2+y_2)y_1$	$-2i(x_2+y_2)x$	$-iJ_2$ / $-3e_N$	0	$-2(x_2-y_2)y_1$	$-2(x_2-y_2)x$	$-J+J_1$	$4xy_1$

(Plenum Press), et également par un grand nombre d'auteurs, toujours dans un cadre empirique.

Cette table est susceptible d'une double interprétation:

- On peut chercher à identifier $\{uf,vf\}$ et $-(uf \circ vf)$ pour tout couple, il faut alors ajouter les termes qui apparaissent dans certaines cases, en bas à droite, on voit que comme plus haut l'identification en ce qui concerne $\{uf,vf\}_0$ n'est pas triviale, elle requiert l'action de e_N dans certains cas.

- On peut ne pas ajouter ces coefficients, alors les éléments de notre algèbre de Lie graduée orthosymplectique sont des transformations associées aux vecteurs de $S \oplus \underline{Q}(1,3) \oplus i\,\mathbb{R} \oplus \mathbb{R}\, e_N$, conctruits à l'aide du principe de trialité affaibli où l'on choisit $F^+ = \underline{Q}(1,3) \oplus \mathbb{R}\, i$.

4. Algèbres de Lie graduées minkowskiennes

Si nous revenons au (I, 2.), on a posé

$$A = F \oplus S = F \oplus S^+ \oplus S^-,$$

mais on pourrait aussi bien envisager un espace A somme directe formelle de F et de deux espaces isomorphes à S, alors A devient

$$F \oplus S \oplus S.$$

Si r est impair, on peut poser:

$$u^+f \circ u'^-f = \beta(x u^+f, u'^-f), \quad x \in F^+,$$

$$x \circ u^+f = xu^+f, \; x \circ u^-f = xu^-f$$

retrouvant ainsi la définition du (I, 2.).

Si r est pair, il serait loisible de poser, aussi bien

$$u^{\mp} f \circ u^{\mp\prime} f = \beta(x u^{\mp}f, u'^{\mp}f), \quad x \in F^+,$$

$$x \circ u^{\mp}f = xu^{\mp}f \,;$$

($\tilde{\beta}$ aussi bien que β peut être envisagée).

Dans cette optique il est possible de construire des algèbres de Lie graduées orthosymplectiques, en particulier si $r = 2$ on va donner une table analogue à (15) et une table analogue à (21).

Prenons l'espace de Minkowski réel E avec

$$(e_1)^2 = 1, \quad (e_2)^2 = (e_3)^2 = (e_4)^2 = -1.$$

Les mêmes calculs qu'au (2.) conduisent à la table, utilisant $\mathcal{B}$:

(22)

0	f	x_1x_2f	x_1f	x_2f
f	$2y_1y_2$	$-\frac{1}{2}(J_1 + J_2)$	0	0
x_1x_2f	$-\frac{1}{2}(J_1 + J_2)$	$2x_1x_2$	0	0
x_1f	0	0	$-2x_1y_2$	$\frac{1}{2}(J_1 - J_2)$
x_2f	0	0	$\frac{1}{2}(J_1 - J_2)$	$-2y_1x_2$

Dans ce cas l'identification de $\{u^+f, v^+f\}$ avec $-(uf^+ \circ v^+f)$ est triviale (resp.: $\{u^-f, v^-f\}$ et $-(u^-f \circ v^-f)$. De même avec $\mathcal{H}$, structure hermitienne, on voit que la structure symplectique associée:

$$\sigma(f, x_1f) = 1, \qquad \mathcal{T}(f) = -ix_2f, \qquad \mathcal{T}(x_2f) = if,$$

$$\sigma(x_1x_2f, x_2f) = 1, \quad \mathcal{T}(x_1x_2f) = ix_1f, \quad \mathcal{T}(x_1f) = -ix_1x_2f,$$

et les compositions se font entre spineurs de parité différente:

$$f \circ x_1f = -\frac{1}{2} J_1,$$

$$f \circ x_2f = -(x_2 - y_2)y_1$$

$$x_1x_2f \circ x_2f = \frac{1}{2} J_1,$$

$$x_1x_2f \circ x_1f = -(x_2 - y_2)x_1,$$

$$if \circ x_1f = -\frac{i}{2} J_2, \quad x_1f \circ ix_1x_2f = -i(x_2 + y_2)x_1,$$

$$ix_1x_2f \circ x_2f = \frac{i}{2} J_2, \quad x_2f \circ if = i(x_2 + y_2)y_1.$$

Finalement on obtient le tableau pour le repère de Majorana symplectique:

(23) Table, voir p. 77.

On complètera les tables (22) et (23) par les relations analogues à (16), mais plus simples.

Remarque. On note que pour le (4.) les identifications entre $-(uf \circ vf)$ et $\{uf, vf\}$ sont toutes triviales et que pour le (3.), peut

(23)

σ	$f-ix_2f$	$x_1f+ix_1x_2f$	$-x_2f+if$	$x_1x_2f+ix_1f$
$f-ix_2f$	$2i(x_2+y_2)y_1$	$-J_1$	$+2(x_2-y_2)y_1$	iJ_2
$x_1f+ix_1x_2f$	$-J_1$	$-2i(x_2+y_2)x_1$	$-iJ_2$	$-2(x_2-y_2)x_1$
$-x_2f+if$	$2(x_2-y_2)y_1$	$-iJ_2$	$-2i(x_2+y_2)y_1$	$-J_1$
$x_1x_2f+ix_1f$	iJ_2	$-2(x_2-y_2)x_1$	$-J_1$	$2i(x_2+y_2)x_1$

intervenir e_N. Cela s'explique aisément: $\{uf,vf\}(wf) = z\,wf$, $z \in C(Q')$, dans tous les cas, car $C(Q')$ représente tous les endomorphismes de S, z représente un endomorphisme symplectique: il doit conserver la parité donc $z \in C^+(Q')$ et $\beta(z) + z = 0$; z ne peut donc contenir que des termes de degré $2, 6$ dans les cas du (3.), de degré 2 dans les cas du (4.), comme on le voit, en utilisant l'identification linéaire de $C(Q')$ avec l'algèbre extérieure.

Le cas général, lorsque $r = 1,2,3$, mod 4. Les considérations du I s'appliquent toujours, ainsi que les remarques faisant l'objet du (II, 4.). Le problème ouvert est d'expliciter une identification entre $\{uf,vf\}$ et $-(uf \circ vf)$ en modifiant au besoin la loi $\circ$ comme cela est fait au (I,2.)* ($r = 3$). Si on renonce à faire cette identification (facultative) point n'est besoin de modifier la loi $\circ$, on obtient encore en toute dimension paire et toute signature une algèbre de Lie graduée orthosymplectique dont les éléments uf de degré 1 sont des spineurs et les éléments z de degré 0 appartiennent à l'algèbre de Clifford et sont tels que $\beta(z) + z = 0$. Si z est homogène de degré k dans une identification linéaire avec l'algèbre extérieure, $k = 2$, mod 4. De même avec $\tilde{\beta}$.

$\beta(z) + z = 0$ exprime que z appartient à l'algèbre de Lie $\mathcal{A}$ des transformations linéaires qui conservent β, transformations qui selon la parité de r peuvent être orthogonales ou symplectiques. Si on note que $\mathcal{A}$ est invariante par Ad, appliqué à tout groupe pinoriel ou spinoriel, on pourra construire une algèbre de Lie graduée en prenant les espaces $\mathcal{A}^-$, S^+ et S^-, ou les espaces $\mathcal{A}^+$, S^+ et S^- selon que r est pair ou impair (cf. (7), 2.) de manière que trivialement:

$$\{uf,vf\} = -(uf \circ vf),$$

* si cela est possible!

$$[uf,a] = -auf, \quad \forall \quad a \in \mathcal{A}^{\overset{+}{-}}, \quad \text{(selon le cas).}$$

Cette algèbre sera alors orthosymplectique dans les conditions données pour la formule (7) du (2.). La valeur des crochets précédents se déduira de l'équivariance pour les éléments g de groupe de Lie associé à tels que $\beta(g)g = 1$.

On peut également adapter la point de vue développé au (3.) et (4.). On retiendra la richesse et l'extrême souplesse de ce procédé.

III. LA LIAISON AVEC LE FORMALISME HEURISTIQUE DE WESS-ZUMINO ET AUTRES

Dans une multitude de publications récentes on introduit, avec des notations que nous respectons la table d'algèbre de Lie graduée:

$$(24) \quad \begin{cases} [P_m, P_n] = 0, \quad [P_m, Q_\alpha] = [P_m, \bar{Q}_{\dot\alpha}] = 0, \\ \{Q_\alpha, \bar{Q}_{\dot\beta}\} = 2\sigma^m_{\alpha\dot\beta} P_m, \\ \{Q_\alpha, Q_\beta\} = \{Q_{\dot\alpha}, Q_{\dot\beta}\} = 0, \end{cases}$$

(cf. [7]), où P_m est un opérateur de translation dans l'espace de Minkowski, les Q_α définissent un spineur, les $\bar{Q}_{\dot\alpha}$ le spineur complexe conjugué. En terme de coordonnées l'espace est décrit par $\{x^m, \theta^\alpha, \bar\theta^{\dot\alpha}\}$, les coordonnées à indices latins commutent, les autres anticommutent ("analyse graduée"), Wess et Zumino posent:

$$(25) \quad \begin{cases} P_m = i\,\partial/\partial x^m, \\ Q_\alpha = \partial/\partial\theta^\alpha - i\,\sigma^m_{\alpha\dot\beta}\,\bar\theta^{\dot\beta}(\partial/\partial x^m), \\ \bar{Q}_{\dot\alpha} = -\partial/\partial\bar\theta^{\dot\alpha} + i\theta^\beta\,\sigma^m_{\beta\dot\alpha}(\partial/\partial x^m). \end{cases}$$

Il existe de nombreuses variantes de ces tables dont le lecteur aimerait connaître une justification logique. Du tableau (15) nous pouvons extraire:

$$(26) \quad \begin{cases} fy \circ x_1 fy = 2y_2 y, \\ fy \circ x_2 fy = -2y_1 y, \\ x_1x_2 fy \circ x_1 fy = 2x_1 y, \\ x_1x_2 fy \circ x_2 fy = 2x_2 y; \end{cases} \qquad (26') \quad \begin{cases} xfy \circ xx_1 fy = -2y_2 x, \\ xfy \circ xx_2 fy = 2y_1 x, \\ xx_1x_2 fy \circ xx_1 fy = -2x_1 x, \\ xx_1x_2 fy \circ xx_2 fy = -2x_2 x, \end{cases}$$

tables isomorphes à celle que l'on a donnée dans [3 d]:

$$(27) \quad \begin{cases} f \circ x_1 f = -2y_2, \\ f \circ x_2 f = 2y_1, \\ x_1 x_2 f \circ x_1 f = -2x_1, \\ x_1 x_2 f \circ x_2 f = -2x_2, \end{cases}$$

à partir d'un principe de trialité affaibli. De (15) et (16) nous lirons donc des tableaux isomorphes à des tableaux de la forme:

$$(28) \quad \begin{cases} [uf,a] = -auf, \\ [a,b] = 0, \\ \{uf,vf\} = -uf \circ vf \end{cases} \quad \text{ou} \quad (28') \quad \begin{cases} [uf,a] = 0, \\ [a,b] = 0, \\ \{uf,vf\} = -(uf \circ vf), \end{cases}$$

où a est xe_k ou ye_k, correspondant à des transformations infinitésimales ou conformes spéciales. On voit l'analogie avec (24). Mais nous pouvons préciser encore.

Selon des résultats classiques et familiers aux physiciens théoriciens, choisissons:

$$\sigma^1_{11\cdot} = \sigma^1_{22\cdot} = -1, \quad \sigma^2_{12\cdot} = -i, \quad \sigma^2_{21\cdot} = i,$$

$$\sigma^3_{11\cdot} = 1, \quad \sigma^3_{22\cdot} = -1, \quad \sigma^4_{12\cdot} = \sigma^4_{21\cdot} = 1.$$

Nous trouvons alors:

$$(29) \quad \begin{cases} \{Q_1, \bar{Q}_{1\cdot}\} = -2e_1 + 2e_3, \\ \{Q_2, \bar{Q}_{2\cdot}\} = -2e_1 - 2e_3, \\ \{Q_1, \bar{Q}_{2\cdot}\} = -2ie_2 + 2e_4, \\ \{Q_2, \bar{Q}_{1\cdot}\} = 2ie_2 + 2e_4, \end{cases}$$

d'où nous déduisons que:

$$(30) \quad \begin{cases} \{(Q_2 - Q_1), \quad (\bar{Q}_{2\cdot} + \bar{Q}_{1\cdot})\} = 8y_2, \\ \{(Q_2 + Q_1), \quad (\bar{Q}_{2\cdot} + \bar{Q}_{1\cdot})\} = -8y_1, \\ \{(Q_2 - Q_1), \quad (\bar{Q}_{2\cdot} - \bar{Q}_{1\cdot})\} = -8x_1, \\ \{(Q_2 + Q_1), \quad (\bar{Q}_{2\cdot} - \bar{Q}_{1\cdot})\} = -8x_2 \end{cases}$$

et en posant

$$Q_1 = x_2 f - x_1 f, \quad Q_2 = x_2 f + x_1 f,$$

$$\bar{Q}_{\dot{1}} = -f - x_1x_2f, \quad \bar{Q}_{\dot{2}} = -f + x_1x_2f,$$

nous obtenons l'identification de (24) et (27), $\bar{Q}_{\dot{1}}$ et $\bar{Q}_{\dot{2}}$ sont les conjugués de change de Q_1 et Q_2, respectivement, avec la conjugaison définie par $\gamma f = x_2 f = ie_2 f$ (différente donc de celle du (4.)). $(Q_1, Q_2, \bar{Q}_{\dot{1}}, \bar{Q}_{\dot{2}})$ est un repère "de Majorana".

Ecrivant

$$G = (u^-f \circ u^+f) = (\theta^1 Q_1 + \theta^2 Q_2) \circ (\bar{\theta}^{\dot{1}} \bar{Q}_{\dot{1}} + \bar{\theta}^{\dot{2}} \bar{Q}_{\dot{2}}) = 2(e_3 - e_1)\theta^1 \bar{\theta}^{\dot{1}}$$
$$- 2(e_1 + e_3)\theta^2 \bar{\theta}^{\dot{2}} + 2(e_4 - ie_2)\theta^1 \bar{\theta}_{\dot{2}} + 2(e_4 + ie_2)\theta^2 \bar{\theta}^{\dot{1}},$$

nous obtenons une forme bilinéaire symétrique à valeurs vectorielles à laquelle il est loisible d'associer une forme bilinéaire F, alternée, interprétant les θ^α, $\bar{\theta}^{\dot{\alpha}}$ comme des formes linéaires:

$$F = 2(e_3 - e_1)\theta^1 \wedge \bar{\theta}^{\dot{1}} - 2(e_1 + e_3)\theta^2 \wedge \bar{\theta}^{\dot{2}} + 2(e_4 - ie_2)\theta^1 \wedge \bar{\theta}^{\dot{2}}$$
$$+ 2(e_4 + ie_2)\theta^2 \wedge \bar{\theta}^{\dot{1}} \qquad (F = G^{\simeq}),$$

$$(31) \quad \begin{aligned}
\frac{\partial F}{\partial \theta^1} &= 2(e_3 - e_1)\bar{\theta}^{\dot{1}} - 2(e_4 - ie_2)\bar{\theta}^{\dot{2}} \simeq Q_1 \circ (u^+f) = \frac{\partial}{\partial \theta^1}(u^-f \circ u^+f)^{\simeq},\\
\frac{\partial F}{\partial \theta^2} &= -2(e_1 + e_3)\bar{\theta}^{\dot{2}} + 2(e_4 + ie_2)\bar{\theta}^{\dot{1}} \simeq Q_2 \circ (u^+f) = \frac{\partial}{\partial \bar{\theta}^2}(u^-f \circ u^+f)^{\simeq},\\
\frac{\partial F}{\partial \bar{\theta}^{\dot{1}}} &= -2(e_3 - e_1)\theta^1 - 2(e_4 + ie_2)\theta^2 \simeq -\bar{Q}_{\dot{1}} \circ (u^-f) = \frac{\partial}{\partial \bar{\theta}^{\dot{1}}}(u^-f \circ u^+f)^{\simeq},\\
\frac{\partial F}{\partial \bar{\theta}^{\dot{2}}} &= +2(e_1 + e_3)\theta^2 - 2(e_4 - ie_2)\theta^1 \simeq -\bar{Q}_{\dot{2}} \circ (u^-f) = \frac{\partial}{\partial \bar{\theta}^{\dot{2}}}(u^-f \circ u^+f)^{\simeq}
\end{aligned}$$

(le signe $\simeq$ désigne une identification évidente).

Dans cette optique est associé biunivoquement au couple de spineurs (u^-f, u^+f) de composantes $(\theta^\alpha, \theta^{-\dot{\alpha}})$, l'ensemble des 2-formes définies par:

$$(32) \quad \begin{aligned}
X_1 &= -2(\theta^1 \wedge \bar{\theta}^{\dot{1}} + \theta^2 \wedge \bar{\theta}^{\dot{2}}), \qquad X_2 = -2i(\theta^1 \wedge \bar{\theta}^{\dot{2}} - \theta^2 \wedge \bar{\theta}^{\dot{1}}),\\
X_3 &= 2(\theta^1 \wedge \bar{\theta}^{\dot{1}} - \theta^2 \wedge \bar{\theta}^{\dot{2}}), \qquad X_4 = 2(\theta^1 \wedge \bar{\theta}^{\dot{2}} + \theta^2 \wedge \bar{\theta}^{\dot{1}}),
\end{aligned}$$

et venant par l'identification $\simeq$ des composantes de $(u^-f \circ u^+f)$. Φ étant une fonction à valeurs scalaires de 4 variables scalaires (x_1, x_2, x_3, x_4) on conviendra de dire que c'est une fonction des X_1, X_2, X_3, X_4, 2-formes associées et de poser:

$$\frac{\partial\phi}{\partial\bar\theta^{\alpha\bullet}} = \frac{\partial\phi}{\partial x^k}\frac{\partial x^k}{\partial\bar\theta^{\alpha\bullet}}, \qquad \frac{\partial\Phi}{\partial\theta^\alpha} = \frac{\partial\Phi}{\partial x^k}\frac{\partial x^k}{\partial\theta^\alpha},$$

cela permet ensuite de définir

$$\frac{\partial^2\Phi}{\partial\theta^\alpha\,\partial\bar\theta^{\beta\bullet}}$$

qui est une 2-forme, et

$$(33) \qquad \frac{\partial^2\Phi}{\partial\theta^\alpha\,\partial\bar\theta^{\beta\bullet}} + \frac{\partial^2\Phi}{\partial\bar\theta^{\dot\beta}\,\partial\theta^\alpha} = 0.$$

On observe que:

$$\frac{\partial^2\Phi}{\partial\theta^\alpha\,\partial\theta^\beta} = \frac{\partial^2\Phi}{\partial\bar\theta^{\alpha\bullet}\partial\bar\theta^{\beta\bullet}} = 0,$$

de sorte que si Φ dépend aussi directement de θ^α, $\bar\theta^{\alpha\bullet}$ on postule cette propriété d'antisymétrie (33) pour les dérivées; (25) est alors formellement obtenue en posant, avec ces conventions:

$$\mathcal{Q}_\alpha(\Phi) = \left(\frac{\partial\Phi}{\partial\theta^\alpha}\right)_{\text{partiel}} + \frac{\partial\Phi}{\partial x^k}\frac{dx^k}{\partial\theta^\alpha}$$

$$(34) \qquad \bar{\mathcal{Q}}_{\alpha\bullet}(\Phi) = -\left(\frac{\partial\Phi}{\partial\bar\theta^{\alpha\bullet}}\right)_{\text{partiel}} + \frac{\partial\Phi}{\partial x^k}\frac{\partial x^k}{\partial\bar\theta^{\alpha\bullet}}$$

$$\mathcal{P}_m = \partial/\partial x^m$$

et prenant "l'anticrochet" de $\mathcal{Q}_\alpha$ et $\bar{\mathcal{Q}}_{\alpha\bullet}$.

R e m a r q u e s. On voit que $\mathcal{Q}_1$ est l'opérateur

$$\frac{\partial}{\partial\theta^1} + 2\left(\frac{\partial}{\partial x^3} - \frac{\partial}{\partial x^1}\right)\bar\theta^{\dot 1} - 2\left(\frac{\partial}{\partial x_4} - i\frac{\partial}{\partial x_2}\right)\bar\theta^{\dot 2},$$

que $\bar{\mathcal{Q}}_{2\bullet}$ est:

$$-\frac{\partial}{\partial\bar\theta^{2\bullet}} + 2\left(\frac{\partial}{\partial x_1} + \frac{\partial}{\partial x^3}\right)\theta^2 - 2\left(\frac{\partial}{\partial x_4} - i\frac{\partial}{\partial x_2}\right)\theta^1 \quad \text{etc.},$$

en relation avec les seconds membres de (31).

On retiendra que la notion de "principe de trialité affaibli" se substitue au formalisme heuristique de Wess-Zumino et autres.

On notera aussi que l'utilisation de $\mathcal{B}$, $\tilde{\mathcal{B}}$, $\mathcal{H}$, $\tilde{\mathcal{H}}$ et de combinaisons linéaires des deux premières formes d'une part, des deux dernières d'autre part permet d'envisager une foule de tables de type Wess-Zumino * ; par exemple avec la forme symplectique σ déduite de H comme au (4.), on peut dans le même cadre construire la table:

$$f \circ f = -2y_1, \quad f \circ x_1x_2f = x_2 - y_2, \quad x_1x_2f \circ x_1x_2f = -2x_1,$$

$$x_1f \circ x_1f = 2x_1, \quad x_2f \circ x_2f = 2y_1, \quad x_1f \circ x_2f = (x_2 - y_2),$$

$$x_2f \circ ix_1f = -i(x_2 + y_2), \quad if \circ x_1x_2f = -i(x_2 + y_2),$$

où les autres compositions sont nulles.

Rien ne semble logiquement justifier la nécessité de l'introduction de relations d'anticommutativité pour les variables ou les dérivées. Si on considère G au lieu de F et des variables scalaires ordinaires θ^α, $\bar{\theta}^{\alpha^\bullet}$, on trouve aisément

$$\frac{\partial G}{\partial \theta^\alpha} = \frac{\partial F}{\partial \theta^\alpha}, \quad \frac{\partial G}{\partial \bar{\theta}^{\alpha^\bullet}} = -\frac{\partial F}{\partial \bar{\theta}^{\alpha^\bullet}}$$

et on peut garder des commutateurs ordinaires pour les opérateurs différentiels. L'algèbre d'interaction suffit, elle est naturellement associée à la donnée de la structure spinorielle et des formes qui s'y attachent. L'analyse "graduée" ne semble pas s'imposer dans ce contexte bien qu'elle soit conceptuellement d'un grand intérêt.

Nous voudrions précisément nous intéresser à l'étude d'une structure de variété graduée attachée à tout fibré spinoriel au sens large, au-dessus d'une variété différentiable au sens classique.

IV. LA STRUCTURE DE VARIÉTÉ GRADUÉE ATTACHÉE A UN FIBRE SPINORIEL AU SENS LARGE ET L'APPROCHE DE KOSTANT

Nous rappelons la définition d'une variété graduée au sens de Kostant et résumons les principales idées exposées dans [5]. On peut au lieu d'algèbres réelles envisages tout aussi bien des algèbres complexes.

M est une variété C^∞, paracompacte de dimension m sur $\mathbb{R}$. U étant un ouvert quelconque de M, appelons $A = \{A(U), \rho_{U,V}\}$ un préfaisceau complet d'algèbres graduées complexes associatives et commutatives (cette dernière propriété prise au sens gradué). $A(U)$ désigne comme à l'usuel l'algèbre "au-dessus de U" et $\rho_{U,V}$ la restric-

* Au-dessus des variétés munies de courbure, l'introduction de fibrations, de connexions, de dérivées de Lie, de dérivations covariantes, etc., pourra limiter les possibilités d'utilisation de certaines algèbres graduées, il faut donc se garder de certaines généralisations.

tion de V à U; l'hypothèse de complétion permet de raisonner indifféremment sur le préfaisceau A ou sur le faisceau $S = F_S(A)$ naturellement associé par limite inductive. En abrégé les hypothèses faites par Kostant sont en substance les suivantes:

a) Pour tout ouvert U il existe un homomorphisme d'algèbres graduées $\phi: A(U) \to C^\infty_{\mathbb{C}}(U)$ qui commute avec les restrictions, $C^\infty_{\mathbb{C}}(U)$ est considérée comme algèbre graduée de degré 0, donc si $(f) = \tilde{f}$, $\tilde{f} = 0$ lorsque $f \in A(U)_1$ ou est engendrée par les éléments de $A(U)_1$.

b) Il existe un "atlas de trivialisations" (U_α) de dimension finie r, étant entendu que U est un ouvert trivialisant (splitting map), s'il existe une algèbre graduée $C(U)$, incluse dans $A(U)_0$, isomorphe à $C^\infty_{\mathbb{C}}(U)$ et une algèbre extérieure $D(U)$, de dimension 2^r sur $\mathbb{C}$, telles que $C(U) \otimes_{\mathbb{C}} D(U)$ soit $\mathbb{C}$-isomorphe à $A(U)$: de manière précise cet isomorphisme associe à

$$\sum_\sigma f_\sigma \otimes \omega^\sigma \in C(U) \otimes D(U),$$

l'élément $\sum_\sigma f_\sigma \omega^\sigma$ de $A(U)$ (σ varie de 1 à 2^r).

Un exemple banal d'une telle situation est fourni par le fibré des formes différentielles C^∞ au-dessus de M.

On établit alors dans [5]:

PROPOSITION A. *Soit $A^1(U)$ l'ensemble des éléments nilpotents de $A(U)$ qui constituent un idéal gradué contenant $A(U)_1$, la suite:*

$$0 \to A^1(U) \xrightarrow[\text{Id}]{} A(U) \xrightarrow[\phi]{} C^\infty_{\mathbb{C}}(U) \to 0$$

est exacte pour tout ouvert U de M.

La propriété, immédiate quand U est trivialisant résulte de la finesse du faisceau S_0 est s'obtient d'une partition de l'unité.

PROPOSITION B. *On peut associer au préfaisceau A, des fibrés complexes de rang C^j_r, $j = 0,1,\dots,r$, notés $F^0(A)$, $F^1(A),\dots,F^r(A)$; $F^r(A)$ est un fibré en droites complexes.*

$A^j(U)$ désignant la $j^{\text{ième}}$ puissance de l'idéal $A^1(U)$, on a la suite d'inclusions:

$$0 = A^{r+1}(U) \subseteq A^r(U) \subseteq \dots \subseteq A^1(U) \subseteq A(U).$$

Il est facile de voir que

$$\tilde{A}^j(U) = A^j(U) / A^{j+1}(U)$$

est un module sur $C^{\infty}_{\mathbb{C}}(U) \simeq A(U)/_{A^1(U)}$; le préfaisceau des $\tilde{A}^j(U)$ étant tel que $\tilde{A}^j(U)$ est un $C^{\infty}_{\mathbb{C}}(U)$-module, pour tout ouvert U, un théorème standard permet de lui associer un fibré vectoriel, ce fibré vectoriel est noté $F^j(A)$.

Il existe un morphisme τ_j de préfaisceaux: $A^j(U) \to \Gamma(U, F^j(A))$, ce dernier déterminé par les sections de $F^j(A)$; si les s_i, $s_i \in A(U)_1$, $i = 1,2,\ldots,r$, sont telles que $\tau_r(s_1 s_2 \ldots s_r)$ est une section partout non nulle du fibré $F^r(A)_{|U}$, on dit que les s_i sont des coordonnées "gauches", les coordonnées "droites" étant les coordonnées usuelles $(x^1, x^2, \ldots, x^m)$ dans M.

L'auteur cité remarque ensuite que l'on peut associer au fibré vectoriel

$$F(A) = \bigoplus_{j=0}^{r} F^j(A)$$

un préfaisceau $G_r A$ en algèbres extérieures, de telle manière que pour tout ouvert U trivialisant $A(U) \cong (G_r(A)(U))$ et que A apparaît linéairement comme le préfaisceau canonique associé au faisceau des germes des sections de $F(A)$, puisque A est complet et que pour tout point $m \in M$ il existe un voisinage U_m sur lequel $A(U_m) \cong G_r A(U_m)$, d'après la définition même de la variété graduée. On note cependant qu'en général A et $G_r A$ sont algébriquement distincts.

C'est cette situation qu'il est possible de retrouver dans notre cadre spinoriel.

Supposons donné au-dessus de M un fibré riemannien ou pseudo-riemannien ξ, de rang réel 2r - avec signature quelconque - ce fibré (ξ) portant lui-même un fibré spinoriel au sens large, de sorte qu'il existe un fibré vectoriel (Σ) de rang complexe 2^r, associé à la représentation spinorielle usuelle d'un groupe spinoriel élargi dans un idéal à gauche de l'algèbre de Clifford complexifiée [3b]. On sait qu'une telle situation est liée à l'existence dans le fibré pseudo-riemannien complexifié d'un champ de sous-espaces totalement isotropes maximaux, ou dans le fibré de Clifford complexifié d'un pseudo-champ de r-vecteurs isotropes (définis modulo un scalaire) et déterminant un fibré en droites complexes et un fibré vectoriel η de rang complexe r.

Appelons $x \mapsto f(x)$ le représentant local d'un tel pseudo-champ, la fibre en $x \in M$ de (Σ) sera désignée par $C(Q')_x f(x)$. Si (U_α) est un ouvert de trivialisation pour (Σ) avec isomorphisme ϕ_x^α de l'algèbre de Clifford $C(Q')$ type, sur $C(Q')_x$, $x \in U_\alpha$, choisissant une base de Witt type (x_i, y_j), $f = y_1 y_2 \ldots y_r$,

$$\phi^\alpha : \{x \to \phi_x^\alpha(x_{i_1} x_{i_2} \ldots x_{i_h} f, \quad x \in U_\alpha, \quad 1 \leq i_1 < i_2 < \ldots < i_h \leq r\}$$

est une section locale repère de (Σ), il sera commode d'écrire ϕ^α sous forme

$$\{x \to (x_{i_1 i_2 \ldots i_h} f)_x^\alpha\}.$$

Un atlas localement fini de telles sections repères (U_α, ϕ^α) permet d'identifier linéairement le préfaisceau en idéaux à gauche minimaux du fibré de Clifford à un préfaisceau A' en algèbres extérieures * puisque sur l'espace des $(x_1, x_2, \ldots, x_r)$ la forme quadratique est identiquement nulle; ce préfaisceau donne par une construction classique un faisceau $S = F_S A'$ dont les fibres sont des algèbres extérieures de dimension 2^r sur $\mathbb{C}$. $P_r S = A''$, P_r étant le foncteur préfaisceau, A'' est un préfaisceau complet en algèbres associatives et graduées commutatives.

Il faut observer qu'en général (Σ) ne peut être considéré comme un fibré en algèbres extérieures.

En effet, soit $x \mapsto \gamma_{\alpha\beta}(x)$ une fonction de transition à valeurs dans un groupe de spinorialité élargi telle que $\gamma_{\alpha\beta}(x)$ conserve globalement l'espace type F' des $(y_1, y_2, \ldots, y_r)$ avec $f = y_1 y_2 \ldots y_r$, selon un résultat facile à établir [2, 3a]:

$$\gamma_{\alpha\beta}(x) = \prod_k (1 + a^{\alpha_k \beta_k}(x)\, y_{\alpha_k} y_{\beta_k}),$$

nécessairement, de sorte que si $x \in U_\alpha \cap U_{\alpha'}$, avec

$$\phi_x^{\alpha'}(x_{i_1} \ldots x_{i_h} f) = x_{i'_1} \ldots x_{i'_h} f_{\alpha'}(x),$$

$$(35) \qquad x_{i'_1} \ldots x_{i'_h} f_{\alpha'}(x) = \lambda(x)\, x_{i_1} \ldots x_{i_h} f_\alpha(x), \qquad \lambda_\alpha(x) \in C^*,$$

modulo des termes de degré inférieur en $x_1, x_2, \ldots, x_r$.

On peut également remarquer que si $\{(x_i, y_j)\}$, $\{(x_{i'}, y_{j'})\}$ constituent deux bases de Witt de l'espace standard, les (y_j) et $(y_{j'})$ déterminant le même sous-espace totalement isotrope maximal, toute

---

* avec la loi déduite de:

$$x_{i_1 i_2 \ldots i_h} f \,\dot{\wedge}\, x_{j_1 j_2 \ldots j_k} f = x_{i_1 i_2 \ldots i_h j_1 \ldots j_k} f.$$

isométrie qui envoie une base sur l'autre est elle-même le produit de deux isométries dont l'une conserve séparément l'espace des (x_i) et celui des (y_i):

$$y_i \to A_i^k y_k = Y_i, \quad x_i \mapsto B_i^k x_k = X_i, \quad {}^tAB = Id,$$

et dont l'autre s'exprime par:

$$Y_i \to Y_i, \quad X_i \to X_i + b_i^k y_k, \quad b_i^k + b_k^i = 0.$$

On peut obtenir un élément particulier σ du préfaisceau des sections de (Σ) au-dessus de tout ouvert U, en munissant U d'une partition de l'unité (ψ_α) associée aux (U_α), si

$$\sigma = \sum_\alpha \sum_{i_1,i_2 \ldots i_h} \psi_\alpha(x)\lambda^{i_1 \ldots i_h}(x)(x_{i_1} x_{i_2} \ldots x_{i_h} f)_x$$

chaque terme obtenu en fixant α est bien identifiable à un élément d'une algèbre extérieure, mais cette identification dépend du choix des trivialisations locales de sorte que σ considéré comme élément de $A''(U)$ n'est pas une section pour un fibré en algèbres extérieures.

Modulo peut être un scalaire non nul, sans importance ici, il est toujours possible de définir sur le fibré spinoriel une forme bilinéaire β telle que: $\beta(uf,vf)f = \beta(uf)vf$, β antiautomorphisme principal, β non dégénéré permet de mettre en dualité les champs de spineurs et co-spineurs. Modulo un facteur scalaire le champ local: $x \mapsto x_{i_1} \ldots x_{i_h} f_\alpha(x)$ peut s'identifier à

$$x \mapsto f_\alpha(x)\,\xi_{k_1} \ldots \xi_{k_{r-h}},$$

où $\xi_{k_1}, \xi_{k_2}, \ldots, \xi_{k_{r-h}}$ est la suite complémentaire de $x_{i_1}, x_{i_2}, \ldots, x_{i_h}$ dans $(x_1, x_2, \ldots, x_r)$.

Soit $A^{r-h}(U_\alpha)$ la composante des degrés supérieurs où égaux à $(r-h)$ de l'algèbre locale ainsi déterminée par le fibré co-spinoriel. Selon (35)

$$A^{r-h}(U_\alpha)/A^{r-h+1}(U_\alpha)$$

est module des sections au-dessus de (U_α) d'un fibré vectoriel G^h de rang C_r^h, $h = 0,1,\ldots,r$ (identifiable d'ailleurs à $\Lambda^{r-h}\eta$).

Les $(\xi_1, \xi_2, \ldots, \xi_r)$ constituent un système de coordonnées gauches pour une variété graduée "au-dessus" de M.

Il est immédiat de voir que l'on a retrouvé la situation décrite par Konstant, dans le cas réel, si on est parti d'une structure pseudo-riemannienne neutre, ce qui permet de définir des spineurs sans complexifier.

Réciproquement. Peut-on attacher à toute variété graduée réelle au sens de [5] le faisceau des sections d'un fibré spinoriel? On part de la situation caractérisée par la donnée:

- d'une variété ordinaire réelle M,
- d'un préfaisceau A en R-algèbres graduées commutatives tel que A soit linéairement identifiable au préfaisceau canonique des germes des sections d'un fibré vectoriel $F(A)$,
- le fibré vectoriel $F(A)$ de base M, dont les fibres ont la structure d'une algèbre extérieure réelle est la somme de Whitney de $(r+1)$ fibrés vectoriels $F^j(A)$, $j = 0,1,\ldots,r$,

$$rg(F^j(A)) = C_r^j.$$

Soit $F^r(A) = \xi$, fibré de rang r, et posons $\eta = \xi^*$ puis $\mathfrak{T} = \xi \circ \eta$. $\mathfrak{T}$ est un fibré pseudo-riemannien réel dont les fibres admettent la décomposition de Witt

$$\xi_x \circ \eta_x = \mathfrak{T}_x$$

pour le métrique de dualité qui est neutre. $\Lambda^r(\xi^*)$ définit selon un critère classique une structure spinorielle au sens large pour $\mathfrak{T}$, cette structure au sens large est d'ailleurs réductible au sens strict [4'].

Le faisceau des germes des sections de $F^{r-h}(A)$ s'identifie au faisceau des germes de sections du fibré $G^h(A)$ associé aux

$$A^{r-h}(U_\alpha)/A^{r-h+1}(U_\alpha)$$

selon notre démonstration antérieure. Les préfaisceaux complets déterminés par $F(A)$ et la somme directe des $G^h(A)$ sont nécessairement identiques, comme préfaisceaux en algèbres extérieures.

Ainsi *à toute variété graduée réelle, au sens de Kostant, se trouve associé le préfaisceau complet* $Spin(A)$ *des sections d'un fibré spinoriel au sens large au-dessus d'un fibré pseudo-riemannien neutre et réciproquement*.

Kostant avait bien remarqué un lien entre ses variétés et les algèbres de Clifford mais seulement lorsque M est réduite à un point [5, p. 296], cette remarque était trop particulière pour conduire à des résultats significatifs en quantification.

Il faut bien noter que cette association n'est pas une identification pour les structures algébriques, il n'y a identification (non canonique) par le choix d'un atlas de trivialisations localement fini que pour les structures linéaires de A et Spin A (et aussi $G_r(A)$). Il apparait néanmoins que dans les applications à la physique mathématique la variété graduée déterminée par Spin A semble une bonne structure à retenir comme le montre la possibilité de construire des "supersymétries" à partir du principe de trialité élargi.

V. LES DERIVATIONS "GAUCHES" $\overleftarrow{\partial}/\partial x^i$ SUR LES VARIÉTÉS GRADUÉES SPINORIELLES

On a considéré dans la littérature une analyse où le rôle des variables usuelles est joué par les éléments d'une algèbre de Grassmann. De manière précise on introduit, par exemple dans [1]:

- Une algèbre $\mathcal{A}$, $\mathbb{Z}_2$-graduée, sur un corps K de caractéristique 0,

- Cette algèbre est une algèbre de séries formelles de générateurs homogènes $X_1, X_2, \ldots, X_n$ tels que

$$X_i X_j = (-1)^{|X_i||X_j|} X_j X_i,$$

- $\overleftarrow{\partial}/\partial X^i$ est une "dérivation":

$$(36)\qquad \overleftarrow{\partial}/\partial X^i(X_{k_1} X_{k_2} \ldots X_{k\ell}) = \begin{cases} 0, \text{ si } k_s \neq i, \text{ pour tout } k_s, \\ (-1)^h X_{k_1} X_{k_2} \ldots X_{k_{s-1}} X_{k_{s+1}} \ldots X_{k_\ell}, \\ \qquad\qquad \text{si } k_s = i, \end{cases}$$

où

$$h = |X_i|(|X_{k_1}| + \ldots + |X_{k_{s-1}}|).$$

Pour retrouver cette situation, prenons un espace E de dimension $n = 2r$, pseudo-euclidien $(p+q=n)$ et introduisons un espace hyperbolique $E_{1,1}(x,y)$, pour construire l'algèbre de Lie du groupe conforme en signature (p,q) (notations du (2.), mais en dimension paire quelconque).

Le principe de trialité affaibli possède la propriété d'équivariance [3 d]:

$$(37)\qquad g(ufy \circ u'fy)g^{-1} = gufy \circ gu'fy, \quad \text{si } g \in G_o^+.$$

Les xx_i, xy_i, $i = 1,\ldots,r$, constituent des éléments de l'algèbre de Lie des translations de l'espace $E'(p,q)$ complexifié de $E(p,q)$. Pour mettre en évidence la translation $\xi \mapsto \xi + a$, on pose

$$w(\xi) = x\xi^2 + \xi - y, \quad \xi \in E'(p,q),$$

et il vient

$$w(\xi + a) = gw(\xi)g^{-1} ,$$

avec $g = 1 + xa = \exp xa$. Ainsi: *L'équivariance de* o *montre qu'un translation de vecteur* a *dans* $E'(p,q)$ *associée à l'action de* $g = 1 + xa$ *sur* $E'(p+1, q+1)$, *est en correspondance naturelle avec le produit à gauche par* g *dans l'espace spineurs*.

Du point de vue infinitésimal à la translation définie par xa est associée la transformation:

$$ufy \mapsto xaufy$$

dans l'espace des spineurs $C'(p+1, q+1)fy$ et, par identification, *à* $a \in E'(p,q)$ *est associée la transformation*: $uf \mapsto auf$ *dans l'espace des spineurs* $C'(p,q)f$.

Plus particulièrement, prenons *a* de la forme: $a_1y_1 + a_2y_2 + \ldots + a_ry_r$, $a_i \in C$, y_i isotropes, éléments d'une base de Witt (x_i,y_j), et:

$$(38) \qquad y_ix_{k_1}x_{k_2}\ldots x_{k_\ell}\, f = \begin{cases} 0, & \text{si } k_s \neq i, \text{ pour tout } k_s, \\ (-1)^h\, x_{k_1}x_{k_2}\ldots x_{k_{s-1}}x_{k_{s+1}}\ldots x_{k_\ell}\, f, & \text{si } k_s = i, \end{cases}$$

où

$$h = |y_i|(|x_{k_1}| + \ldots + |x_{k_{s-1}}|).$$

Ecrivant $y_i = \not{\partial}/\partial x_i$, on retrouve la formule (36); on voit donc que: *l'équivariance de l'interaction* o *conduit à associer naturellement à une dérivation ordinaire une dérivation graduée*.

$\bar{\bar{x}}_i$ étant le conjugué de x_i pour une conjugaison complexe "tenant compte de la signature" on voit que $\not{\partial}/\partial x_i$ est le produit à gauche par $\bar{\bar{x}}_i$, expliquant la propriété d'anticommutation si x_i est isotrope.

Remarque. a) Dans [1] on introduit une dérivation à gauche et une dérivation à droite:

$$X_{k_1}\dots X_{k_\ell}(\partial/\partial X_i) = \begin{cases} 0, & \text{si } k_s \neq i, \text{ pour tout } k_s, \\ (-1)^h X_{k_1} X_{k_2} \dots X_{k_{s-1}} \hat{X}_{k_s} \dots X_{k_\ell}, & \text{si } k_s = i, \end{cases}$$

où

$$h = |X_i|(|X_{k_\ell}| + \dots + |X_{k_{s+1}}|).$$

La signification de cette dérivation à droite apparaîtra si on fait opérer y_i sur l'espace des cospineurs, qui n'est autre que $fC'(p,q)$, par multiplication naturelle à droite.

b) Pour exploiter pleinement la théorie de ces dérivations graduées, il faudrait envisager des générateurs X_i pairs; on y parviendra en introduisant des spineurs orthosymplectiques [3 e].

Avec la table déduite de $\hat{x} * \hat{y} - \hat{y} * \hat{x} = F(\hat{x},\hat{y})$ pour l'algèbre de Clifford symplectique attachée à la forme F, on a pour les éléments d'une base symplectique:

$$(\hat{e}_{\alpha *})\,(\hat{e}_\alpha)^k = (\hat{e}_\alpha)^k\,(\hat{e}_{\alpha *}) - k(\hat{e}_\alpha)^{k-1};$$

le produit à gauche par $\hat{e}_{\alpha *} - 1$ dans l'espace des spineurs symplectiques $C_S(F)\Phi^*$ s'identifie à la dérivée ordinaire $\partial/\partial\hat{x}^\alpha$, $\hat{x}^\alpha$ coordonnées pour le repère $(\hat{e}_\alpha)$ [3 c].

Sur le spineur orthosymplectique

$$(\hat{e}_1)^{k_1}(\hat{e}_2)^{k_2}\dots(\hat{e}_m)^{k_m}\, x_{i_1} x_{i_2} \dots x_{i_\ell}\,(f \otimes \Phi^*)$$

le produit à gauche par $(\hat{e}_{\alpha *} - 1)$ correspond à $\partial/\partial\hat{x}^\alpha$, et le produit à gauche par $\bar{\bar{x}}_i$ à $\not{\partial}/\partial x_i$.

On voit donc que les dérivations usuelles sont liées aux spineurs symplectiques et les dérivations "graduées" aux spineurs orthogonaux.

References

[1] BEREZIN, F.A. et G.I. KAC, Lie groups with commuting and anticommuting parameters, Mat. Sbornik, t. 82 (124), N° 3, 1970.

[2] CHEVALLEY, C., The algebraic theory of spinors, Columbia U.P., New York 1954.

[3] CRUMEYROLLE, A.:

a) Algèbres de Clifford et spineurs, Cours 3ème cycle, Toulouse III, 1974.

b) Fibrations spinorielles et twisteurs généralisés, Periodica

Math. Hungarica 6 (2)(1975), 143-171.

c) Un formalisme de seconde quantification sur les fibrés spinoriels orthogonaux ou symplectiques, C.R.A.S. Paris 282 (1976).

d) Bilinéarite et géométrie affine attachées aux espaces de spineurs ..., Ann. Inst. H. Poincaré 34, N° 3, Sec. A, 1981.

e) Algèbres et faisceaux d'algèbres de Lie graduées associées à des espaces spinoriels par un principe de trialité, Ann. Inst. H. Poincaré 37, N° 2, Sec. A, 1982.

[4] JADCZYK, A. et K. PILCH, Superspaces and supersymmetries, Commun. Math. Phys. 78 (1981), 373-390.

[4'] KARCHENASSE, A., Structures spinorielles neutres, Thèse, Toulouse 1976.

[5] KOSTANT, B., Graded manifold, graded Lie theory and prequantization, in: Differential Geometrical Methods in Math. Phys., Bonn 1975, Proceedings (Lecture Notes in Math. 570), Springer, Berlin-Heidelberg-New York 1977.

[6] VOLKOV, D.V. et V.P. AKULOV, Phys. Lett. 46B (109) (1973).

[7] WESS, J. et B. ZUMINO, Nucl. Phys. 70 (B-39) (1974).

Université Paul Sabatier
Mathématiques
118, route de Narbonne
F-31062 Toulouse Cédex, France

CONSTANTE DE PLANCK ET GÉOMÉTRIE SYMPLECTIQUE

Albert Crumeyrolle (Toulouse)

Table des matières

RÉSUMÉ

L'introduction de la constante de Planck dans la construction des algèbres de Clifford symplectiques et des groupes "spinoriels" symplectiques permet de donner un cadre géométrico-algébrique à des notions qui pourraient être présentées par le biais plus compliqué, de l'analyse fonctionnelle. Il est remarquable que ce point de vue abstrait s'utilise de manière concrète pour construire des déformations de l'algèbre associative et de l'algèbre de Poisson des fonctions C^∞ au-dessus d'une variété symplectique ; le succès de cette méthode repose sur la remarque banale que les algèbres de Clifford symplectiques sont des déformations d'algèbres symétriques , la constante de Planck étant le paramètre de déformation.

Constante de Planck et géométrie symplectique

Nous donnons différentes propriétés des algèbres de Clifford symplectiques et de groupes de revêtements ; nous montrons qu'au-dessus de toute variété symplectique V il existe des déformations pour l'algèbre associative $C^\infty(V, \mathbb{C})$ et pour l'algèbre de Poisson. L'algèbre associative $C^\infty(V, \mathbb{R})$ admet des déformations si V est munie d'un champ global d'espaces lagrangiens. L'algèbre réelle de Poisson admet des déformations quelle que soit la variété symplectique. Les déformations construites présentent un caractère universel.

INTRODUCTION.

La constante de Planck h introduite en 1900 en mécanique statistique à propos de l'étude du spectre du rayonnement thermique, apparaît comme une constante universelle dont le rôle est absolument fondamental en mécanique quantique et dont l'importance ne peut se comparer qu'à celle de la vitesse c de la lumière. L'une et l'autre de ces constantes c et h, dont le produit est relié au carré de la charge de l'électron et à bien d'autres grandeurs, ont été aussi interprétées comme des paramètres variables susceptibles d'être considérés, l'un comme "infiniment" grand, l'autre comme "infiniment" petit ; les développements selon les puissances de $\frac{1}{c^2}$ ont été fréquemment utilisés en relativité générale et dans une série d'articles récents [5]. J. Leray a établi que la constante de Planck intervenait nécessairement dans l'étude des fonctions lagrangiennes. Nous nous proposons de montrer que l'introduction de cette constante en géométrie symplectique permet de donner un cadre géométriquement approprié à l'étude des algèbres de Clifford symplectiques (qui apparaissent comme des déformations d'algèbres symétriques), aux revêtements du groupe symplectique et à tous les problèmes annexes, par exemple le rôle du groupe de Heisenberg en analyse harmonique, abordé par R.Howe dans un esprit bien différent [3].

1°) Généralités : Nous reprenons en principe les notations et les thèmes de nos articles antérieurs [2,a,c]. Il sera commode si E est un espace vectoriel symplectique de dimension $n = 2r$, sur un corps $\mathbb{K}$ commutatif de caractéristique nulle (qui sera soit $\mathbb{R}$ soit $\mathbb{C}$ usuellement) de noter hF la forme symplectique, h étant soit une indéterminée, soit un paramètre à valeurs dans $\mathbb{K}$ qui dans les applications à la mécanique quantique correspond à la constante de Planck (de l'ordre de 10^{-27}C.G.S.). Nous définissons l'algèbre de Clifford symplectique $C_S(hF)$ à l'aide du produit tensoriel $\otimes E$, quotienté par l'idéal bilatère $\mathcal{J}$ engendré par les éléments de la forme :

(1) $\quad x \otimes y - y \otimes x - hF(x,y), \; x, y \in E,$

(algèbre enveloppante d'une algèbre de Lie dite de Heisenberg, de dimension 2r+1).
$C_S(hF)$ peut donc s'interpréter comme une algèbre réelle ou complexe (h est alors un paramètre réel ou complexe), ou comme une algèbre sur l'anneau des séries formelles $\mathbb{K}[[h]]$ (h est alors une indéterminée et E devient un module libre E_H sur $\mathbb{K}[[h]]$ que nous appellerons un module de Planck. Dans ce dernier cas il est facile de voir que $C_S(hF)$ possède les mêmes propriétaires élémentaires que l'algèbre notée $C_S(F)$ dans [2,a] (exemples : propriété universelle, trivialité du centre, isomorphisme linéaire avec une algèbre symétrique...). La propriété universelle entraîne que si $h_1 h_2 \neq 0$, $C_S(h_1 F)$ et $C_S(h_2 F)$ sont isomorphes.

On rappelle que si $(e_\alpha, e_{\beta^*}, \alpha, \beta = 1,2,\ldots r$, est une base symplectique :

$$F(e_\alpha, e_\beta) = F(e_{\alpha^*}, e_{\beta^*}) = 0, \quad F(e_\alpha, e_{\beta^*}) = h\,\delta_{\alpha\beta^*}$$

on a les seules formules non banales pour la loi multiplicative :

$$(2) \quad (e_{\alpha^*})^k (e_\alpha)^\ell = (e_\alpha)^\ell (e_{\alpha^*})^k - h\ell k (e_\alpha)^{\ell-1} (e_{\alpha^*})^{k-1} + \ldots +$$

$$+ (-h)^p C_\ell^p C_k^p p! \; (e_\alpha)^{\ell-p} (e_{\alpha^*})^{k-p} + \ldots$$

avec $p \leqslant \ell$, $p \leqslant k$;

de sorte que beaucoup de vérifications pourront se faire dans des espaces de dimension 2.

2°) Les algèbres de Clifford symplectiques formelles.

h est ici une indéterminée, l'élément général de $C_S(hF)$ est une série formelle "symplectique" $\hat{u}$ dont le terme général s'écrit :

$$\lambda_{h_1,h_2\dots h_r,k_1,k_2,\dots,k_r}(e_1)^{h_1}(e_2)^{h_2}\dots(e_r)^{h_r}$$
$$(e_1{*})^{k_1}(e_2{*})^{k_2}\dots(e_r{*})^{k_r},$$

les h_i, k_i sont des entiers positifs ou nuls, soit symboliquement :

$$\hat{u} = \sum_{H,K^*} \lambda_{HK^*}\, e^H e^{K^*}, \quad (e_\emptyset = 1)$$

où les λ_{HK^*} sont des séries formelles à coefficients dans $\mathbb{K}[[h]]$.

Si nous considérons simplement

$$(3) \quad \underline{\hat{u} = \sum_k \frac{a_k}{k!}(e_1{*})^k}, \quad \underline{\hat{v} = \sum_\ell \frac{b_\ell}{\ell!}(e_1)^\ell},$$

a_k et b_ℓ étant des séries formelles en h, on observera que le coefficient dans $\mathbb{K}$ de $h^p(e_1)^\ell(e_1{*})^k$ qui apparaît dans le développement de $\hat{u}\,\hat{v}$ vient d'après (2) de

$$(4) \quad \frac{1}{k!}\,\frac{1}{\ell!}\left(a_k b_\ell - h\, a_{k+1}\, b_{\ell+1} + \dots + (-h)^s \frac{a_{k+s} b_{\ell+s}}{s!} + \dots\right)$$

et est obtenu :

du coefficient de h^p dans $a_k b_\ell$, de h^{p-1} dans $a_{k+1}b_{\ell+1}$, ... de h^{p-s} dans $a_{k+s}b_{\ell+s}$ donc d'un nombre fini de termes car $s \leqslant p$.

Nous désignerons par $\check{C}_S(hF)$ cette algèbre de Clifford symplectique formelle (algèbre sur $\mathbb{K}[[h]]$).

3°) Les algèbres de Clifford symplectiques larges.

h est ici un paramètre réel ou complexe. Choisissons à partir d'un certain rang, $|H|$ et $|K^*|$ suffisamment grands :

$$(5) \quad |\lambda_{HK^*}| \leqslant \frac{\sigma(\hat{u})(\rho(\hat{u}))^{|H|+|K^*|}}{(H!\ K^*!)^{1/2+\alpha}}$$

$H! = h_1!h_2!\ldots h_r!$, $K^*! = k_1!k_2!\ldots k_r!$, $|H| = \Sigma h_i$, $|K^*| = \Sigma k_i$,

α est un nombre réel, $\alpha > 0$, choisi une fois pour toute (indépendant de $\hat{u}$), $\sigma(\hat{u})$ et $\rho(\hat{u})$ sont des constantes positives. Ces conditions sont indépendantes de la base choisie (utiliser la formule de Stirling).

Notons que dans [2], $\alpha = \frac{1}{2}$.

On voit aisément que selon des règles élémentaires de convergence, on obtient une algèbre sur $\mathbb{K}$ que nous noterons $C_S(hF)^{\alpha}_{\ell}$ et appellerons symplectique élargie (ou large). La seule vérification qui n'est pas immédiate concerne des produits où $\hat{u}$ et $\hat{v}$ sont de la forme (3) avec compte tenu de (5) :

$$|a_k| \leqslant a(k!)^{1/2-\alpha}(t)^k, \quad |b_\ell| \leqslant b(\ell!)^{1/2-\alpha}(t')^\ell,$$

a, b, t, t' constantes positives, pour k et ℓ suffisamment grands.

Compte tenu de l'inégalité binomiale évidente : $(k+s)! \leqslant 2^{k+s}\, k!s!$,

$$\frac{|a_{k+s}b_{\ell+s}|}{s!k!\ell!}|h|^s \leqslant \frac{ab|h|^s\, 2^{(2s+k+\ell)(1/2-\alpha)}(k!\ell!)^{1/2-\alpha}}{k!\ell!(s!)^{2\alpha}}(tt')^s t^k t'^{\ell}$$

et si $M(\alpha)$ est la somme de la série convergente :

$$(6)\quad \sum_{s=0}^{\infty} \frac{2^{2s(1/2-\alpha)}(tt')^s|h|^s}{(s!)^{2\alpha}}\text{, il vient :}$$

$$(6\text{ bis})\quad \sum_{s=0}^{\infty} \frac{|a_{k+s}b_{\ell+s}||h|^s}{s!k!\ell!} \leqslant \frac{ab\ M(\alpha)}{(k!\ell!)^{1/2+\alpha}}(At)^k(At')^{\ell}$$

avec $A = 2^{1/2-\alpha}$.

Ainsi l'élément général de notre algèbre élargie $C_S(hF)^{\alpha}_{\ell}$ se notera :

$$(7)\quad \hat{u}_{\alpha} = \sum_{H,K^*} \frac{M(H,K^*)e^He^{K^*}}{(H!K^*!)^{1/2+\alpha}},\quad \alpha > 0 \quad \text{(on écrira } \hat{u} \text{ par abus}$$

usuellement) avec $|M(H,K^*)| < \sigma(\hat{u})(\rho(\hat{u}))^{|H|+|K^*|}$ pour $|H|$ et $|K^*|$ suffisamment grands.

Proposition 1. Si $\alpha_1 > 0$ et $\alpha_2 > 0$, les algèbres $C_S(hF)^{\alpha_1}_{\ell}$ et $C_S(hF)^{\alpha_2}_{\ell}$ sont homomorphes :

$$C_S(hF)^{\alpha_1}_{\ell} \longrightarrow C_S(hF)^{\alpha_2}_{\ell}.$$

Il suffit d'associer à $\hat{u}_{\alpha_1}$ obtenu avec $\alpha = \alpha_1$, dans (7) $\hat{u}_{\alpha_2} = f_{\alpha_1\alpha_2}(\hat{u}_{\alpha_1})$ obtenu avec $\alpha = \alpha_2$ et le même coefficient $M(H,K^*)$ dépendant éventuellement de α_2 comme le premier de α_1.

Par limite inductive on a évidemment une structure d'algèbre sur $\bigcup_{\alpha>0} C_S(hF)^{\alpha}_{\ell}$, avec la famille $Id_{\alpha_i\alpha_j}$ des applications identiques et la relation d'ordre opposée à l'ordre naturel.

De particulière importance sont les éléments :

$$\exp(ta^2) = 1 + ta^2 + \frac{t^2a^4}{2!} + \ldots \frac{t^na^{2n}}{n!} + \ldots$$

où $a \in E$, $t \in \mathbb{K}$, éléments qui jouent au 4°) ci-dessous un rôle fondamental grâce à la formule (8). Ils donnent lieu à la remarque suivante :

<u>Remarque</u> : Les éléments de $\check{C}_S(hF)$, $\exp t(e_\alpha)^2$ et $\exp t(e_{\alpha^*})^2$, $t \in \mathbb{K}$, satisfont à la condition (5) où l'on fait $\alpha = 0$, en effet :

si $\hat{u} = \exp t(e_{\alpha^*})^2 = \Sigma \frac{t^k}{k!} (e_{\alpha^*})^{2k}$ on a :

$$\frac{a_{2k}}{(2k)!} = \frac{t^k}{k!}, \quad a_{2k+1} = 0,$$

or de $(2k!)^{1/2} \leqslant 2^k.k!$ résulte que :

$$\frac{|a_{2k}|}{(2k)!} = \frac{|t|^k}{k!} \leqslant \frac{(2|t|)^k}{(2k!)^{1/2}} .$$

Si l'on prend $\alpha = 0$, la convergence de (6) requiert une limitation supérieure de $\rho(\hat{u})$. Cette objection signalée dans [4] n'entraîne pas cependant l'impossibilité de définir globalement les groupes de revêtements du groupe symplectique dans ce contexte (cf. plus bas). La série (6) converge si $\mathrm{Sup}(\rho(\hat{u}), \rho(\hat{v})) < \frac{1}{\sqrt{2h}}$. On notera, que si h est la constante de Planck, cette borne supérieure est de l'ordre de 10^{13} C.G.S.

<u>3° bis) Algèbres de Clifford symplectiques larges et complètes.</u>

Nous affinons légèrement la condition (5) en la condition (5 bis).

(5 bis) $\quad |M(H,K^*)| < ST^{(|H|+|K^*|}$

qui exprime que dans leur ensemble nous supposons $\sigma(\hat{u})$,

$\rho(\hat{u})$ bornés respectivement par S et T. S et T peuvent être fixés (mais arbitrairement grands). Cette condition apporte donc une <u>contrainte</u> sur les lois de composition de $C_S(hF)^{\alpha}_{\ell}$, h fixé.

<u>Proposition 3</u>. <u>Dans les conditions ainsi précisées le produit $\hat{u}\,\hat{v}$ est continu séparément par rapport à $\hat{u}$ et par rapport à $\hat{v}$.</u>

Il est entendu que l'on munit l'algèbre d'une norme $\|u_1\|$ avec

$$\|u_1\|^2 = \sum_{H,K^*} |\lambda_{HK^*}|^2$$

et que la topologie est définie par cette norme.
Les seules vérifications non triviales se ramènent au cas où $\hat{u}$ et $\hat{v}$ sont de la forme (3).

Posant $\lambda(k) = \dfrac{a_k}{(k!)^{1/2-\alpha}}$, $\mu(\ell) = \dfrac{b_\ell}{(\ell!)^{1/2-\alpha}}$

$$\|\hat{u}\hat{v}\|_1 \leq \sum_{k=0}^{\infty} \sum_{\ell=0}^{\infty} \sum_{s=0}^{\infty} \frac{|h|^s |\lambda(k+s)| \, |\mu(\ell+s)| \, ((k+s)!(\ell+s)!)^{1/2-\alpha}}{s!\;k!\;\ell!}$$

Tenant compte de l'inégalité

$$k!\;s! \leq (k+s)! \leq 2^{k+s}\;k!\;s!$$

et choisissant $\frac{1}{2} < \alpha' < \alpha + \frac{1}{2}$

$$\|\hat{u}\hat{v}\|_1 \leq \sum_{k,\ell,s=0}^{\infty} \frac{|\lambda(k+s)|}{((k+s)!)^{1/2+\alpha-\alpha'}} \, \frac{|\mu(\ell+s)|}{((\ell+s)!)^{1/2+\alpha-\alpha'}} \, \frac{2^{k+\ell+2s}|h|^s}{(k!\ell!)^{\alpha'}(s!)^{2\alpha'-1}}$$

$$(11) \quad \|\hat{u}\hat{v}\|_1 \leq A \sum_k \frac{|\lambda(k)|2^k}{(k!)^{1/2+\alpha-\alpha'}} \times \sum_\ell \frac{|\mu(\ell)|2^\ell}{(\ell!)^{1/2+\alpha-\alpha'}} ,$$

avec $A = \sum_s \dfrac{|h|^s}{(s!)^{2\alpha'-1}}$.

Si $(\hat{u}_p)_{p\in\mathbb{N}}$ est une suite d'éléments de notre algèbre qui tend vers 0 quand $p \to +\infty$:

$$\hat{u}_p = \sum_k \frac{\lambda^{(p)}(k)}{(k!)^{1/2+\alpha}} (e_1^*)^k$$

$\lambda^{(p)}(k)$, pour k fixé tend nécessairement vers 0.

Ecrivons :

$$(12)\quad \sum_{k=0}^{\infty} \frac{|\lambda^{(p)}(k)|2^k}{(k!)^{1/2+\alpha-\alpha'}} = \sum_{k=0}^{k_o-1} \frac{|\lambda^{(p)}(k)|2^k}{(k!)^{1/2+\alpha-\alpha'}} + \sum_{k=k_o}^{\infty} \frac{|\lambda^{(p)}(k)|2^k}{(k!)^{1/2+\alpha-\alpha'}}$$

k_o étant fixé le premier paquet tend vers 0 quand $p \to +\infty$.

On peut supposer au préalable que k_o a été choisi de manière que

$$\sum_{k=k_o}^{\infty} \frac{|\lambda^{(p)}(k)|2^k}{(k!)^{1/2+\alpha-\alpha'}} < \varepsilon, \text{ pour tout } p$$

car il suffit de faire en sorte que

$$\sum_{k=k_o}^{\infty} \frac{ST^k\, 2^k}{(k!)^{1/2+\alpha-\alpha'}} < \varepsilon \ ,$$

et cela assure la continuité à gauche.

Proposition 3 : Le complété de l'algèbre large avec contrainte (5 bis) admet une structure d'algèbre associative (avec contrainte).

Si $(\hat{u}_p)$, $(\hat{v}_p)$ sont des suites de Cauchy de l'algèbre large avec contrainte, nous appelons $\hat{u}$ la limite de la suite $(\hat{u}_p)$ et $\hat{v}$ la limite de la suite $(\hat{v}_p)$ dans le complété vectoriel.

$(\hat{u}_q - \hat{u}_p)\hat{v}_q - \hat{u}_p(\hat{v}_p - \hat{v}_q) = \hat{u}_q\, \hat{v}_q - \hat{u}_p\, \hat{v}_p$

$$\|\hat{u}_q\, \hat{v}_q - \hat{u}_p\, \hat{v}_p\|_1 \leq \|(\hat{u}_q - \hat{u}_p)\, \hat{v}_q\|_1 + \|\hat{u}_p(\hat{v}_p - \hat{v}_q)\|_1$$

La suite $\hat{u}_p\, \hat{v}_p$ est aussi de Cauchy, de sorte que l'on peut poser :

$\hat{u}\,\hat{v} = \lim_{p\to\infty} \hat{u}_p\,\hat{v}_p$, si la suite $\hat{u}_p\,\hat{v}_p$ satisfait à (5 bis)

Même conclusion pour $\hat{u}_p + \hat{v}_p$, $\lambda\hat{u}_p$, $\lambda \in \mathbb{K}$.

4°) Groupes de Clifford et groupes spinoriels symplectiques.

a) Définition 1.

Nous appellerons groupe de Clifford symplectique formel G_S^H le sous-groupe des éléments $\gamma \in \check{C}_S(hF)^*$ tels que $\gamma x \gamma^{-1} \in E_H$, pour tout $x \in E_H$.

Il est immédiat de voir que G_S^H est un groupe, que $p : \gamma \longrightarrow p(\gamma)$ avec $p(\gamma)(x) = \gamma x \gamma^{-1}$ est un homomorphisme de G_S^H dans $Sp(n, \mathbb{K}[[h]])$, groupe des isomorphismes $\mathbb{K}[[h]]$-linéaires σ de E_H tels que $F(\sigma x, \sigma y) = F(x,y)$, et que le noyau de p est $\mathbb{K}[[h]]^*$.

Les éléments $\exp(ta^2)$, $t \in \mathbb{K}[[h]]$, $a \in E_H$, sont tels que

(8) $\exp(ta^2)x\,\exp(-ta^2) = x+2hF(a,x)ta$

et font partie de G_S^H pour tout $t \in \mathbb{K}[[h]]$. Lorsque $\mathbb{K} = \mathbb{R}$ ou $\mathbb{C}$, nous pouvons considérer h comme un élément (arbitraire mais choisi) de $\mathbb{R}$ ou $\mathbb{C}$. E_H devient un espace vectoriel E et $p(\gamma)$ devient une transformation symplectique de E, $p(\gamma) \in Sp(n,\mathbb{K})$

$p(\exp ta^2)$, $t \in \mathbb{K}$, $a \in E$, est une transvection symplectique et comme $Sp(n,\mathbb{K})$ est engendré par les transvections symplectiques on voit que l'image de p devient $Sp(n,\mathbb{K})$, et est engendrée par l'image d'éléments exponentiels selon (8). Cela nous conduit à la définition (2) :

Définition 2.

Nous appellerons groupe de Clifford symplectique G_S le quotient du sous-groupe des éléments de G_S^H de la forme

$\pi_i \lambda_i(h) \exp(t_i(a_i)^2)$ $\lambda_i(h) \in K [[h]]$, $t_i \in \mathbb{K}$, $a_i \in E$,

par le groupe des éléments $(1+u)$, $u \in K [[h]]$.

On observera d'ailleurs que les produits d'exponentielles formelles ne font apparaître que des séries formelles scalaires convergentes, on pourra considérer que le noyau de p est dans $\mathbb{K}^*$ et on aura la suite exacte :

$$(9) : 1 \longrightarrow \mathbb{K}^* \longrightarrow G_S \underset{p}{\longrightarrow} Sp(n,\mathbb{K}) \longrightarrow 1$$

De plus comme $Sp(n,\mathbb{K})$ est connexe, tout élément de $Sp(n,K)$ est un produit fini de transvections symplectiques appartenant à un voisinage arbitrairement petit de l'identité; on voit donc que l'on peut considérer G_S comme engendré par des exponentielles pour lesquelles la condition de convergence $|\rho(\hat{u})| < \frac{1}{\sqrt{2h}}$ est satisfaite et qui constituent un groupe de Lie local Γ que l'on peut construire à partir d'un voisinage $\mathcal{U}$ de 0 dans l'algèbre de Lie $\check{L}_h$ obtenue de $\check{L}$ en fixant h.

Pour X, Y $\in \check{L}$ on a toujours

$$(10) \exp Y. \exp X. \exp(-Y) = \exp(\sum_0^{\infty}(\frac{1}{n!}(adY)^n(X)))$$

ad est continue dans $\check{L}_h$ algèbre de dimension finie de sorte que :

$$\|(adY)(X)\|_1 \leqslant M\|X\|_1 \quad \|Y\|_1$$

M étant une constante et $\hat{u} \longrightarrow \|\hat{u}\|_1$ une norme euclidienne sur $\check{L}_h$.

Si Y est fixé, posant $\alpha = M\|Y\|_1$, la norme de $\sum_0^{\infty} \frac{1}{n!}$ (ad Y)^n(X) est inférieure à $\|X\| \exp \alpha$ et on voit que l'on peut choisir X suffisamment petit : $X \in \mathcal{V} \subseteq \mathcal{U}$ de manière que le deuxième membre de (10) soit dans Γ et arbitrairement voisin de l'élément neutre.

Comme $g \in G_S$ est un produit fini d'éléments exponentiels, on voit que si $X \in \check{L}_h$ est suffisamment petit $g(\exp X)\, g^{-1}$ sera arbitrairement voisin de e. Ce résultat est essentiel pour démontrer la proposition qui suit :

Proposition 4.

G_S a une structure de groupe de Lie.

Il suffit d'appliquer la proposition 118, page 112 de [1] que nous rappelons en adaptant l'énoncé à nos notations :
G_S est un groupe, si on peut trouver dans le groupe de Lie local Γ qu'il contient un ouvert $\mathcal{U}$ contenant e, muni d'une structure de variété analytique et V ouvert de $\mathcal{U}$ tel que $V = V^{-1}$, $V^2 \subseteq \mathcal{U}$ avec :

(I) $(x,y) \longrightarrow xy^{-1}$ de $V \times V$, dans $\mathcal{U}$ analytique

(II) pour tout $g \in G_S$ il existe un voisinage ouvert V' de e, $V' \subseteq V$ avec $gV'g^{-1} \subseteq \mathcal{U}$ et $x \to gxg^{-1}$ de V' dans $\mathcal{U}$ analytique,
alors G_S admet une structure unique de groupe de Lie.

b) Les groupes spinoriels symplectiques et le revêtement d'ordre 2 de $Sp(n, \mathbb{R})$.

Prenons $\mathbb{K} = \mathbb{C}$, le a) donne un groupe G'_S pour un espace E'; β est l'antiautomorphisme principal de l'algèbre de Clifford tel que $\beta_{|E'} = i\mathrm{Id}$. Il est immédiat de voir, utilisant le centre de l'algèbre et la factorisation de γ en exponentielles sur $\check{L}'$ (ou $\check{L}'h$), que si

$\gamma \in G'^{H}_{S}$, $N(\gamma) = \beta(\gamma)\gamma \in \mathbb{C}[[h]]^*$.

$$N(\gamma\gamma') = N(\gamma)\, N(\gamma')$$

et si $\gamma \in G'_S$

$$N(\gamma) \in \mathbb{C}^*.$$

Définition : E' étant l'espace complexifié de E, E muni de la forme symplectique hF, le sous-groupe des éléments γ de G'_S tels que $\gamma x \gamma^{-1} \in E$ avec $|N(\gamma)| = 1$ s'appelle groupe métaplectique $M_p(r)$.

Il donne lieu à la suite exacte :

(11) $1 \longrightarrow S^1 \longrightarrow Mp\,(r) \longrightarrow Sp(n, \mathbb{R}) \longrightarrow 1$ $\qquad (n=2r).$

(de sorte que l'appellation "toroplectique" serait certainement préférable).

Plus particulièrement : Si on choisit dans les mêmes conditions $N(\gamma) = 1$, on obtient le groupe $Sp_2(r)$ donnant la suite exacte :

(12) $1 \longrightarrow \mathbb{Z}_2 \longrightarrow Sp_2(r) \longrightarrow Sp(n, \mathbb{R}) \longrightarrow 1$

$Sp_2(r)$ est le groupe spinoriel symplectique d'ordre 2.
Pour la construction de $Sp_q(r)$, $Sp_\infty(r)$ le lecteur pourra se reporter à notre précédent article [2,a] formellement inchangé.

c) Les représentations spinorielles des groupes cliffordiens et spinoriels symplectiques.

On a introduit dans [2,a] une limite projective d'une suite d'idéaux à gauche de $C_S(hF)^\alpha_\ell$, (avec $\alpha = \frac{1}{2}$) notée $C_S(hF)^\alpha_\ell \, \Phi^*$, nous conserverons cet espace de représentation pour α quelconque, et ferons opérer $C_S(hF)^\alpha_\ell$ par produit à gauche. Si l'on peut faire agir le groupe G_S et les groupes dérivés, on peut opérer localement. Cependant il semble préférable de garder le point de vue formel et d'introduire l'algèbre

$\check{C}_S(hF)$ quotientée par l'idéal à gauche $\mathcal{J}_m$ engendré par les (e_{α^*}) (dont les éléments s'écrivent $\Sigma\lambda_{HK^*}e^H e^{K^*}$ avec K^* non vide) comme espace de représentation, cette représentation s'obtiendra donc par produit à gauche et passage au quotient ; cet espace de spineurs sera encore noté $\check{C}_S(hF)\Phi^*$; ce choix est lié à celui d'une base symplectique ordonnée, les différents choix se déduisent par extension naturelle de transformations symplectiques et correspondent à des représentations équivalentes.

On notera que sauf pour G_S^H, on devra quotienter à droite par les éléments de la forme $(1+u)$, $u \in \mathbb{K}[[h]]$.

Finalement nous retrouverons, en substance, après avoir précisé et affiné nos définitions et nos développements, la situation décrite dans [2,a]. Nous renvoyons le lecteur à cet article en ce qui concerne l'étude locale des groupes de Lie introduits et les diverses applications que nous en avions déduites.

d) Le groupe de Heisenberg et les revêtements du groupe conforme affine symplectique.

Lorsque la forme bilinéaire hF est dégénérée de rang $r < n$, si hF_1 est la restriction de hF à un supplémentaire E_1 du radical de E (rad. E), on a établi dans [2,b] que

$$C_S(hF) \simeq C_S(hF_1) \otimes \vee(\text{rad } E) \text{ et que le centre de } C_S(hF)$$

est $\vee(\text{rad } E)$, $\vee$ désignant le produit symétrique. L'introduction de h permet de donner un sens aux produits d'exponentielles même si les conditions de convergence évoquées au 3°) ne sont pas satisfaites. Il n'y a donc rien à modifier à ce que nous avons présenté formellement dans [2,b].
En particulier, si $r = (n-1)$ et $\mathbb{K} = \mathbb{C}$, on peut obtenir le revêtement d'un groupe produit semi-direct du groupe symplectique de E_1 et du groupe de Weyl-Heisenberg par introduction d'un espace de dimension n+1, somme directe de E_1, de rad E, et d'un

autre espace de dimension 1 ([2.b, p.248]), et du groupe métaplectique correspondant.

5°) Déformations tronquées de l'algèbre symétrique (à l'ordre p).

Considérons le quotient $\overline{H}$ de l'algèbre $\mathbb{K}[h]$ des polynômes en h par l'idéal des multiples de h^p, p étant un entier fixé (p⩾2). $\overline{H}$ est une algèbre de dimension p sur $\mathbb{K}$ d'élément générique :

$$\sum_{i=0}^{p-1} a_i \overline{h}^i \ , \ a_i \in \mathbb{K},\ \overline{h} = C\ell(h)\ (\mathrm{mod}\ h^p).$$

Le $\overline{H}$-module $\overline{H} \otimes E$ noté encore $E_{\overline{H}}$ est libre et admet une structure d'espace vectoriel sur $\mathbb{K}$ de dimension 2pr : on pourra l'appeler un module de Planck tronqué.

$C_S(hF)$ puis $\check{C}_S(hF)$ se construisent comme plus haut, cette dernière algèbre a des éléments qui sont des "séries formelles symplectiques" mais à coefficient dans $\overline{H}$. Il sera aussi intéressant de la considérer comme une algèbre sur $\mathbb{K}$, de séries formelles symplectiques, selon les puissances des $(\overline{h}, (e_i))$.

La définition d'un groupe de Clifford $G_S^{\overline{H}}$ est identique à celle que l'on donne en 4°, a), $\overline{H}$ remplaçant H, on obtient une suite exacte telle que (9) mais $K[[h]]^*$ est remplacé par $\mathbb{K}[\overline{h}]^*$, cependant il n'est plus possible d'interpréter $\overline{h}$ comme un paramètre à valeurs dans $\mathbb{K}$ puisque $(\overline{h})^p = 0$.

Ordre d'une série formelle symplectique.

$\overline{h}$ sera considérée comme une indéterminée de degré 2, avec cette convention le degré total d'un monôme en $\overline{h}^p\, e^H\, e^{K^*}$ est d'après la formule (2) du 1°) un invariant relativement à tout changement de base symplectique ou autre : on pourra donc parler d'un polynôme homogène à coefficients dans $\mathbb{K}$.

<u>On appellera ordre total $\omega(\hat{u})$ de $\hat{u} \neq 0$ le plus petit des entiers $q \geqslant 0$ tels que la partie homogène de degré q soit non nulle.</u>

On aura alors comme en algèbre commutative :

$\omega(\hat{u} + \hat{v}) \geqslant \mathrm{Min}(\omega(\hat{u}), \omega(\hat{v}))$, si $\hat{u} + \hat{v} \neq 0$

$\omega(\hat{u}\, \hat{v}) \geqslant \omega(\hat{u}) + \omega(\hat{v})$

Il sera utile dans les changements quelconques de repère d'écrire (2) sous la forme :

$$(2\ \text{bis})\quad (e_i)^k(e_j)^\ell = (e_j)^\ell(e_i)^k + h\ell k F(e_i,e_j)(e_j)^{\ell-1}(e_i)^{k-1} + \dots + \dots(h)^p(F(e_i,e_j))^p C_\ell^p C_k^p p!(e_j)^{\ell-p}(e_i)^{k-p} + \dots$$

où lorsque h est remplacé par sa classe (mod h^p), ne figurent que p termes au 2ème membre.

Il existe un principe de substitution : Si les X_i sont des séries formelles dans $\check{C}_S(hF)$, sans terme constant et $f(e_1, e_2,\dots,e_n, \bar{h})$ une série formelle symplectique, $f(X_1, X_2,\dots,X_n, \bar{h})$ est encore une série formelle symplectique : la situation est de ce point de vue analogue au cas commutatif car le deuxième membre de la formule (2 bis) ne contient que p termes, les ordres de $(e_i)^k(e_j)^\ell$ et de $(e_j)^\ell(e_i)^k$ en les (e_i) seuls, diffèrent au plus de $2(p-1)$ de sorte que l'on ne rencontre dans la mise en ordre que des sommes d'un nombre fini de termes.

<u>Proposition 5</u>.

<u>Toute série formelle symplectique de $\check{C}_S(hF)$ est inversible si et seulement si son terme constant dans $\mathbb{K}$ est différent de 0.</u>

Cela résulte de l'inversibilité de $1-X$ et du principe de substitution.

(On remarquera que l'inversibilité d'un élément de $\mathbb{K}[h]$ équivaut à la non nullité de son terme constant en $\mathbb{K}$).

Revenant au cas non tronqué, nous introduisons comme dans (4°,c) l'espace de représentation $\check{C}_S(hF)\Phi^* = \mathcal{J}_g$, on pourra remplacer $\mathbb{K}[[h]]$ par $\mathbb{K}((h))$ dans ce qui suit.

Proposition 6.

La représentation de $\check{C}_S(hF)$ dans $\check{C}_S(hF)\Phi^*$ obtenue par produit à gauche est irréductible.

Soit $w = \Sigma\lambda_{JK^*} e^J \Phi^*$ un élément quelconque de $\mathcal{J}_g$. S'il existe un terme constant dans $\mathbb{K}$ dans le développement de w, w est inversible, donc $\mathcal{J}_g$ est monogène et la représentation est irréductible. S'il ne se trouve pas a priori de tel coefficient dans le développement de w on considère un terme de degré minimal pour l'ordre lexicographique en les (e_α), par exemple si e_1 figure avec coefficient différent de 0, on multiplie à gauche par e_1^*, on répète éventuellement, on reprend si nécessaire avec e_2*...etc... et on aboutit à un terme constant différent de 0 , on peut diviser par h.

On peut aussi donner de la proposition 6, une démonstration légèrement différente :
l'idéal à gauche $\mathcal{J}_m$(4°,c), est maximal, car si on adjoint à $\mathcal{J}_m$ un élément $u = \Sigma\lambda_H e^H$, la formule (2) montre que l'adjonction de u conduit à celle de 1, et à celle de tout élément de l'algèbre ; le quotient de $\check{C}_S(hF)$ par cet idéal maximal est donc un espace de représentation irréductible.

Proposition 7

L'algèbre $\check{C}_S(hF)$ est une algèbre simple.

Il existe un système d'idéaux à gauche maximaux dont l'intersection est nulle.

Si n = 2, r = 1, considérons les idéaux à gauche maximaux $\mathcal{J}_1$, $\mathcal{J}_2$, $\mathcal{J}_3$ tels que $\mathcal{J}_m$, construits sur les bases symplectiques (e_1,e_{1^*}), $(-e_{1^*},e_1)$, $(e_1+\sigma e_{1^*},e_{1^*})$, $\sigma \in \mathbb{K}$, $\sigma \neq 0$. On peut montrer que $\mathcal{J}_1 \cap \mathcal{J}_2 \cap \mathcal{J}_3 = 0$.

Un raisonnement par récurrence assure alors la propriété pour n quelconque. On obtient un système $\mathcal{J}_{a_i}$, i = 1,2,...q, d'idéaux maximaux à gauche d'intersection nulle, naturellement construits sur des bases symplectiques. Il existe une transformation symplectique, donc un automorphisme intérieur qui permet d'affirmer que tous ces idéaux maximaux $\mathcal{J}_{a_i}$ sont isomorphes.

$\check{C}_S(hF)$ étant considérée comme module à gauche sur elle-même, soit φ_{a_i} l'homomorphisme canonique :

$$\check{C}_S(hF) \longrightarrow \check{C}_S(hF)/\mathcal{J}_{a_i} = \hat{\mathcal{J}}_{a_i}.$$

$\prod_1^q(\varphi_{a_i})$ est un homomorphisme injectif de $\check{C}_S(hF)$ sur $\prod_1^q(\varphi_{a_i})$, car

$\prod_1^q(\varphi_{a_i})(u) = \prod_1^q(\varphi_{a_i})(u')$ implique $u' - u \in \mathcal{J}_{a_i}$ pour tout a_i,

donc u = u'.

$\mathcal{J}_{a_i}$ étant maximal, $\hat{\mathcal{J}}_{a_i}$ est un module simple, donc $\prod_1^q(\hat{\mathcal{J}}_{a_i})$ est semi-simple et de même $\check{C}_S(hF)$, mais comme tous les modules simples obtenus sont isomorphes, l'algèbre $\check{C}_S(hF)$ est semi-simple.

Remarque : Si $p_2 > p_1$, il existe un homomorphisme naturel $m_{p_2p_1}$ de l'algèbre tronquée $\check{C}_S(hF)_{p_1}$ sur l'algèbre tronquée $\check{C}_S(hF)_{p_2}$. Cette propriété résulte de la propriété universelle donnée dans [2,a].

Le système d'homomorphisme $m_{p_2p_1}$, $p_2 > p_1$, et d'algèbres tronquées $\check{C}_S(hF)_{p_i}$ définissent un système projectif d'applications et d'algèbres associatives dont la limite projective est $\check{C}_S(hF)$.

Remarque : Comme l'algèbre de Heisenberg engendre l'algèbre de Clifford, l'espace des spineurs symplectiques est espace de représentation inductible de cette algèbre. On retrouve ainsi le théorème de Stone-Neumann : toutes ces représentations sont équivalentes, nous en avons obtenu ainsi un modèle.

6°) Déformations des algèbres associées à une variété symplectique.

Nous reprenons brièvement dans ce contexte les notions déjà présentées en détails dans [2,c].

Avec les notations du 3°) nous prenons $\alpha = \frac{1}{2}$ et $h = 1$, posons pour abréger

$$C_S(F)_\ell^{1/2} = C_S(F)_\ell .$$

Nous considérons à partir d'un repère symplectique (e_α, e_{β^*}) également orthonormé pour une métrique euclidienne adaptée à F, un repère hermitien

$$\mathcal{R} = \{\varepsilon_\alpha, \varepsilon_{\beta^*}\}, \quad \varepsilon_\alpha = \frac{e_\alpha - ie_{\alpha^*}}{\sqrt{2}}, \quad \varepsilon_{\alpha^*} = \frac{e_\alpha + ie_{\alpha^*}}{\sqrt{2}}$$

$i_{\mathcal{R}}$ étant l'isomorphisme linéaire entre $C_S(F_{\mathbb{C}})_\ell$ et l'algèbre symétrique de $E_{\mathbb{C}}$, on note ($\circ$ loi de composition de $C_S(F_{\mathbb{C}})_\ell$) que :

(13) $\underline{i_{\mathcal{R}}(\hat{u}) \circ i_{\mathcal{R}}(\hat{v}) = i_{\mathcal{R}'}(\hat{u}) \circ i_{\mathcal{R}'}(\hat{v})}$, si $\mathcal{R}'$ est un autre repère hermitien, avec changement de base réduit au groupe unitaire ($\mathcal{R}' = A(\mathcal{R})$ avec $A^\alpha_{\beta^*} = A^{\alpha^*}_\beta = 0$).

(V,F) est une variété symplectique de dimension $n = 2r$ et le crochet de Poisson est

(14) $\{f,g\} = \frac{1}{2} F^{ij} (\partial_i f)(\partial_j g)$

pour deux éléments f, g $\in C^\infty(V)$, $C^\infty(V)$ devient ainsi une algèbre de Lie N(algèbre de Poisson).

$C_S(V,F)_\ell$ est le fibré de Clifford symplectique avec groupe $Sp(n,\mathbb{R})$ réductible à $U(r)$, il peut se complexifier en $C_S(V,F_{\mathbb{C}})_\ell$.

Soit $x_o \in V$, $(x^1,x^2,\ldots,x^n)$ des coordonnées locales avec $(x^1,x^2,\ldots,x^n)_{x_o} = (0,0\ldots0)$.

$\tilde{f}$ est le germe de $f \in C^\infty(x_o)$, $\tilde{f} \in \tilde{C}^\infty(x_o)$, notons $\overset{**}{V} T^*_{x_o}(V)$ l'algèbre des séries formelles construites sur l'espace cotangent en x_o à V.

∇ est une connexion symplectique, $\nabla F = 0$, sans torsion.

Nous définissons $\hat{f} \in \overset{**}{V} T^*_{x_o}(V)$ par :

(15) $\hat{f} = f(x_o)+(\nabla_i f)_{x_o} dx^i+\ldots S(\nabla_{\ell_1\ell_2\ldots\ell_k} f)_{x_o} dx^{\ell_1} \vee dx^{\ell_2} \vee\ldots dx^{\ell_k}+\ldots$

où le terme général de degré total k s'écrit :

$$\Sigma \frac{1}{q_1!q_2!\ldots q_n!} (\nabla_{(q_1q_2\ldots q_n)} f)_{x_o} (dx^1)^{q_1} \vee (dx^2)^{q_2} \ldots \vee (dx^i)^{q_i}$$

$q_1,q_2,\ldots q_n$ entiers positifs, $q_1+q_2+\ldots q_n = k$

$$\nabla_{(q_1q_2\ldots q_n)} f = \frac{q_1!q_2!\ldots q_n!}{k!} \Sigma(\nabla_{q_1q_2\ldots q_n} f)$$

$$\nabla_{q_1} = \nabla_{\partial/\partial x^1 \ldots \partial/\partial x^1} \quad (q_1 \text{ fois}).$$

Dans le cas plat :

$$\nabla_{(q_1q_2\ldots q_k)} f = \partial_{q_1q_2\ldots q_k} f.$$

Une formule de Leibnitz donne alors

(16) $\widehat{fg} = \hat{f}.\hat{g}$ (homomorphisme de Taylor),

le produit du deuxième membre étant dans l'algèbre symétrique.

Nous pouvons faire les hypothèses de travail :

$$\left.\begin{array}{ll}(H_1) & (\nabla_{(i_1 i_2 \ldots i_k)} f)_{x_0} \leqslant A a^k, \quad k \in \mathbb{N} \\ (H_2) & \varphi|_{\tilde{C}^H(x_0)} : \tilde{f} \longrightarrow \hat{f} \text{ injective}\end{array}\right\} \qquad (17)$$

A, a, constantes dépendant de x_0, $\tilde{C}^H(x_0) \subseteq \tilde{C}^\infty(x_0)$;

(H_1), (H_2) sont possibles si nous postulons pour V des conditions convenables d'analyticité, elles nous permettront de construire un algorithme de calcul que nous conserverons plus loin, même si (17) n'est pas satisfait.

a) déformations des algèbres associées à $\tilde{C}^\infty(x_0)$ (associatives et de Lie)

Nous utilisons (13) avec des sections de repères hermitiens,

$$\varphi_1 = i_{\mathcal{R}} \circ \varphi, \qquad \varphi : \tilde{f} \longrightarrow \hat{f}, \qquad \varphi_1(\tilde{f}) = \hat{\hat{f}}$$

et posons par définition de la loi * :

(18) $\tilde{f} * \tilde{g} = \varphi_1^{-1}(\hat{\hat{f}} \circ \hat{\hat{g}})$, où F a été changée en hF, $h \in \mathbb{R}$.

* donne une déformation de l'algèbre associative $\tilde{C}^H(x_0, \mathbb{C})$, l'associativité résultant d'un transfert de structure. Le calcul explicite de $\tilde{f} * \tilde{g}$ résulte théoriquement de la formule (2 bis) du (6°) adaptée aux bases hermitiennes, finalement :

$$(19) \quad \tilde{f} * \tilde{g} = \tilde{f}\tilde{g} + hF^{\alpha^*\beta}(\nabla_{\alpha^*}\tilde{f})(\nabla_\beta \tilde{g}) + \frac{h^2}{2!} F^{\alpha_1^*\beta_1} F^{\alpha_2^*\beta_2} (\nabla_{(\alpha_1^*\alpha_2^*)}\tilde{f})(\nabla_{(\beta_1\beta_2)}\tilde{g})$$

$$+ \frac{h^3}{3!} F^{\alpha_1^*\beta_1} F^{\alpha_2^*\beta_2} F^{\alpha_3^*\beta_3} (\nabla_{(\alpha_1^*\alpha_2^*\alpha_3^*}\tilde{f})(\nabla_{(\beta_1\beta_2\beta_3)}\tilde{g}) + \ldots$$

$$+ h^2\alpha_2(\tilde{f},\tilde{g}) + h^3\alpha_3(\tilde{f},\tilde{g}) + \ldots\ldots\ldots$$

α_2, α_3... contenant des termes de courbure.

(19 bis) $h[\tilde{f}, \tilde{g}] = \tilde{f} * \tilde{g} - \tilde{g} * \tilde{f}$ donne une déformation de l'algèbre de Poisson $\tilde{C}^H(x_o, \mathbb{C})$ ·

b) déformation de $C^\infty(V, \mathbb{C})$

(19) donne : $x_1 \longrightarrow (\tilde{f}_{x_o} * \tilde{g}_{x_o})(x_1)$

x_1 "voisin" de x_o.

Si $x_1 = x_o$, variant x_o nous obtenons l'algorithme :

$$(20)\quad (f * g)(x) = (\tilde{f}_x * \tilde{g}_x)(x) = f(x)g(x) + hF^{\alpha^*\beta}(\nabla_{\alpha^*}f)_x(\nabla_\beta g)_x$$

$$+ \frac{h^2}{2!}F^{\alpha_1^*\beta_1}F^{\alpha_2^*\beta_2}(\nabla_{(\alpha_1^*\alpha_2^*)}f)_x(\nabla_{(\beta_1\beta_2)}g)_x$$

$$+ \frac{h^3}{3!}F^{\alpha_1^*\beta_1}F^{\alpha_2^*\beta_2}F^{\alpha_3^*\beta_3}(\nabla_{(\alpha_1^*\alpha_2^*\alpha_3^*)}f)_x(\nabla_{(\beta_1\beta_2\beta_3)}g)_x + \dots$$

Oubliant (H_1) et (H_2) on peut considérer des déformations formelles de $C^\infty(V,\mathbb{C})$ et N selon (19) et (19 bis).

Si θ est une transformation symplectique on peut prouver que :

$$\theta(f * g) = \theta f * \theta g, \quad ((\theta f)(x) = f\circ\theta^{-1}(x))$$

Ainsi nous obtenons :

Théorème : Sur chaque variété symplectique il existe des déformations formelles pour l'algèbre associative $C^\infty(V,\mathbb{C})$ et pour l'algèbre de Poisson N.

c) Le cas réel.

En $f * g - g * f$, les termes impairs sont réels et les termes pairs (en h^{2k}, $k \geqslant 1$) sont imaginaires purs, donc

$f * g - g * f$, calculé mod h^{2k}, $k \geqslant 1$,

donne un crochet réel si f, g $\in C^{\infty}(V, \mathbb{R})$, ainsi nous avons le

<u>Corollaire</u> : <u>L'hypothèse suffisante de Vey [6] $H^3(V, \mathbb{R}) = 0$ pour obtenir une déformation de l'algèbre de Poisson est caduque</u>.

Que peut-on dire de la déformation de l'algèbre associative $\mathbb{C}^{\infty}(V, \mathbb{R})$?

Nous en donnons une condition suffisante d'existence.

Soit (V,F) une variété symplectique qui peut se munir d'un champ global d'espaces lagrangiens: le groupe $Sp(n, \mathbb{R})$ est réductible à $\mathcal{O}(r) \times \mathcal{O}(r)$. Utilisant la formule (2) on aboutit à une formule analogue à (19).

Cette construction réussit sur l'espace $T^*(M)$ (espace des phases).

d) <u>Relation avec le produit de Moyal</u>.

Dans le cas plat on peut montrer [2, c] que notre déformation est isomorphe à celle de Moyal (mais les degrés en h ne sont pas respectés).

Dans le cas non plat, notre déformation présente un caractère universel, il existe un homomorphisme de notre déformation sur toute déformation de type Moyal.

Nous renvoyons pour les détails à l'article indiqué [2, c].

On peut montrer aisément que la méthode s'applique, sur chaque variété symplectique aux déformations de l'algèbre associative $C^{\infty}(V.\mathbb{R})$ en utilisant une correspondance de Kähler-Atiyah, analogue à celle du cas orthogonal.

Références

1. N. BOURBAKI - Groupes et algèbres de Lie. Chapitres 2 et 3 - Hermann. Paris. 1972.

2. A. CRUMEYROLLE - a) Algèbre de Clifford symplectique. Revêtement du groupe symplectique ; indices de Maslov et spineurs symplectiques. J. Math. pures et appl. 56. 1977. p.205 à 230.
b) Algèbres de Clifford dégénérées...etc... Annales de l'I.H.P. Vol.33, n°3, 1980 - p.235-249.
c) Déformations d'algèbres associées à une variété symplectique.... Annales I.H.P. Vol.35, n°3, 1981 - p.175-194.

3. R. HOWE - On the rôle of the Heisenberg group in harmonic analysis. (Bull. of the Am. math. Society - Vol.3, n°2 - Sept.1980).

4. J. HELMSTETTER - Produits intérieurs de séries formelles et algèbres de Clifford symplectiques. (C.R. Acad. Sc. Paris, t.293 - 6-7-81).

5. J. LERAY - . Analyse lagrangienne et mécanique quantique. Séminaire du Collège de France. 1976-77. R.C.P. 25, Strasbourg 1978.
Maslov's asymptotic method : the need of Planck's constant in mathematics (Bull. of Am. Math. Society July 1981 - Vol.5, n°1).

6. J. VEY - Déformation du crochet de Poisson sur une variété symplectique. Comment. Math. Helvetici t.50. 1975. p.421-454.

Université Paul Sabatier
Mathématiques
118, route de Narbonne
F-31062 Toulouse Cédex, France

ALMOST PLURIHARMONIC FUNCTIONS AND SYMPLECTIC MAPPINGS ON GENERALIZED KÄHLER MANIFOLDS

Stančo Dimiev and Tsvetan Pelov (Sofia)

Contents

Summary. In this paper we develop an approach to the symplectic mappings of non-compact generalized Kähler manifolds with non-trivial almost pluriharmonic functions.

Introduction

The development of the generalized Kähler geometry (Koto, Gray, and others) inspires the investigations of an appropriate function theory on generalized Kähler manifolds (almost pluriharmonic and almost holomorphic functions on almost Kähler, nearly Kähler etc. manifolds [3], [4], [5], and [6]).

In order to facilitate the communication between the geometry and analysis on mentioned manifolds here we develop the language of symplectic geometry and related questions. The main point is that on every almost complex manifold (M,J) symplectic 2-forms $\Psi = dJ^*dU$, $U \in \mathcal{C}^\infty(M)$, are considered as being closely related to the fundamental equations of almost pluriharmonic functions (see (1.1), (1.3)) and also to the morphisms in the category of almost complex manifolds, i.e. to the almost holomorphic mappings (see [2]). The manifold (M,J) equipped with a fixed 2-form Ψ is denoted (M,J,Ψ). The almost holomorphic symplectic automorphisms of such manifolds, i.e. $f: M \to M$ with $f^*\Psi = \Psi$, are described in terms of almost pluriharmonic functions.

In the case of almost Kähler manifold (M,g,J), if the fundamental 2-form can be represented as a 2-form of the mentioned kind, we say that (M,g,J) is an APH-convex manifold. Certain domains of almost holomorphy give some examples.

In Sec. 4. we discuss some local proprieties of some more general symplectic automorphisms of (M,J,Ψ). A necessary and sufficient condition for the local reduction of the equality

$$f^*\Psi = \Psi$$

to a simpler one $f^*\Psi = \Psi$, Ψ being 1-form, is described. That is the condition (*) (see (4.1)).

In Sec. 5. the results of the previous paragraph are used to prove that the periodic automorphisms satisfy the condition (*) and also that in the class of the mentioned automorphisms the local results are in fact global ones.

1. Recall of the almost pluriharmonic functions

A real-valued function u defined on the almost complex manifold (M,J) is called an almost pluriharmonic (or more precisely: J-almost-pluriharmonic) function if it satisfies the following equation

$$dJ^*du = 0, \tag{1.1}$$

where J^* is the conjugate fiber linear operator of $J: TM \to TM$.

If $(U, x = (x^k))$ is a local coordinate system on M, we set

$$J_q = \begin{bmatrix} J^1_q \\ \vdots \\ J^{2\mu}_q \end{bmatrix}, \quad \nabla u = (\partial u/\partial x^1, \ldots, \partial u/\partial x^{2\mu}),$$

where the matrix $\|J^p_q(x)\|$ is the local expression of the almost complex structure J and $2\mu = \dim_{\mathbb{R}} M$. Then,

$$J^*du = \sum_{q=1}^{2\mu} (J_q \cdot \nabla u)dx^q$$

and

$$J_q \cdot \nabla u = \sum_{p=1}^{2\mu} J^p_q \, \partial u/\partial x^p.$$

It is clear that u is an J-almost-pluriharmonic function iff J^*du is closed and so

(1.2) $\int_\gamma J^* du = 0$

on any simply connected domain D of M, γ being any closed curve (cycle) in D. We also have

(1.3) $\partial(J_q \cdot \nabla u)/\partial x^s = \partial(J_s \cdot \nabla u)/\partial x^q, \quad s,q = 1,\dots,2\mu.$

It is known [1] and [3] that each almost pluriharmonic function satisfies locally the following equation

$$\Delta_J u = 0,$$

where Δ_J is the so-called J-<u>Laplace operator</u>, i.e.

$$\Delta_J = \sum_{s,p=1}^{2\mu} A_{sp} \partial^2/\partial x^s \partial x^p + \sum_{p=1}^{2\mu} B_p \partial/\partial x^p,$$

where

$$A_{sp} = \sum_{q=1}^{2\mu} (J_q^s J_q^p + \delta_q^s \delta_q^p), \quad B_p = \sum_{s,p=1}^{2\mu} J_q^s (\partial J_q^p/\partial x^s - \partial J_s^p/\partial x^q).$$

Let us set $A_J = \|A_{sp}\|$. Then we have

$$A_J = JJ^* + E_{2\mu};$$

here J is the matrix $\|J_q^p\|$ and J^* is the transposed matrix, $E_{2\mu}$ is the unit $(2\mu \times 2\mu)$-matrix. In the particular case when $J = S$ (the standard almost complex structure), $A_S = 2E_{2\mu}$ and $\Delta_S = 2\Delta$ (Δ is the ordinary Laplace operator of several variables).

It is proved that Δ_J is an elliptic operator [1], [3]. This fact implies the fundamental properties of the almost pluriharmonic functions: regularity and maximum principle. In the case of real analytic M and real analytic J, the almost pluriharmonic functions on (M,J) satisfy also the principle of unique continuation.

The almost pluriharmonic functions are canonically connected with the almost holomorphic functions. By definition, the function $f = u + iv$ is said to be <u>almost holomorphic</u> if

$$du = J^* dv.$$

It is clear that the real resp. imaginary part of an almost holomorphic function is an almost pluriharmonic one.

The notion of the almost holomorphic function is the particular case of the notion of almost holomorphic mapping.

2. Some symplectic structures on non-compact almost complex manifolds and almost holomorphic mappings

Let (M,J) be an almost complex manifold of dimension 2μ ($\dim_{\mathbb{R}} M = 2\mu$). By $\mathcal{E}^{o}(M)$, respectively by $\mathcal{E}^{2}(M)$, we denote the space of smooth functions on M of class $\mathcal{C}^{\infty}$, respectively the space of differential 2-forms on M. We consider the operator

$$(2.1)\qquad dJ^{*}d : \mathcal{E}^{o}(M) \to \mathcal{E}^{2}(M),$$

which is globally defined on the manifold M. For every smooth function U defined globally on M we have the closed differential 2-form $dJ^{*}dU$. The set of all such 2-forms is a vector subspace of $\mathcal{E}^{2}(M)$, this is the image of $\mathcal{E}^{o}(M)$ by the operator $dJ^{*}d$. We will denote this subspace of $\mathcal{E}^{2}(M)$ by $Sp_{J}\mathcal{E}^{2}(M)$.

The vector space of all almost pluriharmonic functions defined globally on M will be denoted by $APH_{J}(M)$. There are different examples of (M,J) when the space $APH_{J}(M)$ is non empty: almost complex vector spaces $(\mathbb{R}^{2\mu}, J)$ for some J, all nearly Kähler manifolds [6], some Lie groups and homogeneous manifolds [11], etc.

Evidently the following sequence is exact:

$$(2.2)\qquad 0 \to APH_{J}(M) \hookrightarrow \mathcal{E}^{o}(M) \overset{dJ^{*}d}{\to} Sp_{J}\mathcal{E}^{2}(M) \to 0.$$

Thus, $Sp_{J}\mathcal{E}^{2}(M)$ is isomorphic to the factor space of $\mathcal{E}^{o}(M)$ by $APH_{J}(M)$. By definition every element of $Sp_{J}\mathcal{E}^{2}(M)$ defines a symplectic structure on M if it is a non degenerate 2-form.

Remark. There are no symplectic structures in the above mentioned sense on a compact manifold M.

PROPOSITION 1. *A closed differential 2-form Ψ on M is an element of $Sp_{J}\mathcal{E}^{2}(M)$ if there exists a 1-form ψ on M such that $d\psi = \Psi$ and $d(J^{*}\psi) = 0$. The converse statement holds locally.*

Proof. Let $\Psi = dJ^{*}dU$. Let us set $\psi = J^{*}dU$. Then $dU = -J^{*}\psi$ and also $d(J^{*}\psi) = 0$. Conversely, let $d(J^{*}\psi) = 0$. The Poincaré lemma implies $dU = -J^{*}\psi$ with a smooth U defined locally and also $\Psi = dJ^{*}dU$.

Recall that a map $F: M \to M$ is called an *almost complex map* of M if the following equality holds

$$(2.3)\qquad F_{*} \circ J = J \circ F_{*},$$

F_* being the differential of F. This means that for every point $m \in M$ we have $F_{*,m} \circ J_m = J_{F(m)} \circ F_{*,m}$. By $AH_J(M)$ the group of all almost complex automorphisms of M is denoted.

Now let us denote by (M,J,Ψ) the almost complex manifold (M,J) equipped with a fixed symplectic 2-form Ψ, $\Psi \in Sp_J \mathcal{E}^2(M)$. Ψ-symplectic mappings $f : M \to M$ are defined as usually as

(2.4) $\quad f^*\Psi = \Psi,$

f^* being the codifferential of f.

PROPOSITION 2. *Given* $(M,J,\Psi = dJ^* dU)$ *and* $f \in AH_J(M)$, *we have that* $f^*\Psi = \Psi$, *i.e.* f *is a symplectic mapping of* M *iff*

$$U - U \circ f \in APH_J(M).$$

Proof. Having in mind that f^* commutes with $df = f_*$ and also (2.3) we conclude that $f^*\Psi = f^*(dJ^* dU) = dJ^* d(U \circ f)$.

Now, if $f^*\Psi = \Psi$, we get $dJ^* d(U \circ f) = dJ^* dU$, i.e. $dJ^*d(U - U \circ f) = 0$. Conversely, if $dJ^*d(U - U \circ f) = 0$, we obtain $dJ^*d(U \circ f) = dJ^* dU$, i.e. $f^*\Psi = \Psi$.

3. Potentials of the fundamental 2-form of almost Kähler manifolds and APH-convex manifolds

Let (M,g,J) be an almost hermitian manifold. In general the Riemann metric is not an Hermitian one, but $h(X,Y) = 1/2(g(X,Y) + g(JX,JY))$ is hermitian, i.e. $h(JX,JY) = h(X,Y)$, $X,Y \in \mathfrak{X}(M)$. By definition the 2-form $\Phi(X,Y) = h(X,JY)$ is the fundamental 2-form of (M,g,J). The local expression of Φ in local coordinate system $(U, x = (x^k))$ on M is the following

(3.1) $$\Phi = \sum_{k<l} (J^k_l - J^l_k) dx^k \wedge dx^l.$$

We suppose that the structural group of (M,g,J) is an orthogonal group, i.e. $g(\partial/\partial x^i, \partial/x^j) = \delta^i_j$ in all local coordinate systems.

It is easy to verify that the condition for Φ to be closed is equivalent to the following system of partial differential equations:

(3.2) $$\partial J^k_l/\partial x^m + \partial J^l_m/\partial x^k + \partial J^m_k/\partial x^l = \partial J^l_k/\partial x^m + \partial J^m_l/\partial x^k + \partial J^k_m/\partial x^l, \quad 1 \leq k < l < 2\mu.$$

Now let (M,g,J) be an almost Kähler manifold, i.e. its fundamental form Φ is closed. We can consider (M,g,J) as a symplectic manifold related to Φ. Denote by $Sp_\Phi(M)$ the group of Φ-symplectic automorphisms of M and by $\mathcal{O}(M)$ the group of orthogonal automorphisms of M.

THEOREM 1. The following equalities are valid:

$$AH_J(M) \cap Sp_\Phi(M) = Sp_\Phi(M) \cap \mathcal{O}(M) = \mathcal{O}(M) \cap AH_J(M).$$

Proof. Firstly, we remark that $f \in Sp_\Phi(M)$ iff

(3.3) $\quad f_*^\tau \circ J \circ f_* = J$

(this means that for every point $m \in M$ we have $(f_{*,m})^\tau \circ J_{f(m)} \circ f_{*,m} = J_m$). The meaning of $f_*^\tau : TM \to TM$ is the following: in local coordinates f_*^τ there is the transposed matrix of the Jacobi matrix of f, but it is not difficult to see that in fact f_* does not depend on the choice of the local coordinate system. The same remark is valid for the equality (3.3).

Secondly, having in mind (3.3) and (2.3) we conclude that $f_*^\tau \circ f_* = 1_{TM}$ (1_{TM} is the identity of TM), i.e. $f \in \mathcal{O}(M)$. Thus we have

$$AH_J(M) \cap Sp_\Phi(M) \subset \mathcal{O}(M).$$

Analogously, we obtain

$$Sp_\Phi(M) \cap O(M) \subset AH_J(M)$$

and also

$$\mathcal{O}(M) \cap AH_J(M) \subset Sp_\Phi(M).$$

The theorem follows by the obtained inclusions.

Remark. In the case of $\mathbb{R}^{2\mu}$ equipped with the standard almost complex structure S, i.e. in the case of $\mathbb{C}^\mu$, Theorem 1 is well known (see [7]).

COROLLARY 1. In the case of compact M the group $U_J(M) = AH_J(M) \cap \mathcal{O}(M)$ equipped with an appropriate topology (we say that this is the group of almost holomorphic isometries of the manifold M) is a Lie group of transformations of M.

Since $U_J(M)$ is a closed subgroup of $AH_J(M)$, then this fact follows from [1], where it is proved that $AH_J(M)$ equipped with the topology of uniform convergence on all compact subsets of M including the derivatives up to the third order is a Lie group of transformations.

Given an almost Kähler manifold (M,g,J), let G be an open subset of M. We say that the fundamental 2-form Φ of (M,g,J) has a potential U on G if $\Phi = dJ^* dU$, i.e. $\Phi \in Sp_J \mathcal{E}^2(G)$.

Since the local expression of $dJ^* dU$ reads

$$dJ^* dU = \sum_{q<p} (\partial(J_p \nabla u)/\partial x_q - \partial(J_q \nabla u)/\partial x_p)\, dx_q \wedge dx_p$$

(see the notations in Section 1), the question of the local existence of a potential of Φ is equivalent to the question of the existence of solutions of the following overdetermined system of partial differential equations:

$$(3.4) \quad \partial(J_p \nabla u)/\partial x_q - \partial(J_q \nabla u)\partial x_p = J^p_q - J^q_p, \quad p,q = 1,\ldots,2\mu, \quad q<p.$$

If $V \subset M$, the restriction of Φ on V is denoted by $\Phi|V$. Let $\iota : V \subset M$ be the embedding; then $\Phi|V = \iota^*\Phi$.

The manifold (M,g,J) is called locally APH-convex if for any point $m \in M$ there exists a neighbourhood V_m of m such that $\Phi|V_m = dJ^* dU_m$, $U_m \in \mathcal{C}^\infty(V_m)$. In general we have $\Phi|V_m = d\Psi_m$, Ψ_m is 1-form on V_m, but Ψ_m is not always equal to $J^* dU_m$ (see Proposition 1).

In the case of the global existence of a potential U of Φ on G we say that G is a APH-convex domain of M.

Since the difference $U_1 - U_2$ of two potentials U_1 and U_2 of Φ on G is an almost-pluriharmonic function on G, in the case of real analytic almost Kähler manifold (M,g,J) we define the notion of a maximal APH-convex domain as follows: G is a maximal APH-convex domain if there exist two potentials of Φ on G whose difference cannot be extended outside G. Thus, the maximal APH-convex domains of M are domains of almost holomorphy ([3] and [5]).

COROLLARY 2. *If $\gamma \subset M$ is a cycle on the APH-convex manifold (M,g,J), the integral $\int_\gamma J^* dU$ is an invariant of any symplectic transformation of M, in particular of any almost holomorphic isometry on M.*

The proof follows by the Stokes' formula.

In the case where U is an almost pluriharmonic function, this integral is equal to zero (see (1.2)).

4. Local properties of some symplectic automorphisms

Let f be a smooth automorphism of the almost complex manifold

(M,J) and Ψ be a closed 2-form on M. Suppose that G is an open set in M satisfying the following three conditions: 1) $\pi_1(G) = 0$, i.e. G is a simply connected domain of M; 2) $\iota^*\Psi = dJ^* dU$, $U \in \mathcal{C}^\infty(G)$, i.e. $\Psi \in Sp_J{}^2(G)$; 3) $f(G) = G$. The set G can be chosen as being holomorphic to an open ball in $\mathbb{R}^{2\mu}$ and as small as necessary ($2\mu = \dim_{\mathbb{R}} M$). We shall consider the set of automorphisms of M satisfying the condition 3).

<u>Definition 1</u>. We say that f <u>satisfies the condition</u> $(*)$ if there exist smooth functions θ and ρ, $\theta, \rho \in \mathcal{C}^\infty(G)$, depending on f, such that

$$(*) \quad \begin{aligned} &\text{a)} \quad f^*(J^* dU) - J^* dU = d\theta, \\ &\text{b)} \quad \rho - \rho f = \theta. \end{aligned}$$

<u>Remark 4.1</u>. It is possible to prove that the definition is independent of the choice of $U \in \mathcal{C}^\infty(G)$.

<u>Remark 4.2</u>. In the case where f is almost holomorphic, the condition $(*)$ implies Proposition 2.

<u>PROPOSITION 3</u>. <u>Let</u> f <u>be a symplectic automorphism of</u> G. <u>Then</u> f <u>satisfies the condition</u> $(*)$ <u>iff there exists a</u> 1-<u>form</u> ψ <u>on</u> G <u>such that</u> $d\psi = \iota^*\Psi$ <u>and</u> $f^*\psi = \psi$ <u>on</u> G.

<u>Proof</u>. Let f satisfy $(*)$ and $\psi = J^* dU + d\rho$. Then $d\psi = dJ^* dU = \iota^* \Psi$ and $f^*\psi = f^*(J^* dU) + d(\rho f) = d^* dU + d\rho = \psi$.

Now let there exist a 1-form ψ such that $d\psi = \iota^*\Psi$ and $f^*\psi = \psi$. Then of course $f^*\Psi = \Psi$, i.e. $d(f^*(J^* dU) - J^* dU) = 0$. As G is simply connected, the universal coefficient formula implies the exactness of any closed 1-form on G. Thus, there exists a smooth function θ on G such that $f^*(J^* dU) - J^* dU = d\theta$. Since $d\psi = dJ^* dU = \Psi$ there exists a smooth function ρ on G such that $\psi - J^* dU + d\rho$. Thus $d\theta = d(\rho - \rho f)$, i.e. $\rho - \rho f = \theta + c$, $c = \text{const}$. Let $\theta_1 = \theta + c$. Then $f^*(J^* dU) - J^* dU = d\theta = d\theta_1$ and $\rho - \rho f = \theta_1$, i.e. f satisfies $(*)$.

The condition $(*)$ is not trivial, i.e. there are smooth symplectic automorphisms which do not satisfy it. In other words, there are smooth automorphisms preserving the symplectic 2-form but preserving none of the local potentials of it. One can prove that for example the following smooth automorphism $f = (f_1, f_2)$ of $\mathbb{R}^2$, considered as a Kähler manifold according to the standard almost complex structure, is symplectic with respect to the standard symplectic 2-form $\Psi = dx \wedge dy$, but $f^*\psi \neq \psi$

for any 1-form ψ, such that $d\psi = \Psi$:

$$f_1(x,y) = x + \sin(x+y) + \cos(x+y),$$

$$f_2(x,y) = y - \sin(x+y) - \cos(x+y).$$

Using Darboux theorem in a convenient way one can generalize this example to a class of almost Kähler manifolds.

A question arises whether the system of partial differential equations $f^*\Psi = \Psi$ is equivalent to the simpler one $f^*\psi = \psi$ (ψ being 1-form) in the set of smooth automorphisms

$$F \subset \mathcal{E}^\infty(G,G), \quad F = \{f\}_{\gamma\in\Gamma}$$

(see [9]).

<u>Definition 2</u>. Let $F = \{f_\gamma\}_{\gamma\in\Gamma}$ be a set of smooth automorphisms on M such that $f_\gamma(G) = G$ for every $\gamma \in \Gamma$. The set F <u>satisfies the condition</u> $(*_\Gamma)$ if there exist smooth functions θ_γ, ρ_γ, and σ_γ on G such that:

$(*_\Gamma)$

1. $f^*J^*dU - J^*dU = d\theta_\gamma, \quad \gamma \in \Gamma;$
2. $\rho_\gamma - \rho_\gamma f_\gamma = \theta_\gamma, \quad \gamma \in \Gamma;$
3. $\sigma_\gamma = \sigma_\gamma f_\gamma + c_\gamma, \quad \gamma \in \Gamma, \quad c_\gamma \in \mathbb{R};$
4. $\sigma_\gamma - \sigma_\delta = \rho_\delta - \rho_\gamma, \quad \delta, \gamma \in \Gamma.$

It is clear that $(*_\Gamma)$ is a generalization of $(*)$ and $(*_\Gamma)$ coincides with $(*)$ in the case where Γ is an one-element-set.

The equality $F^*\omega = \omega$ will be considered later as a set of equalities $\{f_\gamma^*\omega = \omega, \ \gamma\in\Gamma\}$ for any differential form ω on M.

<u>PROPOSITION 4</u>. <u>Let</u> $F^*\Psi = \Psi$. <u>There exists a</u> 1-<u>form</u> ψ_0 <u>such that</u> $d\psi_0 = dJ^*\,dU$ <u>and</u> $F^*\psi_0 = \psi_0$ <u>iff</u> F <u>satisfies</u> $(*_\Gamma)$.

<u>Proof</u>. Let F satisfy $(*_\Gamma)$ and $\psi_0 = J^*\,dU + d(\rho_\gamma + \sigma_\gamma)$ for a fixed $\gamma \in \Gamma$. The 1-form ψ_0 is well-defined because of $(*_\Gamma)$. It is not difficult to verify that $f_\gamma^*\psi_0 = \psi_0$ for any $\gamma \in \Gamma$.

Now, let there exist a 1-form ψ_0 satisfying $d\psi_0 = \Psi$ and $f_\gamma^*\psi_0 = \psi_0$ for any index $\gamma \in \Gamma$. Since $f_\gamma^*\Psi = \Psi$, we have $d(f_\gamma^*J^*dU - J^*dU) = 0$ and the universal coefficient formula implies the exactness of the 1-form $\psi_0 - J^*\,dU$ with respect to the simple-connectness of G. Thus, there are smooth functions θ_γ, $\gamma \in \Gamma$, such that $d\theta_\gamma = f_\gamma^*J^*dU - J^*dU$. As $f_\gamma^*\psi_0 = \psi_0$, according to Corollary 1 there exist $\rho_\gamma \in \mathcal{E}^\infty(G)$ with the property $\rho_\gamma - \rho_\gamma f_\gamma = \theta_\gamma$. It is clear that ρ_γ is determined up to a function $\sigma_\gamma \in \mathcal{C}^\infty(G)$, verifying $\sigma_\gamma = \sigma_\gamma f_\gamma + c_\gamma$, c = const. We can choose σ_γ so that $\psi_0 = J^* dU + d(\rho_\gamma + \sigma_\gamma) = J^* dU + d(\rho_\gamma + \sigma_\gamma)$ for any pair $(\gamma,\delta) \in \Gamma \times \Gamma$, Thus, $\rho_\gamma + \sigma_\gamma = \rho_\delta + \sigma_\gamma + c_{\gamma\delta}$ and $\rho_\gamma - \rho_\delta = \sigma_\gamma - \sigma_\delta + c_{\gamma\delta}$. One can choose $\{\sigma_\gamma\}_{\gamma\in\Gamma}$ so that $c_{\gamma\delta} = 0$. Thus F satisfies the condition $(*_\Gamma)$.

As in the case of $(*)$, the condition $(*_\Gamma)$ is not trivial in the same sense (cf. [9]).

5. Periodic symplectic automorphisms

In this section we are going to consider the set of periodic automorphisms on a locally APH-convex almost Kähler manifold (M,J,Φ). Let us recall that the automorphism $f : M \to M$ is said to be periodic if $f^n = 1_M$ for an integer n (1_M is the identity of M).

PROPOSITION 5. For any periodic automorphism f of M there exists an open covering $U = \{V_\alpha\}_{\alpha\in A}$ such that

1) V_α are simply connected for any $\alpha \in A$;

2) $\iota^*_\alpha \Phi = dJ^* dU_\alpha$, $\alpha \in A$. Here $\iota_\alpha : V_\alpha \hookrightarrow M$ is an embedding and $U_\alpha \in \mathcal{C}^\infty(V_\alpha)$ are determined with respect to M being a locally APH-convex manifold;

3) $f(V_\alpha) = V_\alpha$, $\alpha \in A$.

Proof. In view of the periodicity of f, the set $\{f^i(x) : 1 \leq i \leq n\}$, where x is a fixed point on M and $f^n = 1_M$, is a discrete subset of M. Thus there exists a neighbourhood G_x of x in M so small that $f^i(G_x) \cap f^j(G_x) = \emptyset$ or $f^i(G_x) = f^j(G_x)$ for $i \neq j$, $1 \leq i$, and $j \leq n$. Let $V_x = \bigcup_{j=1}^n f^j(G_x)$. It is clear that $f(V_x) = V_x$. Of course one can choose G_x to be simply connected and $\iota^*_x \Phi = dJ^* dU_x$ ($\iota_x : G_x \hookrightarrow M$). Since f is an automorphism, the corresponding group homeomorphisms

$$f_\# : \pi_1(f^k(G_x)) \to \pi_1(f^{k+1}(G_x)),$$

$$f^* : H^*(f^{k+1}(G_x); \mathbb{R}) \to H^*(f^k(G_x); \mathbb{R})$$

are isomorphisms. Since V_x consists of a finite number of simply connected components, $\pi_1(V_x) = 0$ and $j^*_x \Phi$ is exact, i.e. $j^*_x \Phi = dJ^* dU_x$ ($j_x : V_x \hookrightarrow M$). Thus there exists an open covering $\mathcal{V}_x = \{V_x\}_{x\in M}$ of M such that 1) $\pi_1(V_x) = 0$, $x \in M$; 2) $j^*_x \Phi = dJ^* dU_x$, $x \in M$; 3) $f(V_x) = V_x$.

Let $\mathcal{V} = \{V_\alpha\}_{\alpha \in A}$ be a minimal covering of M with the elements of $\mathcal{V}_x$. Let α_o be a fixed index in A and $V = V_{\alpha_o}$. In the following two theorems Φ is considered to be restricted to V.

THEOREM 2. *The periodic automorphism* $f : V \to V$ *is symplectic* ($f^*\phi = \phi$) *iff there exists a* 1-*form* ψ *on* V, *such that* $d\psi = \phi$ *and* $f^*\psi = \psi$.

P r o o f. Evidently, if such ψ exists then f is symplectic ($f^*\phi = f^*\, d\psi = df^*\, \psi = d\psi = \phi$). Now let f be a symplectic periodic automorphism. Having in mind (3.3) it is enough to prove that f satisfies (*). Since $\phi = dJ^*\, dU$ and $f^*\phi = \phi$, then $d(f^*J^*\, dU - J^*\, dU) = 0$. Since $\pi_1(V) = 0$, there exists $\theta \in \mathcal{E}^\infty(V)$ such that $f^*J^*\, dU - J^*\, dU = d\theta$. Thus the first equality of (*) is satisfied. Let

$$\tau = r_o\theta + \sum_{j=1}^{n-1} r_j\theta \circ f^j, \quad r_j = \text{const}, \ \theta\ , \quad f^j \in \mathcal{E}^\infty(V).$$

Then

$$\tau f = r_{n-1}\,\theta + \sum_{j=1}^{n-1} r_{j-1}\,\theta\, f^j .$$

Hence

$$\tau - \tau f = (r_o - r_{n-1})\theta + \sum_{j=1}^{n-1} (r_j - r_{j-1})\theta\, f^j .$$

Let us note that

$$d(\theta + \sum_{j=1}^{n-1} \theta\, f^j) = 0$$

in view of $f^n = 1_M$.

Thus

$$\sum_{j=1}^{n-1} \theta\, f^j = \theta + c, \quad c = \text{const}.$$

Let $r_j = -j$, $0 \leq j \leq n-1$. Then $\tau - \tau f = (1-n)\theta - \theta = -n\theta$. Let $\rho = -1/n\tau$ and $\theta_1 = \theta + c$. Then $\rho - \rho f = \theta_1$ and $f^*J^*dU - J^*dU = d\theta = d\theta_1$, i.e. f satisfies the condition (*).

THEOREM 3. *Let* $F = \{f_\gamma\}_{\gamma\in\Gamma}$ *satisfy the condition* $(*_\Gamma)$ *and let* f_o *be a periodic symplectic automorphism on* V. *We suppose that* $f_o f_\gamma = f_\gamma f_o$ *for every* $\gamma \in \Gamma$. *Then* $F_1 = F \cup \{f_o\}$ *satisfies* $(*_{\Gamma_1})$, *where* $\Gamma_1 = \Gamma \cup \{o\}$, *i.e. there exists a* 1-*form* ψ_o *on* V *such that* $d\psi_o = dJ^*dU$ *and* $f_\delta^*\psi_o = \psi_o$ *for any* $\delta \in \Gamma_1$.

Proof. Since F satisfies $(*_\Gamma)$, according to (3.4), there exists a 1-form ψ on V such that $d\psi = \phi$ and $f_\Gamma^*\psi = \psi$ for any $\gamma \in \Gamma$. Since $\pi_1(V) = 0$ and $d\psi = \phi$, there exists $\theta_0 \in \mathcal{C}^\infty(V)$ such that $f_0^*\psi - \psi = d\theta_0$. According to Theorem 2 there is $\rho \in \mathcal{C}^\infty(V)$ with the following properties:

$$\rho - \rho f_0 = \theta_0$$

and

$$\rho = -1/n \sum_{j=0}^{n-1} j\, \theta_0 f_0^j.$$

Since $f_\gamma f_0 = f_0 f_\gamma$, the equality $\rho = \rho f_\gamma + c_\gamma$ holds for any $\gamma \in \Gamma$, $c_\gamma = \text{const}$. Let $\sigma_\gamma = \rho$ for $\gamma \in \Gamma$ and $\sigma_0 = 0$. Then: 1) $\sigma_\gamma - \sigma_\delta = 0 = \rho_\gamma - \rho_\delta$, $\delta, \gamma \in \Gamma$; 2) $\sigma_\gamma - \sigma_0 = \rho = \rho_0 - \rho_\gamma$, $\gamma \in \Gamma$; 3) $\sigma_\gamma f_\gamma = \sigma_\gamma + c_\gamma$, $\sigma_0 f_0 = \sigma_0 = 0$, $\gamma \in \Gamma$. Here $\rho = \rho_0$, $\rho_\gamma = 0$ since $f_\gamma^*\psi = \psi$. Thus F_1 satisfies $(*_{\Gamma_1})$.

The theorem proved shows that in the case when the equality $f^*\phi = \phi$ is equivalent in the set F to the equality $f^*\psi = \psi$, ψ being a proper 1-form, the same property is valid when a periodic commuting automorphism is joined to the set F.

Let $\mathcal{V} = \{V_\alpha\}_{\alpha \in A}$ be the open covering described in Proposition 4 and $\{\eta_\alpha\}_{\alpha \in A}$ be a partition of unity connected with $\mathcal{V}$. Since $f(V_\alpha) = V_\alpha$ for any index $\alpha \in A$, the set of smooth functions $\{\sigma_\alpha\}_{\alpha \in A}$, defined by the equality $\sigma_\alpha = 1/n \sum_{j=1}^n \eta_\alpha f^j$ (n is the period of f) is also a partition of unity connected with the covering $\mathcal{V}$. Moreover, $\sigma_\alpha = \sigma_\alpha f$ for every $\alpha \in A$.

THEOREM 4. *The periodic automorphism f of $(M,g,J;\phi)$ is ϕ-symplectic iff there exist an open covering $\{V_\alpha\}_{\alpha \in A}$ of M and 1-forms ψ and ψ_α, $\alpha \in A$, such that:*

1) $j_\alpha^* d\psi_\alpha = j_\alpha^*\phi$, $j_\alpha : V_\alpha \hookrightarrow M$;

2) $f^* j_\alpha^* \psi_\alpha = j_\alpha^* \psi_\alpha$, $\alpha \in A$;

3) $d\psi - \phi = \Sigma_\alpha\, d\sigma_\alpha \wedge \psi_\alpha$;

4) $f^*\psi = \psi$.

Proof. We take the covering constructed in Proposition 4. One can easily find 1-forms ψ'_α which verify 1), $\psi'_\alpha = dJ^* dU|V_\alpha$. According to Theorem 2 there exist 1-forms ψ_α, $\alpha \in A$, verifying 1) and 2). Let $\psi = \Sigma_\alpha \sigma_\alpha \psi_\alpha$. Since $\operatorname{supp} \sigma_\alpha \subset V_\alpha$ in view of 2), the following equalities hold: $f^*\psi = \psi$ and $d\psi - \phi = \Sigma_\alpha\, d\sigma_\alpha \wedge \psi_\alpha$, i.e. 3) and 4).

Let us note that ψ and ψ_α depend on f. The following theorem proves that one can choose ψ and ψ_α "independently" of f in a certain way.

THEOREM 5. *Let M be a paracompact almost Kähler manifold and* $F = \{f_1, f_2, \ldots, f_k\}$, $f^{k_i} = 1_M$, $1 \leq i \leq k$, $f_i f_j = f_j f_i$, $1 \leq k, j \leq k$. *Then we have* $F^*\phi = \phi$ *iff there exist* 1-*forms* ψ_α *and* ψ *such that*:

1) $j^*_\alpha d\psi_\alpha = j^*_\alpha \phi$, $\alpha \in A$;

2) $F^* j^*_\alpha \psi_\alpha = j^*_\alpha \psi_\alpha$, $\alpha \in A$;

3) $d\psi - \phi = \Sigma_\alpha \, d\eta_\alpha \wedge \psi_\alpha$;

4) $F^*\psi = \psi$.

Proof. It is easy to find an open covering $\mathcal{V} = \{V_\alpha\}_{\alpha \ A}$ such that $\pi_1(V_\alpha) = 0$, $j^*_\alpha \phi$ is exact and $f_i(V_\alpha) = V_\alpha$, $1 \leq i \leq k$, $\alpha \in A$. One can also find a partition of unity connected with the covering $\mathcal{V}$, $\{\eta_\alpha\}_{\alpha \in A}$ such that $\eta_\alpha = \eta_\alpha f_i$, $\alpha \in A$, $1 \leq i \leq k$. The rest of the proof is analogous to that of Theorem 4.

Let us note that the commutativity of the elements of F is important. In the case where $f_i f_j \neq f_j f_i$, in general the simultaneous preservation of an 1-form is not possible.

References

[1] BOOTBY, W., S. KOBAYASHI, and H. WANG, A note on mappings and automorphisms of almost complex manifolds, Ann. of Math. 77 (1963), 329-334.

[2] HELGASON, S., Differential geometry and symmetric spaces, Academic Press, New York 1962.

[3] DIMIEV, S., Fonctions presque holomorphes, Banach Center Publ. 11, Warsaw 1983, pp. 61-75.

[4] ——, Propriétés locales des fonctions presque-holomorphes, Analytic Functions, Błażejewko 1982 (Lecture Notes in Math. 1039), Springer-Verlag, Berlin-Heidelberg-New York-Tokyo 1983, pp. 102-117.

[5] ——, Systèmes de type Cauchy-Riemann à coefficients non-constants de plusieurs variables, Intern. Conf. Compl. Analysis and Applications, Varna 1981.

[6] MUCHKAROV, O., Existence of almost holomorphic functions on nearly Kähler manifolds, Intern. Conf. Compl. Analysis and Applications, Varna 1981.

[7] ARNOLD, V.I., Méthodes mathématiques de la mécanique classique, Mir, Moskow 1977.

[8] PELOV, Ts.R., A class of mappings preserving a closed differential form on a smooth manifold, C.R. Bulg. Acad. of Sci. 35 (1983).

[9] ——, A condition for simultaneous preservation of a potential of a smooth 2-form on manifolds of a certain type, C.R. Bulg. Acad. of Sci. 36 (1983).

[10] ——, A class of mappings preserving the fundamental form on an almost Kähler manifold, Intern. Conf. Compl. Analysis and Applications, Varna 1981.

[11] DIMIEV, S., N. MILEV, Structures presque-complexes analytiques, to appear.

Institute of Mathematics of the
Bulgarian Academy of Sciences
BG-1090 Sofia, P.O.Box 373, Bulgaria

ON OKA'S ANALYTIC SET-VALUED FUNCTIONS AND SPECTRAL THEORY

Arturo Vaz Ferreira (Bologna)

Contents page

Summary. Some results concerning analytic families of operators (in particular Vesentini's subharmonicity theorems and Słodkowski's local representation theorem) are extended to general spectral theory.

1. Introduction

In his classical memoir [12] Waelbroeck founded analytic calculus on the concept of bounded structure (b-structure) or bornology, giving it a natural setting. In this way, not only did he extend enormously the range of its application but also produced a deeper insight into spectral theory. The consideration of b-structures cannot be dispensed with, also e.g. in studying rings of meromorphic functions because in general these rings cannot be given locally multiplicatively convex topologies.

Analytic calculus and other results (cf. for instance [1], [2], [4], and [5]) give evidence of the idea that spectral properties of b-algebras are not essentially different from those of locally multiplicatively convex topological algebras. (Of course, representation theoretical tools such as Gel'fand-Mazur theorem are not applicable here, and frequently this fact introduces some trouble and forces us to a more subtle approach.)

The present paper is just devoted to checking that idea one more time. Therefore we will be dealing not with single elements of a b-algebra, but with analytic families of such elements.

We shall extend Vesentini's subharmonicity theorems (cf. [10] and [11]) and Słodkowski's local representation of analytic set-valued functions as holomorphic operator-valued functions (cf. [9]). From our result several conclusions can immediately be drawn as in the case of Banach algebras. We will not burden our exposition with a long list of such interesting consequences; instead, we refer the reader to detailed papers by Aupetit, Słodkowski, and Vesentini, and to the items mentioned there.

In those papers, as in ours, questions of essentially local character are dealt with; this has determined the form in which we have presented our statements. It turns out that the possibility of a global formulation requires the use of b-structures! (We intend to dedicate a further paper to some global questions.) Let us point out also that our extension of the theorem on the subharmonicity of the logarithm of hyperbolic spectral radius is not without throwing some light on the importance of the recent introduction of complex-differential tools in spectral theory by Vesentini. As is often the case within bornology theory, our proofs sometimes are made to rely on Banach algebra facts by more or less elaborated techniques but the results so obtained are of an extreme generality and are applied also to operators (in our case, to analytic families of operators) with unbounded point spectrum.

In our statements the supplementary upper semicontinuity hypothesis for spectral set-valued functions is unavoidable and is due to high spectral instability phenomena that arise in so wide a context. (Of course, the upper semicontinuity conditions are superfluous for Banach algebras and more generally for complete locally multiplicatively convex algebras in which the set of invertible elements is open.)

Thus, having situated the contents of the paper we finish our introduction by recalling some notions and facts; we fix also the notation.

The key notion here is the concept due to Oka [7]: let G be an open subset of $\mathbb{C}, \Lambda \subset \mathbb{P}_1(C)$ open and let

$$K : \zeta \to K(\zeta) \subset \Lambda$$

be a set-valued function defined on G with compact values $K(\zeta) \neq \emptyset, \mathbb{P}_1(\mathbb{C})$. Put

$$U_K^\Lambda = \{(\zeta, z) \in G \times \Lambda : z \notin K(\zeta)\}.$$

We will say that K is <u>analytic</u> iff K is upper semicontinuous (usc), so that U_K^Λ is open, and U_K^Λ is a domain of holomorphy. (A study of the usc functions, set-valued, is to be found in [6], ch. I, § 18.)

In the sequel $\mathbb{P}_1(\mathbb{C})$ will be identified to the Riemann sphere $\mathbb{C}\cup\{\infty\}$; doing so we coordinate $\mathbb{P}_1(\mathbb{C})$.

Throught the paper G will denote an open subset of $\mathbb{C}$ and A a unital complete $\mathbb{C}$-b-algebra with continuous inversion (cf. below for this notion); $\mathcal{B}$ will be the class of all completant bounded absolutely convex subsets of A containing 1. If $B\in\mathcal{B}$, its Minkowski functional $\|\cdot\|_B$, given by $\|a\|_B = \inf\{\rho\in[0,+\infty] : a\in\rho B\}$, $a\in A$, is a norm on $\mathbb{C}B$ and $(\mathbb{C}B, \|\cdot\|_B)$ is a Banach space.

For results concerning b-algebras we refer the reader to [8] and [12]. Here we only recall briefly what is essential to understand the text.

Let $a\in A$. A subset Ω of $\mathbb{P}_1(\mathbb{C})$ on which the resolvent function of a (given by $(\lambda - a)^{-1}$ for $\lambda\in\mathbb{C}$) is bounded, is called a <u>cospectral set</u> of a; $S\subset\mathbb{P}_1(\mathbb{C})$ is a <u>spectral set</u> of a iff $\mathbb{P}_1(\mathbb{C})\setminus S$ is cospectral for a; the resolvent function is holomorphic on each open cospectral set.

By Liouville theorem the closed spectral sets constitute a filter basis $\mathcal{F}(a)$; the open cospectral sets constitute a basis for an ideal of sets. We will denote by $\bar{\sigma}(a)$ the complementary set in $\mathbb{P}_1(\mathbb{C})$ of the union of the open cospectral sets of a. (In the case of a Banach algebra A, $\bar{\sigma}(a)$ is just the point spectrum $sp_A(a)$.) The complementary in $\mathbb{P}_1(\mathbb{C})$ of a closed neighbourhood of $\bar{\sigma}(a)$ is clearly cospectral for a.

Let h be a homography and suppose $h(a)$ makes algebraically sense for the element a of A. Then, if Ω is an open cospectral set of a, $h(\Omega)$ is an open cospectral set for $h(a)$ and we have $\bar{\sigma}(h(a)) = h(\bar{\sigma}(a))$. (This is clear for h linear; for h an inversion, it is enough to take into account the axioms of a b-algebra and the relations:

$$\lambda(\lambda - a)^{-1} = 1 + a(\lambda - a)^{-1}, \quad \lim_{\substack{\lambda\to\infty \\ \lambda\in\Omega}}(\lambda - a)^{-1} = 0,$$

whenever ∞ belongs to a cospectral set Ω of a.)

Let us say that a subset K of A is <u>compact</u> iff $K\subset B$ for some $B\in\mathcal{B}$ and K is compact in the Banach space $(\mathbb{C}B, \|\cdot\|_B)$.

The continuity of the inversion $a\mapsto a^{-1}$ in A is to be understood in the following sense: if $K\subset A$ is compact and there is some open neighbourhood of zero in $\mathbb{C}$ cospectral for each $a\in K$, then K^{-1} is bounded in A. Note that in the case when the bornology of A comes from a complete locally multiplicatively convex topological algebra structure, the inversion in A is automatically continuous in our sense. (Therefore, the results of this paper contain all the correspondent existing ones which are formulated within the narrower realm of (locally multiplicatively convex) topological algebra theory.)

In the sequel the following notation will be used. If $S\subset\mathbb{C}$, we put $|S| = \sup_{t\in S}|t|$. If D is an open domain in $\mathbb{P}_1(\mathbb{C})$, c_D will be the Caratheodory pseudodistance on D, and for a point $z_o\in D$ and compact $S\subset D$ we will put

$$\tau_D(z_o,S) = \sup_{t\in S} c_D(z_o,t).$$

Also, for $a\in A$ and $\nu\in\mathbb{P}_1(\mathbb{C})$ we will put $\ell_a(\nu) = a(1-\nu a)^{-1}$ if $\nu\in\mathbb{C}$ and $(1-\nu a)^{-1}$ exists in A, $\ell_a(\nu) = 0$ if $\nu=\infty$ and a^{-1} exists in A. It is clear that whenever S is a closed spectral set of $a\in A$ and $\Omega = \mathbb{P}_1(\mathbb{C})\setminus S$, then $\ell_a(\nu)$ is holomorphic on Ω^{-1} and the set $\{\ell_a(\nu) : \nu\in\Omega^{-1}\}$ is bounded in A.

2. The results

We can state now the following results where $a:\zeta\mapsto a(\zeta)$ denotes a holomorphic function from G into A, $F=\bigcup_{\zeta\in G}\bar{\sigma}(a(\zeta))$ and $\bar{F}$ the closure of F in $\mathbb{P}_1(\mathbb{C})$. We will suppose throughout that the set-valued function $\zeta\mapsto\bar{\sigma}(a(\zeta))$ is usc.

THEOREM 1. Suppose $\infty\notin F$. Then we have $\bar{\sigma}(a(\zeta))\subset\mathbb{C}$, $\zeta\in G$, and the function

$$\zeta\mapsto\log|\bar{\sigma}(a(\zeta))|$$

is subharmonic in G.

THEOREM 2. Suppose $F\neq\mathbb{P}_1(\mathbb{C})$. Then $\zeta\mapsto\bar{\sigma}(a(\zeta))$ is analytic in the sense of Oka.

COROLLARY 1. Let G be supposed to be bounded, and h to be some homography for which $\infty\notin h(\bar{F})$. Then, there is a separable Hilbert space E and a holomorphic function f of G into the Banach algebra $\mathcal{L}(E)$ of continuous linear endomorphisms of E such that we have

$h^{-1}(sp_{\mathcal{L}(E)}(f(\zeta))) = \bar{\sigma}(a(\zeta))$ *for all* $\zeta \in G$.

In the next statement $\sigma(b)$, where b is an element of A, denotes the complementary in $\mathbb{C}$ of the union of open sets $\Omega \subset \mathbb{C}$, cospectral for b.

COROLLARY 2. *Suppose* $\bigcup_{\zeta \in G} \sigma(a(\zeta)) \neq \mathbb{C}$. *Then the set*

$$U_a^o = \{(\zeta, z) \in G \times \mathbb{C} : z \in \mathbb{C} \setminus (\sigma(a(\zeta)))\}$$

is a domain of holomorphy in $\mathbb{C}^2$. *If* $\bigcup_{\zeta \in G} \sigma(a(\zeta)) = \mathbb{C}$, *but* $\infty \notin F$, *the same result holds.*

Remark. The statement of Corollary 2 cannot be reworded by assuming $\zeta \mapsto \sigma(a(\zeta))$ is usc instead of $\zeta \mapsto \bar{\sigma}(a(\zeta))$ as the following example shows. Take as A the field of germs at zero of functions defined and meromorphic in punctured discs centered at zero with its natural bornology, and as a the function $\zeta \mapsto \zeta(\text{germ of } z^{-1})$ on $\mathbb{C}$. We have $\zeta \mapsto \sigma(a(\zeta))$ usc whereas $U_a^o = \mathbb{C}^2 \setminus \{0\}$ which is not a domain of holomorphy by the Hartogs theorem.

THEOREM 3. *Let* D *be any domain in* $\mathbb{P}_1(\mathbb{C})$ *containing* F, *and let* z_o *be a point in* D. *Then, the function*

$$\zeta \mapsto \log \tau_D(z_o, \bar{\sigma}(a(\zeta)))$$

is subharmonic in G.

3. The proofs

For convenience we shall rely on several lemmata in proving our results.

LEMMA 1. *Let* V *be an open set in* $\mathbb{C}$ *and* $p : \zeta \to p(\zeta)$ *be a holomorphic function from* V *into* A *such that the set-valued function* $\zeta \mapsto \bar{\sigma}(p(\zeta))$ *is* usc. *Suppose* $0 \notin \bigcup_{\zeta \in V} \bar{\sigma}(p(\zeta))$ *and put* $m(\zeta) = (p(\zeta))^{-1}$ *for each* $\zeta \in V$. *Then* $\zeta \mapsto \bar{\sigma}(m(\zeta))$ *is* usc, *the function* $\zeta \mapsto m(\zeta)$ *is holomorphic in* V *and* $\infty \notin \bigcup_{\zeta \in V} \bar{\sigma}(m(\zeta))$.

LEMMA 2. *Let* V *be an open set in* $\mathbb{C}$ *and* $p^* : \zeta \mapsto p^*(\zeta)$ *be a holomorphic function from* V *into* A *such that the set-valued function* $\zeta \mapsto \bar{\sigma}(p^*(\zeta))$ *is* usc. *Suppose* $\infty \notin \bigcup_{\zeta \in V} \bar{\sigma}(p^*(\zeta))$. *Define* $W = \{\xi : \zeta \in V, \zeta \notin \bar{\sigma}(p^*(\zeta))\}$, *and put* $m^*(\zeta) = (p^*(\zeta) - \xi)^{-1}$, $\zeta \in W$. *Then* $\zeta \mapsto \bar{\sigma}(m^*(\zeta))$

is usc, the function $\zeta \mapsto m^*(\zeta)$ is holomorphic in the open set W and $\infty \notin \bigcup_{\zeta \in W} \bar{\sigma}(m^*(\zeta))$.

Proof of Lemma 1. Let i denote the homography $z \mapsto z^{-1}$ of $\mathbb{P}_1(\mathbb{C})$. As i is a homeomorphic transformation of $\mathbb{P}_1(\mathbb{C})$, we see that $\zeta \mapsto i(\bar{\sigma}(p(\zeta))) = \bar{\sigma}(i(p(\zeta))) = \bar{\sigma}(m(\zeta))$ is usc. Obviously, we also get $\infty \notin \bigcup_{\zeta \in V} \bar{\sigma}(m(\zeta))$.

To verify that m is holomorphic, we can do it locally. Let $\zeta_0 \in V$ and Ξ be an open neighbourhood of $\bar{\sigma}(p(\zeta_0))$, the closure of which does not contain 0. Put $\Omega = \mathbb{P}_1(\mathbb{C}) \setminus \bar{\Xi}$; fix also $\eta > 0$ for which $\bar{\Delta}(\zeta_0,\eta) \subset V$ and $\bar{\sigma}(p(\zeta)) \subset \Xi$ for every $\zeta \in \bar{\Delta}(\zeta_0,\eta)$. The open neighbourhood Ω of 0 will be cospectral for each $p(\zeta)$, $\zeta \in \bar{\Delta}(\zeta_0,\eta)$. Therefore we see that $K = p(\bar{\Delta}(\zeta_0,\eta))$ and $K^{-1} = m(\bar{\Delta}(\zeta_0,\eta))$ are bounded.

Fix $B_1 \in \mathcal{B}$ for which p is holomorphic on $\Delta(\zeta_0,\eta)$ with values in $(\mathbb{C}B_1, \|.\|_{B_1})$ and $B_1 \supset K \cup K^{-1}$; fix next $B_2 \in \mathcal{B}$ with $B_2 \supset B_1^{\,3}$. Suppose $\zeta_1,\zeta_2 \in \Delta(\zeta_0,\eta)$. From the algebraic relation

$$m(\zeta_1) - m(\zeta_2) = -m(\zeta_1)[p(\zeta_1) - p(\zeta_2)]\, m(\zeta_2)$$

we deduce at once that m is continuous on $\Delta(\zeta_0,\eta)$ as a function with values in $(\mathbb{C}B_2, \|.\|_{B_2})$. Therefore, if we take now $B_3 \in \mathcal{B}$ with $B_3 \supset B_2^{\,3}$, it follows that the quotient

$$(\zeta_1 - \zeta_2)^{-1}(m(\zeta_1) - m(\zeta_2))$$

converges just in $(\mathbb{C}B_3, \|.\|_{B_3})$ to $-m(\zeta_2)(p'(\zeta_2))m(\zeta_2)$ as $\zeta_1 \to \zeta_2$. The proof is complete now.

Proof of Lemma 2. The fact that W is open is clear from the upper semicontinuity of $\zeta \mapsto \bar{\sigma}(p^*(\zeta))$.

Put $p(\zeta) = p^*(\zeta) - \zeta$, $\zeta \in W$; the function p is obviously holomorphic in W. Let us verify that $\zeta \mapsto \bar{\sigma}(p(\zeta))$ is usc. Indeed, given $\zeta_0 \in W$ and a neighbourhood Ξ of $\bar{\sigma}(p(\zeta_0))$, it will be possible to choose $\varepsilon > 0$ so small that $\Delta(\zeta_0,\varepsilon) \subset W$, $\bar{\sigma}(p(\zeta_0)) + \Delta(0,\varepsilon) \subset \Xi$, and η, $0 < \eta < \varepsilon/2$, so that $\bar{\sigma}(p^*(\zeta)) \subset \bar{\sigma}(p^*(\zeta_0)) + \Delta(0,\varepsilon/2)$ for $\xi \in \Delta(\xi_0,\eta)$, in such a way that for those ζ we have

$$\bar{\sigma}(p(\zeta)) = \bar{\sigma}(p^*(\zeta) - \xi) = \bar{\sigma}(p^*(\zeta)) - \zeta \subset \bar{\sigma}(p^*(\zeta_0)) + \Delta(0,\varepsilon/2) - \zeta$$

$$\subset \bar{\sigma}(p^*(\zeta_0)) - \zeta_0 + \Delta(0,\varepsilon/2) + \Delta(0,\varepsilon/2) = \bar{\sigma}(p(\zeta_0)) + \Delta(0,\varepsilon) \subset \Xi.$$

Now, as by the definition of W, $0 \quad \bigcup_{\zeta \in W} \bar{\sigma}(p(\zeta))$, we see that the function p satisfies the hypothesis of Lemma 1 on W from which it follows that the asserted conclusions hold for the function m*.

LEMMA 3. Suppose $m \in A$ is such that for some closed spectral set S of m, $\infty \in \Omega = \mathbb{P}_1(\mathbb{C}) \setminus S$. Let $B_S \in \mathcal{B}$ be chosen in such a way that

(*) $\ell_m(\nu) \in B_S$ for $\nu \in \Omega^{-1}$ and $\ell_m(\nu)$ is holomorphic on Ω^{-1} with values in $(\mathbb{C} B_S, \|.\|_{B_S})$.

Then, for $\nu \in \Omega^{-1}$, we have

$$(\mathrm{dist}(\nu, \partial_{B_S}(\ell_m)))^{-1} = \overline{\lim_{n\to\infty}} \|(m(1-\nu m)^{-1})^n\|_{B_S}^{1/n} \le (\mathrm{dist}(\nu, S^{-1}))^{-1},$$

where $\partial_{B_S}(\ell_m)$ stands for the boundary of regularity of ℓ_m considered as a $(\mathbb{C}B_S, \|.\|_{B_S})$-valued holomorphic function.

In particular,

$$|\bar{\sigma}(m)| \le (\mathrm{dist}(0, \partial_{B_S}(\ell_m)))^{-1} = \overline{\lim_{n\to\infty}} \|m^n\|_{B_S}^{1/n} \le |S|.$$

We have also,

$$\inf_{B\in\mathcal{B}} \overline{\lim_{n\to\infty}} \|m^n\|_B^{1/n} = \lim_{B\in\mathcal{B}} \overline{\lim_{n\to\infty}} \|m^n\|_B^{1/n} = |\bar{\sigma}(m)|.$$

Besides, if for each $k \in \mathbb{N}$, $N_k(m)$ denotes the set of points in $\mathbb{C}$ at distance $\le 2^{-k}$ from $\bar{\sigma}(m)$, each $N_k(m)$ is a spectral set for m, and

$$|\bar{\sigma}(m)| \le \overline{\lim_{n\to\infty}} \|m^n\|_{B_{N_k(m)}}^{1/n} \le |N_k(m)| = |\bar{\sigma}(m)| + 2^{-k},$$

where $B_{N_k}(m) \in \mathcal{B}$ is chosen in such a way that it satisfies the condition (*) for the spectral set $N_k(m)$.

Proof. We have, for $\nu \in \Omega^{-1}$ and $n \in \mathbb{N}$,

$$\left(\frac{\partial}{\partial\nu}\right)^n \ell_m(\nu) = n!\ (m(1-\nu m)^{-1})^{n+1}.$$

Hence, by Cauchy integral formula (for Banach space-valued holomorphic functions),

$$(m(1-\nu m)^{-1})^{n+1} = (1/2\pi i) \int_{\partial\Delta(\nu,r)} (\xi-\nu)^{-n-1}\, \ell_m(\xi)\, d\xi,$$

where $\partial\Delta(\nu,r)$ is the oriented boundary of the disc of radius r centred at ν, $\Delta(\nu,r) \subset\subset \Omega^{-1}$, so that

$$\|(m(1-\nu m)^{-1})^{n+1}\|_{B_S} \leq r^{-n} \sup_{\xi\in\partial\Delta(\nu,r)} \|\ell_m(\xi)\|_{B_S},$$

$$(\mathrm{dist}(\nu,\partial_{B_S}(\ell_m)))^{-1} = \overline{\lim_{n\to\infty}}\|(m(1-\nu m)^{-1})^n\|_{B_S}^{1/n} \leq (\mathrm{dist}(\nu,S^{-1}))^{-1}.$$

From this the other stated relations follow by taking into account that $\mathbb{P}_1(\mathbb{C})\setminus\bar{\sigma}(m)$ is the union of all open sets cospectral for m.

<u>Proof of Theorem 1</u>. For the sake of convenience we write

$$\rho_a(\zeta) = |\bar{\sigma}(a(\zeta))|.$$

First of all we observe that the upper semicontinuity of $\zeta\mapsto\bar{\sigma}(a(\zeta))$ forces ρ_a to be upper semicontinuous. (Indeed, if for $\zeta_0\in G$, $c\in\mathbb{R}$, we have $\rho_a(\zeta_0)<c$, necessarily $\rho_a(\zeta)<c$ in a neighbourhood of ζ_0 such that for each point ζ in it, $\bar{\sigma}(a(\zeta))\subset\Delta(0,c)$.)

Now, the proof that ρ_a is subharmonic leans on the following two assertions that we shall establish first:

(A) <u>Let</u> $f : G\to[-\infty,+\infty[$ <u>be an upper semicontinuous function satisfying the condition</u>: <u>for every</u> $\zeta\in G$ <u>there exists</u> r, $0<r<\mathrm{dist}(\zeta,\mathbb{C}\setminus G)$, <u>such that for each</u> r', $0<r'\leq r$, <u>and every holomorphic polynomial</u> q, <u>the inequality</u> $f\leq\mathrm{Re}\,q$ <u>on</u> $\partial\Delta(\zeta,r')$ <u>implies</u> $f(\zeta)\leq\mathrm{Re}\,q(\zeta)$. <u>Then</u> f <u>is subharmonic</u>.

(B) <u>Given</u> $a : G\to A$ <u>satisfying the hypothesis in the statement of Theorem</u> 1, <u>we have that for every</u> $\zeta\in G$, <u>if</u> $0<r<\mathrm{dist}(\zeta,\mathbb{C}\setminus G)$, <u>the inequality</u> $\rho_a\leq 1$ <u>on</u> $\partial\Delta(\zeta,r)$ <u>implies</u> $\rho_a(\zeta)\leq 1$.

<u>Proof of assertion (A)</u>. By Littlewood's criterion it is enough to verify that for each ζ where $f(\zeta)>-\infty$,

$$f(\zeta)\leq\int_0^{2\pi} f(\zeta+\eta e^{i\theta})\frac{d\theta}{2\pi}$$

for η conveniently small.

By the hypothesis of our assertion, we are sure that this inequality does hold for $\eta\leq r$, just by approximating uniformly on $\partial\Delta(\zeta,\eta)$ every continuous function g which is a majorant of $f(\zeta+\eta e^{i\theta})$ with trigonometrical polynomials $\geq g$.

<u>Proof of assertion (B)</u>. We proceed by absurd. Thus, suppose $a : G\to A$ satisfies the hypothesis of Theorem 1, there exist $\zeta_0\in G$ and r_0 with $0<r_0<\mathrm{dist}(\zeta_0,\mathbb{C}\setminus G)$ such that $\rho_a\leq 1$ on $\partial\Delta(\zeta_0,r_0)$

and $\rho_a(\zeta_o) > 1$.

Fix $k \in \mathbb{N}$ large enough and such that $\rho_a(\zeta_o) - 1 > 2^{-k+1}$. Then for each $\zeta \in \bar{\Delta}(\zeta_o, r_o)$ choose $\eta(\zeta)$, $0 < \eta(\zeta) < (1/2)\min\{r_o, \mathrm{dist}(\zeta_o, \mathbb{C}\setminus G) - r_o\}$ such that for every $\zeta' \in \bar{\Delta}(\zeta, \eta(\zeta))$, $\bar{\sigma}(a(\zeta')) \subset \mathrm{int}(N_{k+2}(\bar{\sigma}(a(\zeta))))$ (notation of Lemma 3).

Given $\zeta \in \bar{\Delta}(\zeta_o, r_o)$, the image in A of $\bar{\Delta}(0, |N_k(a(\zeta))|^{-1}) \times \bar{\Delta}(\zeta, \eta(\zeta))$ by the function $(\nu, \zeta') \mapsto (1 - \nu a(\zeta'))$ is a compact set K constituted by invertible elements in A and for each of them $\Delta(0, (2+2^{k+1}|\bar{\sigma}(a(\zeta))|)^{-1})$ is cospectral. Hence K^{-1} is a bounded set. Let us denote by L the bounded set $a(\bar{\Delta}(\zeta, \eta(\zeta)))$ and fix $B' \in \mathcal{B}$ such that $a : \Delta(\zeta, \eta(\zeta)) \to (CB', \|.\|_{B'})$ is holomorphic and $B' \supset L \cup (K^{-1}L) \cup (K^{-1}L)^2$; next denote by $B''(\zeta)$ an element of $\mathcal{B}$ which contains $(B')^2$. We have:

(1) $\ell_{a(\zeta')}(\nu) \in B''(\zeta)$ for $(\nu, \zeta') \in \Delta(0, |N_k(a(\zeta))|^{-1}) \times \bar{\Delta}(\zeta, \eta(\zeta))$;

(2) for each $\zeta' \in \bar{\Delta}(\zeta, \eta(\zeta))$, $\ell_{a(\zeta')}$ is holomorphic on $\Delta(0, |N_k(a(\zeta))|^{-1})$ with values in $(CB''(\zeta), \|.\|_{B''(\zeta)})$.

Therefore we are in a position that enables us to apply Lemma 3 in order to conclude that the following estimates hold for each $\zeta' \in \bar{\Delta}(\zeta, \eta(\zeta))$:

$$(i^\zeta) \qquad \rho_a(\zeta') \leq \overline{\lim_{n\to\infty}} \|a(\zeta')^n\|_{B''(\zeta)}^{1/n} \leq |\bar{\Delta}(0, |N_k(a(\zeta))|)| = |N_k(a(\zeta))| = \rho_a(\zeta) + 2^{-k}.$$

At this point let us pick up from the family of open discs $\Delta(\zeta, \eta(\zeta))$, $\zeta \in \partial\bar{\Delta}(\zeta_o, r_o)$, a finite covering $\Delta(\zeta_1, \eta(\zeta_1)), \ldots, \Delta(\zeta_s, \eta(\zeta_s))$ for $\partial\bar{\Delta}(\zeta_o, r_o)$, and choose a finite number of discs

$$\Delta(\zeta_{s+1}, \eta(\zeta_{s+1})), \ldots, \Delta(\zeta_{s+t}, \eta(\zeta_{s+t}))$$

in such a way that

$$\Delta(\zeta_o, \eta(\zeta_o)),\ \Delta(\zeta_1, \eta(\zeta_1)), \ldots, \Delta(\zeta_s, \eta(\zeta_s)),\ \Delta(\zeta_{s+1}, \eta(\zeta_{s+1})), \ldots, \Delta(\zeta_{s+t}, \eta(\zeta_{s+t}))$$

constitutes a covering of $\bar{\Delta}(\zeta_o, r_o)$. Fix also $\bar{B} \in \mathcal{B}$ which contains $B''(\zeta_o) \cup B''(\zeta_1) \cup \ldots \cup B''(\zeta_s) \cup \ldots \cup B''(\zeta_{s+t})$.

Moreover, denote for each $\zeta' \in \bar{\Delta}(\zeta_o, r_o)$ by $j(\zeta')$ the first integer j, for which $\zeta' \in \Delta(\zeta_j, \eta(\zeta_j))$. We have, for every $\zeta' \in \bar{\Delta}(\zeta_o, r_o)$,

$(i^{\bar{\Delta}})$ $\quad \rho_a(\zeta') \leq \overline{\lim_{n\to\infty}} \|a(\zeta')^n\|_{\bar{B}}^{1/n} \leq \rho_a(\zeta_{j(\zeta')}) + 2^{-k}$,

and - in particular - for $\zeta' \in \partial\bar{\Delta}(\zeta_o, r_o)$,

$(i^{\partial\bar{\Delta}})$ $\quad \rho_a(\zeta') \leq \overline{\lim_{n\to\infty}} \|a(\zeta')^n\|_{\bar{B}}^{1/n} \leq 1 + 2^{-k}$.

Furthermore we have: a is a $(\mathbb{C}\bar{B}, \|\cdot\|_{\bar{B}})$-valued holomorphic function in the neighbourhood $\Delta(\zeta_o, \eta(\zeta_o)) \cup \ldots \cup \Delta(\zeta_{s+t}, \eta(\zeta_{s+t}))$ of $\bar{\Delta}(\zeta_o, r_o)$.

That being, define recursively a sequence B_n, $n \geq 1$, of elements in $\mathfrak{B}$ in such a way that $B_1 = \bar{B}$, and - if $n > 1$ and $B_1, \ldots, B_{n-1}$ have been already chosen - B_n is taken so that is satisfies $B_n \supset B_{n-1}$, whereas $\zeta \mapsto (a(\zeta))^n$ is holomorphic in a neighbourhood of $\bar{\Delta}(\zeta_o, r_o)$ with values in $(\mathbb{C}B_n, \|\cdot\|_{B_n})$.

Then, for each $n \geq 1$, $\zeta \mapsto \|a(\zeta)^n\|_{B_n}^{1/n}$ is continuous and subharmonic (because $\log\|a(\zeta)^n\|_{B_n}$ is subharmonic) in a neighbourhood of $\bar{\Delta}(\zeta_o, r_o)$.

On the other hand, by $(i^{\partial\bar{\Delta}})$,

$$\rho_a(\zeta_o) - 2^{-k} > \int_0^{2\pi} \overline{\lim_{n\to\infty}} \|a(\zeta_o + r_o e^{i\theta})^n\|_{B_n}^{1/n} \frac{d\theta}{2\pi}$$

$$\geq \overline{\lim_{n\to\infty}} \int_0^{2\pi} \|a(\zeta_o + r_o e^{i\theta})^n\|_{B_n}^{1/n} \frac{d\theta}{2\pi},$$

from which, for some $n_o \geq 1$,

$$\|a(\zeta_o)^{n_o}\|_{B_{n_o}}^{1/n_o} > \int_0^{2\pi} (\|a(\zeta_o + r_o e^{i\theta})^{n_o}\|_{B_{n_o}})^{1/n_o} \frac{d\theta}{2\pi},$$

that is a contradiction! Thus, assertion (B) is established.

Let us finish the p r o o f of Theorem 1. Let a be as in the statement; to see that $\log \rho_a$ is subharmonic we have to apply assertion (A).

Let $\zeta \in G$ and suppose r satisfies $0 < r < \mathrm{dist}(\zeta, \mathbb{C} \setminus G)$. If q is a holomorphic polynomial such that $\log \rho_a \leq \mathrm{Re}\, q$ on $\partial\Delta(\zeta, r)$, we have to obtain $\rho_a \leq |e^q|$, i.e. $|e^{-q}|\rho_a \leq 1$ on $\partial\Delta(\zeta, r)$. Now, a routine verification shows that the function $a_o = e^{-q} a$ satisfies also the hypothesis of Theorem 1 on G, and we have $\rho_{a_o} = |e^{-q}|\rho_a$. It follows by assertion (B) that $\rho_{a_o}(\zeta) \leq 1$, i.e. $\log \rho_a(\zeta) \leq \mathrm{Re}\, q(\zeta)$, as we wanted to prove.

Proof of Theorem 2. We are going to show that for a fixed $u \in \mathbb{P}_1(\mathbb{C}) \setminus F$, if we put $\Lambda = \mathbb{P}_1(\mathbb{C}) \setminus \{u\}$ and

$$U_a^\Lambda = \{(\zeta, z) : \zeta \in G, \quad z \in \Lambda \setminus \bar{\sigma}(a(\zeta))\},$$

U_a^Λ is a domain of holomorphy.

Suppose ι is a homography of $\mathbb{P}_1(C)$ such that $\iota(a(\zeta))$ exists for each $\zeta \in G$, and let $\iota(a)$ denote the mapping $\zeta \mapsto \iota(a(\zeta))$. Since the transformation $\phi_\iota : (\zeta, z) \to (\zeta, \iota(z))$ is a biholomorphism of the complex manifold $G \times \mathbb{P}_1(\mathbb{C})$, we can see that U_a^Λ is a domain of holomorphy iff

$$\phi_\iota(U_a^\Lambda) = \{(\zeta, \iota(z)) : \zeta \in G, \quad z \in \Lambda \setminus \bar{\sigma}(a(\zeta))\}$$

$$= \{(\zeta, w) : \zeta \in G, \quad w \in \iota(\Lambda) \setminus \iota(\bar{\sigma}(a(\zeta)))\} = U_{\iota(a)}^{\iota(\Lambda)}$$

is itself a domain of holomorphy. If, moreover, $\iota(u) = \infty$, in the non-trivial case when $u \neq \infty$, we can clearly invoke Lemma 1 to assert that $\iota(a)$ is holomorphic in G and $\zeta \to \bar{\sigma}(\iota(a(\zeta)))$ is usc. Therefore, as $\iota(\Lambda)$ is $\mathbb{C}$ for ι with $\iota(u) = \infty$, there will be no loss of generality if we suppose henceforth $\infty \notin F$ and we take $\Lambda = \mathbb{C}$, i.e. $U_a^{\mathbb{C}} \subset G \times \mathbb{C}$.

That being, by [9], Th. 3.2. condition (iii), to verify that $\zeta \mapsto \bar{\sigma}(a(\zeta))$ is an Oka's analytic set-valued function, it suffices to prove that for every $\alpha, \beta \in \mathbb{C}$ the function

$$\delta_{\alpha,\beta;a} : \zeta \mapsto \log \max\{|z - \alpha\zeta - \beta|^{-1} : z \in \bar{\sigma}(a(\zeta))\}$$

is subharmonic on the following open subset of G:

$$G_{\alpha,\beta;a} = \{\zeta \in G : \alpha\zeta + \beta \notin \bar{\sigma}(a(\zeta))\}.$$

To see this we note that, for $\zeta \in G_{\alpha,\beta;a}$,

$$\delta_{\alpha,\beta;a}(\zeta) = \log|\bar{\sigma}((a(\zeta) - (\alpha\zeta + \beta))^{-1})|.$$

Thus, if $\alpha = 0$,

$$\delta_{0,\beta;a}(\zeta) = \log|\bar{\sigma}((p_{\beta;a}(\zeta))^{-1})| = \log|\bar{\sigma}(m(\zeta))|,$$

where $p_{\beta;a}(\zeta) = a(\zeta) - \beta$, $m(\zeta) = (p_{\beta;a}(\zeta))^{-1}$. Then Lemma 1 can be applied to the function $p_{\beta;a}$ on $G_{0,\beta;a}$ and Theorem 1 tells us that $\log|\bar{\sigma}(m(\zeta))|$ is subharmonic in this open set.

For $\alpha \neq 0$, put $p^*_{\alpha,\beta;a}(\zeta) = \alpha^{-1} a(\zeta) - \alpha^{-1}\beta$ on G. We have

$$G_{\alpha,\beta;a} = G_{1,0;p^*_{\alpha,\beta;a}}$$

and

$$\delta_{\alpha,\beta;a}(\zeta) = -\log|\alpha| + \delta_{1,0;p^*_{\alpha,\beta;a}},$$

so that we can restrain ourselves to verify that $\delta_{1,0;p^*_{\alpha,\beta;a}}$ is subharmonic in the open set $G_{\alpha,\beta;a}$.

This is done analogously as above: Lemma 2 can be applied to the function $p^*_{\alpha,\beta;a}$ and Theorem 1 warrants us that the function

$$\delta_{1,0;p^*_{\alpha,\beta;a}}(\zeta) = \log|\bar{\sigma}(m^*(\zeta))|,$$

where $m^*(\zeta) = (p^*_{\alpha,\beta;a}(\zeta) - \zeta)^{-1}$, is subharmonic in $G_{\alpha,\beta;a}$. This ends the proof.

<u>Proof of Corollary 1</u>. Let the function a and the homography h be as stated. By Theorem 2,

$$U_a = U_a^{\mathbb{P}_1(\mathbb{C})\setminus\{h^{-1}(\infty)\}}$$

is a domain of holomorphy. Then,

$$\phi_h(U_a) = \{(\zeta,z) : \zeta \in G, \quad z \in \mathbb{C} \setminus h(\bar{\sigma}(a(\zeta)))\}$$

is also a domain of holomorphy. As we have $h(\bar{F})$ bounded, we may apply [9], Th. IV to conclude that there is a separable Hilbert space E and a holomorphic function $f : G \to \mathcal{L}(E)$ for which

$$sp_{\mathcal{L}(E)}(f(\zeta)) = h(\bar{\sigma}(a(\zeta))), \quad \zeta \in G,$$

as we wanted to prove.

<u>Proof of Corollary 2</u>. Let us denote by X the Stein manifold $G \times \mathbb{C}$. Fix $u \in \mathbb{C} \setminus \bigcup_{\zeta \in G} \sigma(a(\zeta))$. Theorem 2 tells us that

$$U_a^{\mathbb{P}_1(\mathbb{C})\setminus\{u\}}$$

is a Stein open submanifold of $G \times \mathbb{P}_1(\mathbb{C})$. Hence

$$D = U_a^{\mathbb{P}_1(\mathbb{C})\setminus\{u\}} \cap (G \times \mathbb{C})$$

is a Stein open submanifold of X.

On the other hand, U_a^o is open in X because $U_a^o = U_a^{\mathbb{P}_1(\mathbb{C})} \cap (G \times \mathbb{C})$ (note that $\sigma(b) \subset \bar{\sigma}(b) \subset \sigma(b) \cup \{\infty\}$!), and moreover we have $U_a^o = D \cup (G \times \{u\})$, so that $G \times \{u\}$ is a closed subset of X contained in the open subset U_a^o.

Now, the fact that U_a^o is Stein follows easily from the Docquier-Grauert theorem. Indeed, for any point x of the boundary of U_a^o in X there will be an open polydisc P_x centred at x such that $P_x \cap (G \times \{u\}) = \emptyset$; it follows that $P_x \cap U_a^o = P_x \cap D$ which shows $P_x \cap U_a^o$ is Stein. Therefore U_a^o is Stein.

Theorem 2 implies directly the second part of the statement.

Proof of Theorem 3. If c_D is identically 0 we have nothing to prove. Suppose $c_D \neq 0$. We can fix a point $u \in \mathbb{P}_1(\mathbb{C}) \setminus D$ and a homography h such that $h(u) = \infty$.

Put $\kappa(\zeta) = \log \tau_D(z_o, \bar{\sigma}(a(\zeta)))$, $\zeta \in G$. To prove that κ is subharmonic we may proceed locally. Given $\zeta_o \in G$, we fix a compact neighbourhood $H \subset D$ of $\bar{\sigma}(a(\zeta_o))$ and some open disc $\Delta(\zeta_o, \varepsilon) \subset\subset G$, $\varepsilon > 0$, such that $\bar{\sigma}(a(\zeta)) \subset \operatorname{int}(H)$, $\zeta \in \Delta(\zeta_o, \varepsilon)$. As we have for $z \in D$, $c_D(z_o, z) = c_{h(D)}(h(z_o), h(z))$, by [11], Th. II, and Corollary 1, we conclude that

$$\zeta \mapsto \log \sup_{w \in h(\bar{\sigma}(a(\zeta)))} c_{h(D)}(h(z_o), w) = \log \sup_{z \in \bar{\sigma}(a(\zeta))} c_D(z_o, z) = \kappa(\zeta)$$

is subharmonic on $\Delta(\zeta_o, \varepsilon)$. This establishes Theorem 3.

References

[1] AROSIO, A., A.V. FERREIRA, Une caractérisation des fonctionnelles linéaires multiplicatives dans les algèbres topologiques générales, C.R. Acad. Sci. Paris 287A (1978), 327-330.

[2] ——, ——, Caractérisation spectrale des fonctionnelles linéaires multiplicatives, C.R. Acad. Sci. Paris 287A (1978), 1085-1088.

[3] AUPETIT, B., Propriétés spectrales des algèbres de Banach (Lecture Notes in Math. 735), Springer-Verlag, Berlin 1979.

[4] FERREIRA, A.V., G. GIGANTE, A. VENTURI, A theorem of the Gleason-Kahane-Żelazko type for general topological algebras, Bollettino U.M.I. 5 15B (1978), 897-905.

[5] HOUZEL, Ch., Espaces analytiques rélatifs et théorèmes de finitude, Math. Ann. 205 1973 , 13-54.

[6] KURATOWSKI, K., Topology, Vol. I, Academic Press, New York 1966.

[7] OKA, K., Note sur les familles de fonctions analytiques multiformes..., J. Sci. Hiroshima Univ. Ser. A 4 (1934), 93-98.

[8] SEBASTIAO, J., E. SILVA, Sur le calcul symbolique d'opérateurs permutables à spectre vide ou non-borné, Ann. Math. Pura Appl. (4) 58 (1962), 219-276.

[9] SŁODKOWSKI, Z., Analytic set-valued functions and spectra, Math. Ann. 256 (1981), 363-386.

[10] VESENTINI, E., Maximum theorems for spectra, Essays on topology and related topics, Mémoires dédiés à Georeges de Rham, pp. 111-117, Springer-Verlag, Berlin 1970.

[11] ——, Carathéodory distances and Banach algebras, Adv. in Math. 47 (1983), 50-73.

[12] WAELBROECK, L., Etude spectrale des algèbres complètes, Acad. Roy. Belg. Cl. Sci. Mem. Coll. in 8° (2) 31 , 7 (1960),1-142.

Dipartimento di Matematica
Università di Bologna
Piazza di Porta S. Donato, 5
I-40127 Bologna, Italy

SUR L'ÉQUATION DE DIRAC AVEC CHAMP ÉLECTROMAGNÉTIQUE QUELCONQUE

Guy Laville (Paris)

Sommaire

1 - Introduction et notations.

Nous nous proposons ici de construire explicitement des solutions de l'équation de Dirac en présence de champs électromagnétiques très généraux. Il est inutile d'insister sur l'importance d'un tel problème, toute l'électrodynamique quantique étant basée sur cette équation. D'autre part, les équations du type Pauli ou Schrödinger peuvent être étudiées en partant de l'équation de Dirac.

On peut résumer la méthode de la façon suivante.

A certaines transformées de Fourier près, l'équation de Dirac est équivalente à la restriction sur une sous-variété d'un système d'équations aux dérivées partielles à coefficients constants ; on cherche d'abord des solutions de ce système, puis par restriction et transformée de Fourier inverse, on revient à des solutions de l'équation de Dirac. Bien entendu, la difficulté, et c'est en fait la seule difficulté de l'étude de l'équation de Dirac, est la non-commutativité des matrices γ . Cette difficulté est contournée à l'aide d'une idée simple : quand on a qu'une seule matrice au cours d'un calcul, on peut calculer comme dans le cas commutatif.

En poussant un peu plus loin l'interprétation, on peut considérer la présence du champ électromagnétique comme une déformation d'une sous-variété et même les "corrections radiatives" sont une nouvelle déformation. Remarquons que, dans tout le présent travail, il n'y a pas de "corrections", de "petites" (ou "grandes") perturbations, des séries convergentes en un sens "généralisé". Il n'y a pas de termes grands ou petits.

Nous noterons x^1, x^2, x^3 les coordonnées d'espace, $x^4 = t$ le temps (nous n'utiliserons pas ici le temps complexifié) $\hbar = 1$, $c = 1$, $\gamma^1, \gamma^2, \gamma^3, \gamma^4$ les matrices de Dirac, la forme explicite n'a pas d'importance, en fait, seule compte les règles de multiplication de ces matrices. Soit A l'algèbre de Clifford du complexifié de l'espace de Minkovoski (c'est-à-dire en fait les matrices 4x4 à coefficients complexes, c'est l'algèbre sur le corps $\mathbb{C}$ dont la base, en tant qu'algèbre est $\gamma^1, \gamma^2, \gamma^3, \gamma^4$). Pour simplifier les écritures, posons :

$$\gamma^{\mu\nu} = \gamma^\mu \gamma^\nu$$
$$\gamma^{\mu\nu\rho} = \gamma^\mu \gamma^\nu \gamma^\rho \tag{1}$$

A^1, A^2, A^3, A^4 sera le 4-potentiel du champ electromagnétique, champ de composantes $F^{\mu\nu}$, ces composantes formant un tenseur antisymétrique $(1 \leq \mu \leq 4,\ 1 \leq \nu \leq 4)$. Le 4-potentiel satisfait à la condition de Lorentz

$$\frac{\partial A^1}{\partial x^1} + \frac{\partial A^2}{\partial x^2} + \frac{\partial A^3}{\partial x^3} - \frac{\partial A^4}{\partial x^4} = 0$$

La particule sera de masse m et de charge e. En général, nous utiliserons la convention de sommation.

2 - Les équations de Maxwell et de Dirac.

Nous allons écrire les équations de Maxwell dans le même formalisme que l'équation de Dirac. Si l'on écrivait les équations du champ électromagnétique par exemple dans le formalisme du calcul différentiel extérieur, ceci compliquerait beaucoup tous les calculs, donnerait un formalisme différent de celui de l'équation de Dirac rendrait très compliqué l'introduction éventuelle de monopoles magnétiques.

Les équations définissant le potentiel :

(2) $$\frac{\partial A^{\mu}}{\partial x_{\nu}} - \frac{\partial A^{\nu}}{\partial x_{\mu}} = F^{\mu\nu}$$

s'écrivent aussi :

(3) $$\gamma_{\nu} \frac{\partial}{\partial x_{\nu}} (\gamma_{\mu} A^{\mu}) = \sum_{\mu<\nu} \gamma_{\mu\nu} F^{\mu\nu}$$

Définissons le "deuxième potentiel" P, comme une solution de :

(4) $$\gamma_{\mu} \frac{\partial}{\partial x_{\mu}} P = e\gamma_{\nu} A^{\nu}$$

(le facteur e est introduit dans P seulement pour condenser les écritures). On peut comparer P au tenseur de Hertz, bien que celui-ci ne coïncide pas, en général avec P.

$$P = \sum_{\mu<\nu} \gamma_{\mu\nu} P^{\mu\nu} .$$

On peut trouver P à l'aide de ses composantes : comme

$$\gamma_{\rho} \frac{\partial}{\partial x_{\rho}} \gamma_{\mu} \frac{\partial}{\partial x_{\mu}} = - \frac{\partial^2}{\partial x_1^2} - \frac{\partial^2}{\partial x_2^2} - \frac{\partial^2}{\partial x_3^2} + \frac{\partial^4}{\partial x_4^2} = - \Box$$

une dérivation dans (4) donne :

(5) $$-\Box \sum_{\mu<\nu} \gamma_{\mu} \gamma_{\nu} P^{\mu\nu} = e \sum_{\mu<\nu} \gamma_{\mu} \gamma_{\nu} F^{\mu\nu} - \frac{\partial A^1}{\partial x_1} - \frac{\partial A^2}{\partial x_2} - \frac{\partial A^3}{\partial x_3} + \frac{\partial A^4}{\partial x_4}$$

tenant compte de la condition de jauge de Lorentz :

(6) $$\Box P^{\mu\nu} = -eF^{\mu\nu}$$

Soit N une solution fondamentale de l'équation des ondes, on pourra prendre :

(7) $$P^{\mu\nu} = -eN * F^{\mu\nu}$$

sous réserve d'un bon comportement de $F^{\mu\nu}$ à l'infini. Bien entendu, on peut prendre n'importe quelle autre solution de (6) en imposant d'autres conditions : par exemple en enfermant le champ dans une boîte et en prenant des conditions au bord périodiques, ou en partant d'un champ polynomial ou développable en série entière et en résolvant par la méthode des coefficients indéterminés.

Dans ces conditions, l'équation (4) n'est plus satisfaite qu'à une transformation de jauge près. Ceci n'est pas gênant, il suffit de définir la nouvelle jauge pour cette égalité (4). La condition de Lorentz est encore satisfaite puisque (4) implique (5), et puisque (5) et (6) impliquent

$$\frac{\partial A^1}{\partial x_1} + \frac{\partial A^2}{\partial x_2} + \frac{\partial A^3}{\partial x_3} - \frac{\partial A^4}{\partial x_4} = 0 \quad .$$

Un changement de jauge s'interprête de la façon suivante : soit P^{00} une fonction telle que

$$\square\, P^{00} = 0$$

effectuons la transformation de jauge :

$$A_\mu \longmapsto A_\mu + \frac{\partial P^{00}}{\partial x_\mu}$$

le nouveau second potentiel sera :

$$P = P^{00} + \sum_{\mu<\nu} \gamma_{\mu\nu}\, P^{\mu\nu}$$

puisque l'on a encore :

$$\gamma_\rho \frac{\partial}{\partial x_\rho} (P^{00} + \sum_{\mu<\nu} \gamma_{\mu\nu}\, P^{\mu\nu}) = \gamma_\mu (A_\mu + \frac{\partial P^{00}}{\partial x_\mu}) \; .$$

L'équation de Dirac sous la forme habituelle s'écrit :

(8) $$\gamma_\mu (i \frac{\partial}{\partial x_\mu} - e\, A^\mu)\psi - m\psi = 0$$

tenant compte de l'égalité (4) :

(9) $$(\gamma_\mu\, i \frac{\partial}{\partial x_\mu} - \gamma_\mu \frac{\partial P}{\partial x_\mu})\psi - m\psi = 0$$

effectuons maintenant une transformée de Fourier relativement à la masse m, notons x_0 la variable conjuguée et $\tilde{\psi}$ la transformée de Fourier partielle correspondante :

$$\tilde{\psi}(x_0,x_1,x_2,x_3,x_4) = \int_{\mathbb{R}} e^{-imx_0}\, \psi(m,x_1,x_2,x_3,x_4)\; dm \; .$$

L'égalité (9) devient :

(10) $$(-i \frac{\partial}{\partial x_0} + i \gamma_\mu \frac{\partial}{\partial x_\mu})\tilde{\psi} - \gamma_\mu \frac{\partial P}{\partial x_\mu} \tilde{\psi} = 0$$

Introduisons l'opérateur de dérivation matriciel

(11) $$D_x = \frac{\partial}{\partial x_0} - \gamma_\mu \frac{\partial}{\partial x_\mu} \quad .$$

L'équation de Dirac à résoudre est en fait :

(12) $$(D_x - iD_x P)\tilde{\psi} = 0.$$

3 - Les fonctions à plusieurs variables cliffordiennes.

Remarquons tout d'abord une façon d'engendrer des solutions de l'équation (12) dans le cas sans champ (cas trivial). On a $P = 0$, il faut résoudre

(13) $$D_x \Phi = 0$$

Cette équation possède comme solution l'exponentielle :

$$\exp[(\lambda_1 x_1 + \ldots + \lambda_4 x_4) + (\lambda_1 \gamma_1 + \ldots + \lambda_4 \gamma_4) x_0]$$

où $\lambda_1, \ldots, \lambda_4$ sont des paramètres réels.
Par superposition, on obtient la famille de solutions

$$\Phi(x_0, x_1, x_2, x_3, x_4) = \int \exp i[(\lambda_1 x_1 + \ldots + \lambda_4 x_4) + (\lambda_1 \gamma_1 + \ldots + \lambda_4 \gamma_4) x_0] \varphi(\lambda_1, \ldots, \lambda_4) d\lambda_1 \ldots d\lambda_4$$

Si $\varphi : \mathbb{R}^4 \longrightarrow A$ est une fonction L^2 en chacun des paramètres $\lambda_1, \ldots, \lambda_4$, Φ sera L^2 en $x_1, \ldots, x_4$.

On peut cependant obtenir aussi L^2 en la variable x_0, ceci grâce à l'hyperbolicité de l'opérateur. Si l'on était en elliptique, ceci serait impossible.

Prenons φ telle que

$$\text{support de } \varphi \subset \{(\lambda_1, \ldots, \lambda_4) \in \mathbb{R}^4 : \lambda_4^2 \geq \lambda_1^2 + \lambda_2^2 + \lambda_3^2\}$$

Φ est bien une fonction L^2 en les 4 variables $x_1,\dots,x_4$. Pour x_0, remarquons que :

$$[i(\lambda_1\gamma_1+\dots+\lambda_4\gamma_4)]^{2k} = (-1)^k\,(-\lambda_1^2-\lambda_2^2-\lambda_3^2+\lambda_4^2)^k$$

donc

$$e^{i(\lambda_1\gamma_1+\dots+\lambda_4\gamma_4)x_0} = \cos(\sqrt{-\lambda_1^2-\lambda_2^2-\lambda_3^2+\lambda_4^2}\,x_0)$$

$$+\ i(\lambda_1\gamma_1+\dots+\lambda_4\gamma_4)\ \sin(\sqrt{-\lambda_1^2-\lambda_2^2-\lambda_3^2+\lambda_4^2}\ x_0)\ .$$

En passant en coordonnées sphériques on trouve des intégrales du type $\int\cos(r\,x_0)\,f(r)\,dr$ et $\int_r \sin(r\,x_0)\,f(r)\,dr$ qui sont L^2 en x_0 si $rf(r)$ est L^2.

D'une autre façon, on peut engendrer des solutions de (13) en posant :

$$W = \lambda_1x_1+\dots+\lambda_4x_4 + (\lambda_1\gamma_1+\dots+\lambda_4\gamma_4)x_0$$

toute fonction

$$f(W) = \sum_{n=o}^{\infty} W^n\,C_n$$

avec C_n matrice 4x4 à coefficients constants sera encore solution, quels que soient les paramètres $\lambda_1,\dots,\lambda_4$.

Passons maintenant à plusieurs variables cliffordiennes. Il y a environ un demi-siècle, après un premier départ dû à G. Moisil [9] , R. Fueter et son école [4], développèrent systématiquement la théorie des fonctions quaternionniennes, généralisant la théorie des fonctions d'une variable complexe. De façon moins systèmatique, étant donné l'espace $\mathbb{R}^n$ euclidien et son algèbre de Clifford associée les fonctions "monogènes" sur cette algèbre ont été étudiées (pour une bibliographie détaillée voir [5] et [6]). Notons aussi les travaux plus récents de l'école italienne, mais a tendance plus "géométrique" [8],[10]. Cependant deux points de vue peuvent être étudiés. Prenons l'exemple simple de $\mathbb{R}^4$, on peut le considérer comme $\mathbb{C}^2$ ou bien comme $\mathbb{H}$ (quaternions). La structure algébrique sous-jacente est totalement différente. De façon plus générale on on peut considérer les espaces à plusieurs variables cliffordiennes. Ce domaine qui, à la connaissance de l'auteur, n'a pas été exploré mais il se révèle être le cadre naturel dans lequel il faut placer l'équation de Dirac avec champ, si l'on veut comprendre la structure de cette équation et espérer trouver des solutions explicites.

Considérons l'espace $\mathbb{R}^{5n}$ (n entier strictement positif). Les variables seront notées $(x_{ko}, x_{k1}, x_{k2}, x_{k3}, x_{k4})$ pour $1 \leq k \leq n$. Soit A l'algèbre de Clifford de l'espace $\mathbb{C}^4$ muni de la métrique de Minkowski. Cette algèbre est réalisée habituellement à l'aide des matrices γ de Dirac, nous le ferons aussi dans la présente étude A est donc considérée comme l'ensemble des matrices 4x4 à coefficients complexes. Etudions les fonctions :

$$f : \mathbb{R}^{5n} \longrightarrow A$$

Posons

(13) $$D_k = \frac{\partial}{\partial x_{ko}} - \gamma_\mu \frac{\partial}{\partial x_{k\mu}} \quad \text{pour tout} \quad 1 \leq k \leq n$$

(rappelons que la convention de sommation est toujours utilisée).

<u>Définition</u> : Une fonction $f : \mathbb{R}^{5n} \longrightarrow A$, de classe $\mathcal{C}^1$ sera dite monogène quand

(14) $$D_k f = 0 \quad \text{pour tout} \quad 1 \leq k \leq n$$

On peut comparer ceci à la définition d'une fonction de plusieurs variables complexes $\frac{\partial f}{\partial \bar{z}_k} = 0$ pour tout $1 \leq k \leq n$. Cependant les deux théories seront profondément différentes car la non-commutativité fait que le produit de deux fonctions monogènes n'est pas forcément monogène. Ceci est, bien sûr, analogue au cas des fonctions quaternionniennes : (cf. : C. Deavours [3]). Nous avons mis l'opérateur à gauche, si on l'avait mis à droite nous obtiendrions l'équation de Dirac conjuguée.

La construction de solutions polynomiales pour le système (14), qui est un système à coefficients constants, n'est pas immédiate : il y a en tout 4n équations. Le plus simple est d'introduire les expressions :

$$(15)\quad Z_k = \lambda_1 x_{k1}+\lambda_2 x_{k2}+\lambda_3 x_{k3}+\lambda_4 x_{k4}+(\lambda_1\gamma_1+\lambda_2\gamma_2+\lambda_3\gamma_3+\lambda_4\gamma_4)x_{ko}$$

où $\lambda_1,\lambda_2,\lambda_3,\lambda_4$ sont quatre paramètres réels.

On remarque que les Z_k sont monogènes quels que soient les λ_j. Soient $\alpha_1,\ldots,\alpha_n$, n entiers, posons $d=\alpha_1+\ldots+\alpha_n$. Développons suivant les puissances des λ_j :

$$(16)\qquad Z_1^{\alpha_1} Z_2^{\alpha_2}\ldots Z_n^{\alpha_n} = \sum \lambda_1^{m_1}\lambda_2^{m_2}\lambda_3^{m_3}\lambda_4^{m_4}\ P_{\alpha_1\ldots\alpha_n}^{m_1\ldots m_4}$$

la somme portant sur tous les entiers m_j tels que $m_1+m_2+m_3+m_4 = d$, degré total en λ ou en x.

Dans ces calculs, seule intervient la matrice constante

$$(17)\qquad M = \lambda_1\gamma_1+\lambda_2\gamma_2+\lambda_3\gamma_3+\lambda_4\gamma_4$$

donc, on peut calculer comme dans le cas commutatif. En particulier (16) satisfait encore aux équations (14), donc tous les polynômes $P_{\alpha_1\ldots\alpha_4}^{m_1\ldots m_4}$ sont monogènes. Ceci donne des solutions polynômiales des équations (14).

Nous pouvons aussi trouver des fonctions monogènes L^2 en les variables $x_{j\mu}$, $1\leq j\leq n$, $1\leq\mu\leq 4$: posons :

$$(18)\qquad F = \exp\left\{\sum_{j=1}^{n} a_j\left[\left(\sum_{\mu=1}^{4}\lambda_\mu x_{j\mu}\right)^2 + (-\lambda_1^2-\lambda_2^2-\lambda_3^2+\lambda_4^2)\,x_{jo}^2 + 2\,M\left(\sum_{\mu=1}^{4}\lambda_\mu x_{j\mu}\right)x_{jo}\right.\right\}$$

les a_j étant des réels strictement négatifs.

Dans cette expression, il n'y a qu'une seule matrice, M, donc nous pouvons calculer comme dans le cas commutatif : Pour $1\leq\nu\leq 4$

$$\frac{\partial F}{\partial x_{p\nu}} = a_p \left[2\lambda_\nu \left(\sum_{\mu=1}^{4} \lambda_\mu x_{j\mu}\right) + 2\left(\sum_{\mu=1}^{4} \lambda_\mu \gamma_\mu\right)\lambda_\nu x_{po}\right]$$

$$\frac{\partial F}{\partial x_{po}} = a_p \left[2x_{po}(-\lambda_1^2-\lambda_2^2-\lambda_3^2+\lambda_4^2) + 2\left(\sum_{\mu=1}^{4} \lambda_\mu \gamma_\mu\right)\left(\sum_{\mu=1}^{4} \lambda_\mu x_{j\mu}\right)\right]$$

$$\frac{\partial F}{\partial x_{po}} - \sum_{\nu=1}^{4} \gamma_\nu \frac{\partial F}{\partial x_{p\nu}} = 0$$

puisque $\sum_{\mu=1}^{4} \lambda_\mu \gamma_\mu = -\lambda_1^2-\lambda_2^2-\lambda_3^2+\lambda_4^2$ on voit facilement que F est L^2 en $x_{j\mu}$, $1 \leq j \leq n$; $1 \leq \mu \leq 4$ si l'on prend λ_4 tel que $\lambda_4^2 < \lambda_1^2+\lambda_2^2+\lambda_3^2$:

$$F = \exp\left\{ \sum_{j=1}^{n} a_j \left[\left(\sum_{\mu=1}^{4} \lambda_\mu x_{j\mu}\right)^2 + (-\lambda_1^2-\lambda_2^2-\lambda_3^2+\lambda_4^2) x_{jo}^2 \right]\right\} \exp\left\{ 2 \sum_{j=1}^{4} a_j M \left(\sum_{\mu=1}^{4} \lambda_\mu x_{j\mu}\right) x_{jo} \right.$$

Comme $M^2 = (-1)(\lambda_1^2+\lambda_2^2+\lambda_3^2-\lambda_4^2)$

$$\exp\left\{ 2 M a_j \left(\sum_{\mu=1}^{4} \lambda_\mu x_{j\mu}\right) x_{jo} \right\} = \cos\left[2 a_j \left(\sum_{\mu=1}^{4} \lambda_\mu x_{j\mu}\right) x_{jo} \sqrt{\lambda_1^2+\lambda_2^2+\lambda_3^2-\lambda_4^2} \right]$$

$$+ \frac{M}{\sqrt{\lambda_1^2+\lambda_2^2+\lambda_3^2-\lambda_4^2}} \sin\left[2 a_j \left(\sum_{\mu=1}^{4} \lambda_\mu x_{j\mu}\right) x_{jo} \sqrt{\lambda_1^2+\lambda_2^2+\lambda_3^2-\lambda_4^2} \right]$$

Remarquons que par les dérivées partielles successives en les paramètres a_j et λ_μ, on engendre une famille de fonctions monogènes ayant les mêmes propriétés.

On peut aussi trouver une fonction monogène L^2 en les variables $x_{j\mu}$, $1 \leq j \leq n$, $0 \leq \mu \leq 3$

$$(19) \qquad G = \exp\left\{ \sum_{j=1}^{n} a_j \left[\left(\sum_{\mu=0}^{3} \lambda_\mu x_{j\mu}\right)^2 - \sum_{\mu=0}^{3} \lambda_\mu^2 x_{j4}^2 + 2\left(\sum_{\mu=0}^{3} \lambda_\mu \gamma_\mu \gamma_4\right)\left(\sum_{\mu=0}^{3} \lambda_\mu x_{j\mu}\right) x_{j4} \right]\right\}.$$

La démonstration est en tous points analogue.

Nous pouvons engendrer encore d'autres fonctions monogènes d'une façon qui rappelle les théorèmes de Paley-Wiener ou Plancherel-Polya : soient

$$\xi_{k\rho} \quad , 1 \leq k \leq n \quad , \quad 1 \leq \rho \leq 4$$

4n variables réelles,

$$\zeta_k = \lambda_\rho \, \xi_{k\rho}$$

$f : \mathbb{R}^{4n} \longrightarrow A$ une fonction, continue à support compact pour simplifier l'exposé ; considérons :

$$(20) \qquad \int_{\mathbb{R}^{4n}} e^{M(Z_1\zeta_1 + \ldots + Z_n\zeta_n)} \, f(\xi_{11}, \ldots, \xi_{n4}) \, d\xi$$

les Z_k étant toujours définis par (15). Comme

$$Z_k \, \zeta_k = (\lambda_\rho \, x_{k\rho} + M \, x_{ko})(\lambda_\nu \, \xi_{k\nu})$$

$$MZ_k \, \zeta_k = M(\lambda_\rho \, x_{k\rho})(\lambda_\nu \, \xi_{k\nu}) + (-\lambda_1^2 - \lambda_2^2 - \lambda_3^2 + \lambda_4^2) \, x_{ko} \lambda_\nu \, \xi_{k\nu}$$

$$\exp[M(Z_1\zeta_1 + \ldots + Z_n\zeta_n)] = \exp\left[(-\lambda_1^2 - \lambda_2^2 - \lambda_3^2 + \lambda_4^2) x_{ko} \lambda_\nu \, \xi_{k\nu}\right] \exp[M(\lambda_\rho x_{k\rho})(\lambda_\nu \xi_{k\nu})]$$

et si $\lambda_1^2 + \lambda_2^2 + \lambda_3^2 - \lambda_4^2 \geqslant 0$

$$\exp[M(\lambda_\rho \, x_{k\rho})(\lambda_\nu \, \xi_{k\nu})] = \cos\left[(\lambda_\rho \, x_{k\rho})(\lambda_\nu \xi_{k\nu}) \sqrt{\lambda_1^2 + \lambda_2^2 + \lambda_3^2 - \lambda_4^2}\right]$$

$$+ \frac{M}{\sqrt{\lambda_1^2 + \lambda_2^2 + \lambda_3^2 - \lambda_4^2}} \sin\left[(\lambda_\rho \, x_{k\rho})(\lambda_\nu \, \xi_{k\nu}) \sqrt{\lambda_1^2 + \lambda_2^2 + \lambda_3^2 - \lambda_4^2}\right]$$

4 - Interprétation de l'équation de Dirac.

Pour retrouver l'équation de Dirac, nous allons particulariser les considérations du paragraphe précédent en prenant $n=7$, donc travailler avec 35 variables réelles, ces variables étant notées $x'_{k\rho}$. Le système (14), par exemple, s'écrira :

$$(21) \qquad D_{k'} \, f = 0 \qquad \text{pour tout} \qquad 1 \leq k \leq 7 \quad .$$

Effectuons le changement de variables :

$x_p = x'_{1p}$ pour $0 \leq p \leq 4$

$x_{2p} = x'_{2p}$ pour $p = 0, 2, 3, 4$; $x_{21} = x'_{21} - P^{12}$

$x_{3p} = x'_{3p}$ pour $p = 0, 1, 3, 4$; $x_{32} = x'_{32} - P^{23}$

$x_{4p} = x'_{4p}$ pour $p = 0, 1, 2, 4$; $x_{43} = x'_{43} - P^{31}$

$x_{kp} = x'_{kp}$ pour $5 \leq k \leq 7$ et $0 \leq p \leq 3$

$x_{54} = x'_{54} - P^{14}$; $x_{64} = x'_{64} - P^{24}$; $x_{74} = x'_{74} - P^{34}$

Dans ce changement de variables, on a, pour $0 \leq j \leq 4$

$$\frac{\partial}{\partial x'_{1j}} = \frac{\partial}{\partial x_p}\frac{\partial x_p}{\partial x'_{1p}} + \frac{\partial}{\partial x_{2p}}\frac{\partial x_{2p}}{\partial x'_{1p}} + \frac{\partial}{\partial x_{3p}}\frac{\partial x_{3p}}{\partial x'_{1j}} + \ldots$$

$$\frac{\partial}{\partial x'_{1j}} = \frac{\partial}{\partial x_j} + \frac{\partial P^{12}}{\partial x'_{1j}}\frac{\partial}{\partial x_{21}} + \frac{\partial P^{23}}{\partial x'_{1j}}\frac{\partial}{\partial x_{32}} + \frac{\partial P^{31}}{\partial x'_{1j}}\frac{\partial}{\partial x_{43}} + \frac{\partial P^{14}}{\partial x'_{1j}}\frac{\partial}{\partial x_{54}} + \frac{\partial P^{24}}{\partial x'_{1j}}\frac{\partial}{\partial x_{64}} + \frac{\partial P^{34}}{\partial x'_{1j}}\frac{\partial}{\partial x_{74}}$$

$$D_{1'} = D_1 + D_1\, P^{12}\frac{\partial}{\partial x_{21}} + D_1\, P^{23}\frac{\partial}{\partial x_{32}} + D_1\, P^{31}\frac{\partial}{\partial x_{43}} + D_1\, P^{14}\frac{\partial}{\partial x_{54}} + D_1\, P^{24}\frac{\partial}{\partial x_{64}} + $$

$$+ D_1\, P^{34}\frac{\partial}{\partial x_{74}}$$

avec $D_1 = \frac{\partial}{\partial x_0} - \sum_{\mu=1}^{4} \gamma_\mu \frac{\partial}{\partial x_\mu}$.

Remarquons que $D_{1'}\, P^{ij} = D_1\, P^{ij}$.

Toute fonction monogène de 7 variables cliffordiennes se trouvera dans le noyau de l'opérateur :

$$D_{1'} + (D_{1'}\, P^{12})\gamma_1\, D_{2'} + (D_{1'}\, P^{23})\gamma_2\, D_{3'} + (D_1\, P^{31})\gamma_3\, D_{4'} - (D_{1'}\, P^{14})\gamma_4\, D_{5'} - (D_{1'}\, P^{24})\gamma_4\, D_{6'} - (D_{1'}\, P^{34})\gamma_4\, D_{7'} \; . \tag{22}$$

Effectuons le changement de variables, l'opérateur (22) devient :

$$\begin{aligned} D_1 &+ (D_1\, P^{12})\gamma_1 \left(\frac{\partial}{\partial x_{20}} + \gamma_2 \frac{\partial}{\partial x_{22}} + \gamma_3 \frac{\partial}{\partial x_{23}} + \gamma_4 \frac{\partial}{\partial x_{24}}\right) + \\ &+ (D_1\, P^{23})\gamma_2 \left(\frac{\partial}{\partial x_{30}} + \gamma_1 \frac{\partial}{\partial x_{31}} + \gamma_3 \frac{\partial}{\partial x_{32}} + \gamma_4 \frac{\partial}{\partial x_{34}}\right) + \\ &+ (D_1\, P^{31})\gamma_3 \left(\frac{\partial}{\partial x_{40}} + \gamma_1 \frac{\partial}{\partial x_{41}} + \gamma_2 \frac{\partial}{\partial x_{42}} + \gamma_4 \frac{\partial}{\partial x_{44}}\right) - \\ &- (D_1\, P^{14})\gamma_4 \left(\frac{\partial}{\partial x_{50}} + \gamma_1 \frac{\partial}{\partial x_{51}} + \gamma_2 \frac{\partial}{\partial x_{52}} + \gamma_3 \frac{\partial}{\partial x_{53}}\right) - \end{aligned} \tag{23}$$

$$- (D_1 P^{24})\gamma_4 (\frac{\partial}{\partial x_{60}} + \gamma_1 \frac{\partial}{\partial x_{61}} + \gamma_2 \frac{\partial}{\partial x_{62}} + \gamma_3 \frac{\partial}{\partial x_{63}}) -$$

$$- (D_1 P^{34})\gamma_4 (\frac{\partial}{\partial x_{70}} + \gamma_1 \frac{\partial}{\partial x_{71}} + \gamma_2 \frac{\partial}{\partial x_{72}} + \gamma_3 \frac{\partial}{\partial x_{73}}) \ .$$

Cet opérateur est indépendant des variables $x_{21}, x_{32}, x_{43}, x_{54}, x_{64}, x_{74}$, donc, l'opérateur (22) est tangent à la sous-variété d'équations :

$$x'_{21} - P^{12} = 0 \ , \ x'_{32} - P^{23} = 0 \ , \ x'_{43} - P^{31} = 0$$

$$x'_{54} + P^{14} = 0 \ , \ x'_{64} + P^{24} = 0 \ , \ x'_{74} + P^{34} = 0 \ .$$

Effectuons une transformée de Fourier en les variables

$\{x_{pq} \ , \ 2 \le p \le 7, \ 0 \le q \le 4, \ pq \notin \{(21),(32),(43),(54),(64),(74)\}\}$

soient ξ_{pq} les variables transformées .

Prenons $\xi_{20} = \xi_{23} = \xi_{24} = \xi_{30} = \xi_{31} = \xi_{34} = \xi_{40} = \xi_{42} = \xi_{44} = \xi_{50} = \xi_{52} = \xi_{53} =$

$= \xi_{60} = \xi_{61} = \xi_{63} = \xi_{70} = \xi_{71} = \xi_{72} = 0$

$$\xi_{22} = \xi_{32} = \xi_{41} = \xi_{51} = \xi_{62} = \xi_{73} = 1$$

L'opérateur (23) devient

(24) $$D_1 - iD_1(P^{12}\gamma_{12} + P^{23}\gamma_{23} + P^{13}\gamma_{13} + P^{14}\gamma_{14} + P^{24}\gamma_{24} + P^{34}\gamma_{34})$$

qui est exactement l'opérateur apparaissant dasn (12), c'est-à-dire une forme de l'équation de Dirac, dans le cas où l'on a choisi une jauge pour laquelle $P^{00} = 0$.

Pour une jauge quelconque (mais satisfaisant à la condition de Lorentz), il suffira de multiplier une solution éventuelle Φ par :

$$e^{iP^{00}} \Phi$$

car, dans le cas :

$$\left[D_1 - iD_1(P^{00} + \sum_{\mu<\nu} P^{\mu\nu}\gamma_{\mu\nu})\right] e^{iP^{00}} \Phi$$

$$= iD_1P^{00} e^{iP^{00}}\Phi + e^{iP^{00}}\left[D_1 - iD_1(P^{00} + \sum_{\mu<\nu} P^{\mu\nu}\gamma_{\mu\nu})\right]\Phi$$

$$= iD_1P^{oo} e^{iP^{oo}}\Phi - e^{iP^{oo}} iD_1 P^{00}\Phi + e^{iP^{oo}}\left[D_1 - iD_1(\sum_{\mu<\nu} P^{\mu\nu}\gamma_{\mu\nu})\right]\Phi$$

$$= 0$$

Théorème : Une fonction monogène, restreinte à la sous-variété définie ci-dessus, puis transformée de Fourier donne des solutions de l'équation de Dirac avec champ electromagnétique.

Corollaire : La fonction G définie par (19) donne par restriction et transformation de Fourier des solutions de l'équation de Dirac avec un champ purement électrique.

Remarque : Par transformation de Lorentz on peut ramener beaucoup de champs électromagnétiques à ce cas.

Bibliographie

[1] J. Bjorken, S. Drell. Relativistic quantum mechanics. Mac Graw-Hill.

[2] Bracks, R. Delanghe, F. Sommen. Clifford analysis Pitman.

[3] C. Deavours. The quaternion calculus Am. Math. Monthly, 1973, p.995-1008.

[4] R. Fueter. Die Funktionentheorie der Differential-gleichungen $\Delta u = 0$ und $\Delta\Delta u = 0$ mit vier reelen variablen. Comm. Math. Helvetici, 7, 1934, p.307.

[5] H. Haefeli : Hypercomplexe Differentiale, Comm. Math. Heln. 20 (1947) p.382-4

[6] V. Iftimie. Fonctions hypercomplexes. Bull. Soc. Math. de Roumanie, 9, 1965.

[7] G. Laville. Une famille de solutions de l'équation de Dirac avec champ electromagnétique quelconque C.R. Acad. Sc. de Paris, t.296 (1983), p.1029-1032.

[8] E. Martinelli. Variétés à structure quaternionniennes généralisées. Revue Roumaine de Math. pures et appl. tome X, n°7, (1965) p.915-922

[9] G. Moisil. Sur les quaternions monogènes Bull. Sci. Math. Paris (2), 55, p.168, (1931).

[10] G.B. Rizza. Contributi rencenti alla tearia delle funzioni nelle algebre "Rend. del Sem. Mat. e Fisico di Milano, vol XLIII, p.45-54, 1973.

Mathématiques, L.A. 213 du C.N.R.S.
Université Pierre et Marie Curie
4, Place Jussieu
F-75230 Paris Cédex 05, France

RÉGULARISATION SUR UNE VARIÉTÉ

Christine Laurent-Thiébaut (Paris)

Table des matières

Résumé

L'objet de ce travail est l'étude de la régularisation des distributions sur une variété réelle $\mathcal{C}^\infty$ (ce qui peut être appliqué en particulier à une variété analytique complexe). On donne des conditions nécessaires et suffisantes sur des familles de noyaux $(K_\varepsilon)_{\varepsilon \in \mathbb{R}^+}$ généralisant les noyaux de convolution pour que les familles $(T_\varepsilon)_{\varepsilon \in \mathbb{R}^+}$ des régularisées d'une distribution T convergent vers cette distribution pour les topologies faible et forte de l'espace vectoriel des distributions.

On exprime ensuite une condition suffisante pour que la convergence se fasse pour la topologie de l'espace des distributions à front d'onde dans un cône donné.

On montre finalement que les noyaux associés aux opérateurs de de Rham vérifient ces conditions.

Cela nous permet de donner une démonstration simple d'une condition suffisante d'existence de l'indice de Kronecker de deux courants.

Christine Laurent-Thiébaut

Introduction

L'objet de ce travail est l'étude de la régularisation des fonctions, des distributions et des courants sur une variété réelle $\mathcal{C}^\infty$ et a fortiori sur une variété analytique complexe puisque cette dernière est naturellement muni d'une structure réelle $\mathcal{C}^\infty$.

Si nous nous plaçons dans $\mathbb{R}^n$, la méthode habituelle de régularisation est liée à la convolution : en effet on considère une fonction θ, $\mathcal{C}^\infty$ à support compact au voisinage de 0 , vérifiant $\int_{\mathbb{R}^n} \theta(x)dx = 1$, on pose $\theta_\varepsilon(x) = \frac{1}{\varepsilon^n}\theta(\frac{x}{\varepsilon})$ et $K_\varepsilon(x,y) = \theta_\varepsilon(x-y)$ définie sur $\mathbb{R}^n \times \mathbb{R}^n$

Si u est une fonction sur $\mathbb{R}^n$, les régularisées u_ε de u sont définies par $u_\varepsilon(x) = \int_{\mathbb{R}^n} K_\varepsilon(x,y)u(y)dy$.

Exhibons les principales propriétés de la fonction K_ε définie sur $\mathbb{R}^n \times \mathbb{R}^n$

1) K_ε est une fonction $\mathcal{C}^\infty$

2) K_ε est à support dans une bande d'ordre ε contenant la diagonale de $\mathbb{R}^n \times \mathbb{R}^n$

3) $\int_{\mathbb{R}^n} K_\varepsilon(x,y)dy = \int_{\mathbb{R}^n} K_\varepsilon(x,y)dx = 1$

4) $\int_{\mathbb{R}^n}(\frac{\partial}{\partial x_i} + \frac{\partial}{\partial y_i})K_\varepsilon(x,y) = 0$ pour $i = 1 , \dots , n$

Une variété n'étant, en général, pas munie d'une loi de groupe , nous ne pouvons pas utiliser la convolution. L'idée principale est d'utiliser des noyaux $(K_\varepsilon)_{\varepsilon\in\mathbb{R}^+}$ qui seront des fonctions définies sur $V \times V$, si V est la variété considérée vérifiant des propriétés analogues aux quatre propriétés citées précédemment.

Dans un premier paragraphe nous définissons l'analogue des propriétés 1, 2 et 3 et nous montrons au paragraphe 2 que les noyaux ainsi définis donnent une bonne notion de régularisation dans les espaces $\mathcal{L}^p_{loc}(V)$, c'est-à-dire que si $f \in \mathcal{L}^p_{loc}(V)$ la famille f_ε des régularisées de f converge vers f dans $\mathcal{L}^p_{loc}(V)$ quand ε tend vers 0 .

Au paragraphe 3 nous donnons tout d'abord des conditions nécessaires et suffisantes sur la famille des noyaux $(K_\varepsilon)_{\varepsilon\in\mathbb{R}^+}$, qui seront des généralisations de la condition 4) , pour que les régularisées T_ε d'une distribution T convergent vers cette distribution pour les topologies faible et forte de l'espace des distributions sur V . On exprime ensuite une condition suffisante pour que la convergence se fasse pour la topologie de l'espace des distributions à front d'onde dans un cône donné (cf. [1] et [3]) .

Nous prouvons au paragraphe 4 l'existence de noyaux vérifiant les conditions trouvées au paragraphe 3 . Pour cela nous montrons que les noyaux associés aux opérateurs régularisants de de RHAM [6] vérifient ces conditions.

Dans le paragraphe 5 nous étendons la notion de régularisation au cas des courants en remplaçant les noyaux $K_\varepsilon(x,y)$ par des formes différentielles doubles sur $V \times V$ (cf. [6], § 17) et nous en déduisons une démonstration extrêmement simple d'un des **principaux résultats de** [5] .

Finalement nous considérons le cas où la distribution donnée est dans $\mathcal{H}^s_{loc}(V)$, l'espace de Sobolev d'ordre s sur V et nous en déduisons une nouvelle condition suffisante pour que les régularisées T_ε d'une distribution T convergent faiblement vers T .

1 . Notations et définitions.

On désigne par V une variété $\mathcal{C}^\infty$, orientée , paracompacte de dimension k .

Définition 1-1. Soient π_1 et π_2 les deux projections de $V \times V$ sur V . On dira qu'un sous-ensemble A de $V \times V$ est propre si pour tout compact K de V les ensembles $\pi_1(\pi_2^{-1}(K) \cap A)$ et $\pi_2(\pi_1^{-1}(K) \cap A)$ sont relativement compacts dans V .

On considère sur $V \times V$ un système fondamental d'entourages emboités de la diagonale Δ de $V \times V$ noté $(U_\varepsilon)_{\varepsilon \in \mathbb{R}^+}$ filtré par les voisinages de 0 dans $\mathbb{R}^+$.

Soit enfin dy une forme différentielle $\mathcal{C}^\infty$, positive de degré maximal sur V , ne s'annulant pas .

Définition 1-2. On appelle famille de noyaux régularisants sur $V \times V$, toute famille $(K_\varepsilon(x,y))_{\varepsilon \in \mathbb{R}^+}$ de fonctions $\mathcal{C}^\infty$, positives sur $V \times V$, telles que pour tout $\varepsilon > 0$, le support de K_ε soit propre, contenu dans U_ε et contienne la diagonale Δ de $V \times V$, que

$$m_{1,\varepsilon} = \sup_{y \in V} \int_V K_\varepsilon(x,y)dx$$

soit borné indépendamment de ε et que la famille de fonctions $(x \longmapsto \int_V K_\varepsilon(x,y)dy)_{\varepsilon \in \mathbb{R}^+}$ converge uniformément sur tout compact de V vers la fonction constante égale à 1 quand ε tend vers 0 dans $\mathbb{R}^+$.

Définition 1-3. Soit f une fonction définie presque partout sur V telle que pour tout $x \in V$ l'expression $\int_V K_\varepsilon(x,y)f(\;)dy$ ait un sens et soit finie. On appelle famille des régularisées de f la famille des fonctions $(f_\varepsilon)_{\varepsilon \in \mathbb{R}^+}$ définies par

$$f_\varepsilon(x) = \int_V K_\varepsilon(x,y)f(y)dy$$

pour tout $x \in V$.

On notera $\mathcal{D}(V)$ l'espace vectoriel des fonctions $\mathcal{C}^\infty$ à support compact dans V et $\mathcal{D}'(V)$ l'espace des distributions sur V.

Définition 1-4. Soit $T \in \mathcal{D}'(V)$ une distribution sur la variété V, on définit la famille $(T_\varepsilon)_{\varepsilon \in \mathbb{R}^+}$ des régularisées de T de la manière suivante : pour toute $\varphi \in \mathcal{D}(V)$ on pose $\langle T_\varepsilon, \varphi\, dx \rangle = \langle T, \varphi_\varepsilon\, dx \rangle$ où $\varphi_\varepsilon(x) = \int_V K_\varepsilon(x,y)\, \varphi(y) dy$.

2 . Régularisation dans $L^p_{loc}(V)$ $(1 \leq p \leq +\infty)$.

Nous allons montrer dans ce paragraphe que si $f \in \mathcal{L}^p_{loc}(V)$, les régularisées f_ε de f sont aussi dans $\mathcal{L}^p_{loc}(V)$ et convergent vers f dans $\mathcal{L}^p_{loc}(V)$ quand ε tend vers 0 dans $\mathbb{R}^+$.

Rappelons tout d'abord un cas particulier du théorème d'interpolation de Riesz.

LEMME 2.1. Soient (X,μ_X) et (Y,μ_Y) deux espaces munis d'une mesure positive dénombrablement additive, K un opérateur intégral donné par la formule

$$(K\varphi)(x) = \int_Y k(x,y)\varphi(y)d\mu_Y \quad \text{si } \varphi \in \mathcal{L}^p(Y,\mu_Y)$$

où $k \in \mathcal{C}(X \times Y)$ vérifie

$$m_1 = \sup_{y \in Y} \int_X |k(x,y)| d\mu_X < +\infty$$

et

$$m_\infty = \sup_{x \in X} \int_Y |k(x,y)| d\mu_Y < +\infty$$

alors

$$\| K\varphi \|_{\mathcal{L}^p(X,\mu_X)} \leq m_1^{1/p}\, m_\infty^{1-\frac{1}{p}} \|\varphi\|_{\mathcal{L}^p(Y,\mu_Y)} .$$

Démonstration. Ier cas : $p = +\infty$:

$$\| K\varphi \|_{\mathcal{L}^\infty(X,\mu_X)} = \sup_{x \in X} \left| \int_Y k(x,y)\, \varphi(y) d\mu_Y \right|$$

$$\leq \sup_{x \in X} \int_Y |k(x,y)|\, |\varphi(y)| d\mu_Y$$

$$\leq m_\infty \|\varphi\|_{\mathcal{L}^\infty(Y,\mu_Y)} .$$

2ème cas : $p = 1$:

$$\| K\varphi \|_{\mathcal{L}^1(X,\mu_X)} = \int_X \left| \int_Y k(x,y)\varphi(y) d\mu_Y \right| d\mu_X$$

$$\leq \int_{X \times Y} |k(x,y)|\, |\varphi(y)|\, d\mu_X \otimes d\mu_Y$$

$$\leq \int_Y |\varphi(y)| (\int_X |k(x,y)| d\mu_X) d\mu_Y$$

$$\leq m_1 \|\varphi\|_{\mathcal{L}^1(Y,\mu_Y)} .$$

<u>3ème cas</u> : $p \in]1, +\infty[$:

$$\|K\varphi\|_{\mathcal{L}^p(X,\mu_X)} = [\int_X |\int_Y k(x,y)\varphi(y) d\mu_Y|^p d\mu_X]^{1/p}$$

$$\leq [\int_X (\int_Y |k(x,y)| |\varphi(y)| d\mu_Y)^p d\mu_X]^{1/p}$$

Soit q le réel positif tel que $\frac{1}{p} + \frac{1}{q} = 1$

$$\int_Y |k(x,y)| |\varphi(y)| d\mu_Y = \int_Y |k(x,y)|^{1/q} |k(x,y)|^{1/p} |\varphi(y)| d\mu_Y$$

$$\leq (\int_Y |k(x,y)| d\mu_Y)^{1/q} (\int_Y |k(x,y)| |\varphi(y)|^p d\mu_Y)^{1/p}$$

d'après l'inégalité de Hölder

$$\|K\varphi\|_{\mathcal{L}^p(X,\mu_X)} \leq [\int_X (\int_Y |k(x,y)| d\mu_Y)^{p/q} (\int_Y |k(x,y)| |\varphi(y)|^p d\mu_Y) d\mu_X]^{1/p}$$

$$\leq m_\infty^{1/q} [\int_{X\times Y} |k(x,y)| |\varphi(y)|^p d\mu_X \otimes d\mu_Y]^{1/p}$$

$$\leq m_\infty^{1-1/p} [\int_Y |\varphi(y)|^p (\int_X |k(x,y)| d\mu_X) d\mu_Y]^{1/p}$$

$$\leq m_1^{1/p} m_\infty^{1-1/p} \|\varphi\|_{\mathcal{L}^p(Y,\mu_Y)} .$$ q.e.d.

PROPOSITION 2.2. <u>Soit</u> $f \in \mathcal{L}^p_{loc}(V)$, <u>on pose</u>

$$f_\varepsilon(x) = \int_V K_\varepsilon(x,y) f(y) dy .$$

<u>Alors</u> $f_\varepsilon \in \mathcal{E}^\infty(V) \cap \mathcal{L}^p_{loc}(V)$ <u>et si</u> $\varphi \in \mathcal{E}^0_0(V)$ <u>et</u> $\psi \in \mathcal{E}^0_0(V)$ <u>sont des fonctions continues positives à support compact dans</u> V <u>telles que</u>

$$\psi|_{\pi_2(\pi_1^{-1}(\operatorname{supp}\varphi) \cap U_1)} \equiv 1$$

<u>où</u> π_1 <u>et</u> π_2 <u>désignent les deux projections de</u> $V \times V$ <u>sur</u> V , <u>on a pour tout</u> $p \in [1, +\infty]$ <u>et pour tout</u> $\varepsilon < 1$

$$\|\varphi f_\varepsilon\|_p \leq m_{1,\varepsilon}^{1/p} m_{\infty,\varepsilon}^{1-1/p} \|\varphi\|_\infty \|\psi f\|_p$$

<u>où</u> $m_{1,\varepsilon} = \sup_{y\in V} \int_V K_\varepsilon(x,y) dx$ <u>et</u> $m_{\infty,\varepsilon} = \sup_{x\in V} \int_V K_\varepsilon(x,y) dy$.

<u>D é m o n s t r a t i o n</u> .

Montrons tout d'abord que $f_\varepsilon \in \mathcal{E}^\infty(V)$.

Soit $x_0 \in V$ et W_{x_0} un voisinage relativement compact de x_0 dans V . Pour tout $\varepsilon > 0$, $\operatorname{supp} K_\varepsilon \cap \pi_1^{-1} W_{x_0}$ est relativement compact car le support de K_ε est propre, et,

K_ε étant continue, elle est bornée par une constante M_ε sur $\operatorname{supp} K_\varepsilon \cap \pi_1^{-1} W_{x_0}$.

Soit θ_ε une fonction $\mathcal{C}^\infty$ à support compact dans V , positive telle que $\theta_\varepsilon|_{\pi_2(\operatorname{supp} K_\varepsilon \cap \pi_1^{-1} W_{x_0})} \equiv M_\varepsilon$

Alors pour tout $y \in V$ et tout $x \in W_{x_0}$ on a

$$|K_\varepsilon(x,y)f(y)| \leq \theta_\varepsilon(y)|f(y)|$$

où $\theta_\varepsilon|f|$ est une fonction intégrable car $f \in \mathcal{L}^p_{loc}(V)$ et est $\mathcal{C}^\infty$ à support compact.

On déduit alors du théorème de Lebesgue que f_ε est continue sur V .

Des coordonnées étant choisies sur W_{x_0} , en remplaçant $K_\varepsilon(x,y)$ par ses dérivées successives par rapport à x et en faisant un raisonnement analogue on montre que $f_\varepsilon \in \mathcal{C}^\infty(V)$

$$\|\varphi f_\varepsilon\|_p = [\int_V \varphi^p(x)|\int_V K_\varepsilon(x,y)f(y)dy|^p\, dx]^{1/p}$$

$$= [\int_V \varphi^p(x)|\int_V K_\varepsilon(x,y)\,\psi(y)f(y)dy|^p\, dx]^{1/p}$$

dès que $\varepsilon < 1$, grâce à l'hypothèse faite sur ψ .

De plus φ étant continue à support compact, elle est bornée dans V et donc

$$\|\varphi f_\varepsilon\|_p \leq \|\varphi\|_\infty \; \|\int_V K_\varepsilon(x,y)\psi(y)f(y)dy\|_p .$$

Nous pouvons alors appliquer le lemme 2.1 car $m_{1,\varepsilon}$ est borné par hypothèse et $m_{\infty,\varepsilon}$ aussi puisque

$$\sup_{x \in V} |1 - \int_V K_\varepsilon(x,y)dy| \xrightarrow[\varepsilon \to 0]{} 0$$

et on a donc

$$\|\varphi f_\varepsilon\|_p \leq \|\varphi\|_\infty \; m_{1,\varepsilon}^{1/p} \; m_{\infty,\varepsilon}^{1-1/p} \; \|\psi f\|_p . \qquad \text{q.e.d.}$$

PROPOSITION 2.3. <u>La famille</u> $(f_\varepsilon)_{\varepsilon \in \mathbb{R}^+}$ <u>des régularisées de</u> $f \in \mathcal{L}^p_{loc}(V)$ <u>converge vers</u> f <u>dans</u> $\mathcal{L}^p_{loc}(V)$.

<u>Démonstration</u>. Soient φ et ψ des fonctions appartenant à $\mathcal{C}^\bullet_0(V)$ et vérifiant les hypothèses de la proposition 2.2.

Par ailleurs pour tout $\eta > 0$ il existe $g \in \mathcal{C}^0_0(V)$ telle que

$$\|\psi f - \psi g\|_p < \eta .$$

Considérons une telle fonction g :

$$\|\varphi f - \varphi f_\varepsilon\|_p \leq \|\varphi f - \varphi g\|_p + \|\varphi g - \varphi g_\varepsilon\|_p + \|\varphi f_\varepsilon - \varphi g_\varepsilon\|_p .$$

On a tout d'abord , grâce à l'hypothèse faite sur le support de ψ

$$\begin{aligned}\|\varphi f - \varphi g\|_p &= \|\varphi(\psi f - \psi g)\|_p \qquad (\psi|_{\mathrm{supp}\,\varphi} \equiv 1)\\ &\leq \|\varphi\|_\infty \|\psi f - \psi g\|_p \\ &\leq \|\varphi\|_\infty \eta \quad \text{grâce au choix de } g .\end{aligned}$$

De plus $\varphi f_\varepsilon - \varphi g_\varepsilon = \varphi(f - g)_\varepsilon$ et d'après l'inégalité de la proposition 2.2

$$\begin{aligned}\|\varphi f_\varepsilon - \varphi g_\varepsilon\|_p &\leq \|\varphi\|_\infty\, m_{1,\varepsilon}^{1/p}\, m_{\infty,\varepsilon}^{1-1/p} \|\psi f - \psi g\|_p \\ &\leq \|\varphi\|_\infty\, m_{1,\varepsilon}^{1/p}\, m_{\infty,\varepsilon}^{1-1/p}\, \eta .\end{aligned}$$

Quant au dernier terme on a

$$\|\varphi g - \varphi g_\varepsilon\|_p \leq \|\varphi\|_\infty \|g - g_\varepsilon\|_p$$

et la majoration cherchée se déduit du lemme suivant :

LEMME 2.4. <u>Soit</u> f <u>une fonction continue à support compact sur</u> V . <u>On pose</u> $f_\varepsilon(x) = \int_V K_\varepsilon(x,y)f(y)dy$, <u>alors</u> f_ε <u>est une fonction</u> $\mathcal{C}^\infty$, <u>à support compact dans</u> V <u>et la famille</u> $(f_\varepsilon)_{\varepsilon \in \mathbb{R}^+}$ <u>converge uniformément vers</u> f <u>dans</u> V <u>quand</u> ε <u>tend vers</u> 0 .

<u>Démonstration</u>. La fonction K_ε étant $\mathcal{C}^\infty$ sur $V \times V$ et f étant continue à support compact dans V , on obtient grâce au théorème de dérivation sous le signe somme que f_ε est $\mathcal{C}^\infty$ sur V . De plus le support de f_ε est contenu dans $\pi_1(\pi_2^{-1}(\mathrm{supp}\, f) \cap \mathrm{supp}\, K_\varepsilon)$ qui est compact puisque K_ε est supposé à support propre.

Par ailleurs les entourages U_ε de la diagonale Δ de $V \times V$ étant emboités et $\mathrm{supp}\, K_\varepsilon \subset U_\varepsilon$, il existe un compact K_f contenant les supports de f et des f_ε pour ε assez petit.

Etudions maintenant la convergence de la famille $(f_\varepsilon)_{\varepsilon \in \mathbb{R}^+}$ quand ε tend vers 0

$$\begin{aligned}f_\varepsilon(x) - f(x) &= \int_V K_\varepsilon(x,y)f(y)dy - f(x)\\ &= \int_V K_\varepsilon(x,y)(f(y) - f(x))dy + f(x)\Big(\int_V K_\varepsilon(x,y)dy - 1\Big)\end{aligned}$$

$$\|f_\varepsilon(x) - f(x)\|_\infty \leq \Big\| \int_V K_\varepsilon(x,y)(f(y)-f(x))dy \Big\|_\infty + \|f\|_\infty \Big\| \int_V K_\varepsilon(x,y)dy - 1 \Big\|_{\infty,\mathrm{supp}\, f}$$

Or f étant continue à support compact, elle est uniformément continue et donc

$(\forall\alpha>0)\ (\exists\varepsilon_0>0)\ (\forall\varepsilon<\varepsilon_0)\ ((x,y)\in U_\varepsilon \Rightarrow |f(x)-f(y)|<\alpha)$

donc si ε est assez petit

$$\left\|\int_V K_\varepsilon(x,y)(f(y)-f(x))dy\right\|_\infty \leq \alpha\ m_{\infty,\varepsilon}$$

de plus par hypothèse sur K_ε, $m_{\infty,\varepsilon}$ est borné et

$(\forall\alpha>0)\ (\exists\varepsilon_0'>0)(\forall\varepsilon<\varepsilon_0')\ \left(\left\|\int_V K_\varepsilon(x,y)dy-1\right\|_{\infty,\mathrm{supp}\, f}<\alpha\right)$

par conséquent pour tout α et pour ε assez petit

$$\|f_\varepsilon - f\|_\infty \leq K\alpha + \|f\|_\infty\,\alpha$$

q.e.d.

Fin de la démonstration de la proposition 2.3.

D'après le lemme 2.4 $g-g_\varepsilon$ converge unformément vers 0 dans V et est à support compact dans V on a donc pour tout $p\in[1,+\infty]$, $\lim\limits_{\varepsilon\to 0}\|g-g_\varepsilon\|_p=0$ et par conséquent pour ε assez petit $\|g-g_\varepsilon\|_p<\eta$.

On a donc montré que pour tout $\eta>0$ et pour ε assez petit on a

$$\|\varphi f-\varphi f_\varepsilon\|_p \leq \|\varphi\|_\infty (2+K)\,\eta$$

où $K\geq m_{1,\varepsilon}^{1/p}\, m_{\infty,\varepsilon}^{1-1/p}$ et donc f_ε converge vers f dans $\mathcal{L}^p_{loc}(V)$.

Remarque. Dans tout ce paragraphe on peut remplacer la forme différentielle dy par une mesure positive, dénombrablement additive sur V et considérer les espaces $\mathcal{L}^p_{loc}$ relatifs à cette mesure.

3 . Régularisation des distributions.

Dans ce paragraphe on s'intéresse à la convergence de la famille T_ε des régularisées d'une distribution T pour les différentes topologies de l'espace des distributions : topologie faible, topologie forte, topologie des distributions à front d'onde dans un cône donné.

Donnons tout d'abord quelques propriétés des régularisées

PROPOSITION 3.1. Soit $T \in \mathcal{D}'(V)$ une distribution sur V, les régularisées T_ε de T sont des fonctions $\mathcal{C}^\infty$ sur V et de plus pour tout $y \in V$, $T_\varepsilon(y) = \langle T , K_\varepsilon(x,y)dx\rangle$.

Démonstration. Soit $\varphi \in \mathcal{D}(V)$ on a par définition (cf. § 1)

$$\langle T_\varepsilon, \varphi\, dx\rangle = \langle T, \varphi_\varepsilon\, dx\rangle = \langle T, (\int_V K_\varepsilon(x,y)\varphi(y)dy)dx\rangle .$$

La fonction $x \mapsto K_\varepsilon(x,y)$ est une fonction $\mathcal{C}^\infty$ à support compact dans V pour chaque y fixé et dépend de y de manière $\mathcal{C}^\infty$, de plus dy est une forme différentielle $\mathcal{C}^\infty$ sur V et donc

$$\langle T_\varepsilon , \varphi\, dx\rangle = \langle\langle T, K_\varepsilon(x,y)dx\rangle\, \varphi(y)\, dy\rangle$$

par conséquent $T_\varepsilon(y) = \langle T , K_\varepsilon(x,y)dx\rangle$ et c'est une fonction $\mathcal{C}^\infty$ de y sur V.

3.1. Convergence faible des régularisées.

Rappelons que la topologie faible de $\mathcal{D}'(V)$ est la topologie de la convergence simple des distributions sur les formes différentielles $\mathcal{C}^\infty$ à support compact dans V de degré maximum (cf. [7]) .

Pour toute forme différentielle ψ, $\mathcal{C}^\infty$ à support compact de degré maximum dans V considérons l'expression

$\langle T - T_\varepsilon , \psi \rangle$.

Puisque par hypothèse la forme différentielle dx ne s'annule pas sur V, il existe $\varphi \in \mathcal{D}(V)$ telle que $\psi = \varphi\, dx$ et donc

$$\begin{aligned}\langle T - T_\varepsilon , \psi \rangle &= \langle T - T_\varepsilon , \varphi\, dx\rangle \\ &= \langle T , (\varphi - \varphi_\varepsilon)dx\rangle \quad \text{par définition de}\end{aligned}$$

T_ε (déf. 1.4) .

Pour obtenir la convergence faible de T_ε vers T il faut donc étudier la convergence de φ_ε vers φ dans $\mathcal{D}(V)$.

D é f i n i t i o n 3.1.1. On appelle opérateur différentiel (linéaire) à coefficient $\mathcal{C}^\infty$, d'ordre fini toute application linéaire $P : \mathcal{C}^\infty(V) \to \mathcal{C}^\infty(V)$ telle que, pour toute carte $\Phi : V_\Phi \to \tilde{V}_\Phi \subset \mathbb{R}^k$ de V , il existe un opérateur différentiel $P^{\mathcal{H}}$ à coefficients $\mathcal{C}^\infty$ sur l'ouvert $\tilde{V}_\Phi$ de $\mathbb{R}^k$ tel que, pour toute $u \in \mathcal{C}^\infty(V)$

$$(Pu) \circ \Phi^{-1} = P^\Phi (u \circ \Phi^{-1}) \quad \text{dans} \quad \tilde{V}_\Phi \ .$$

Un tel opérateur s'étend aux distributions ([2], 1.8, p. 29).

Dans la suite le terme opérateur différentiel désignera toujours un opérateur différentiel à coefficients $\mathcal{C}^\infty$ d'ordre fini .

R e m a r q u e 3.1.2. Soient $(f_\varepsilon)_{\varepsilon \in \mathbb{R}^+}$ et f des fonctions de $\mathcal{D}(V)$ à supports dans un même compact de V , la famille $(f_\varepsilon)_{\varepsilon \in \mathbb{R}^+}$ converge vers f dans (V) pour la topologie usuelle quand ε tend vers 0 si et seulement si pour tout opérateur différentiel P

$$\lim_{\varepsilon \to 0} \| P(x,D_x)f(x) - P(x,D_x)f_\varepsilon(x)\|_\infty = 0$$

PROPOSITION 3.1.3. Soit $f \in \mathcal{D}(V)$, on pose $f_\varepsilon(x) = \int_V K_\varepsilon(x,y)f(y)dy$.
Pour que la famille $(f_\varepsilon)_{\varepsilon \in \mathbb{R}^+}$ converge vers f dans $\mathcal{D}(V)$ quand ε tend vers 0 , il faut et il suffit que pour tout opérateur différentiel P sur V

(*) $\quad \| \int_V ((P(x,D_x) - P^*(y,D_y))K_\varepsilon(x,y))f(y)dy\|_\infty \to 0$ quand $\varepsilon \to 0$

où P^* est l'adjoint de P pour le produit scalaire sur $\mathcal{D}(V)$: $(f/g) = \int_V f(y)g(y)dy$.

D é m o n s t r a t i o n . 1/ Condition nécessaire. On suppose que $(f_\varepsilon)_{\varepsilon \in \mathbb{R}^+}$ converge vers f dans $\mathcal{D}(V)$ quand ε tend vers 0

$$\int_V ((P(x,D_x) - P^*(y,D_y))K_\varepsilon(x,y))f(y)dy$$
$$= \int_V (P(x,D_x)K_\varepsilon(x,y))f(y)dy - \int_V K_\varepsilon(x,y)(P(y,D_y)f(y))dy$$
$$= P(x,D_x)f_\varepsilon(x) - (P(y,D_y)f(y))_\varepsilon(x)$$ (en dérivant le premier terme sous le signe somme, dans chaque carte)

$$= P(x,D_x)f_\varepsilon(x) - P(x,D_x)f(x) + P(x,D_x)f(x) - (P(y,D_y)f(y))_\varepsilon(x),$$

or d'après le lemme 2.4

$$\|P(x,D_x)f(x) - (P(y,D_y)f(y))_\varepsilon(x)\|_\infty \to 0 \text{ quand } \varepsilon \to 0$$

et grâce à la convergence de $(f_\varepsilon)_{\varepsilon\in\mathbb{R}^+}$ vers f dans $\mathcal{D}(V)$

$$\|P(x,D_x)f(x) - P(x,D_x)f_\varepsilon(x)\|_\infty \to 0 \quad \text{quand} \quad \varepsilon \to 0 .$$

On a donc

$$\lim_{\varepsilon\to 0} \left\| \int_V (P(x,D_x) - P^*(y,D_y))K_\varepsilon(x,y))f(y)dy \right\|_\infty = 0 .$$

2/ <u>Condition suffisante</u>. Pour obtenir la convergence de la famille $(f_\varepsilon)_{\varepsilon\in\mathbb{R}^+}$ vers f dans $\mathcal{D}(V)$ quand ε tend vers 0, il suffit de démontrer que, pour tout opérateur différentiel P sur V, on a

$$\lim_{\varepsilon\to 0} \|P(x,D_x)f_\varepsilon(x) - P(x,D_x)f(x)\|_\infty = 0$$

$$P(x,D_x)f(x)-P(x,D_x)f_\varepsilon(x)=P(x,D_x)f(x)-(P(y,D_y)f(y))_\varepsilon(x)+(P(y,D_y)f(y))_\varepsilon(x)-P(x,D_x)f_\varepsilon$$

or d'après le lemme 2.4

$$\|P(x,D_x)f(x) - (P(y,D_y)f(y))_\varepsilon(x)\|_\infty \to 0 \quad \text{quand} \quad \varepsilon\to 0$$

et $\|(P(y,D_y)f(y))_\varepsilon(x) - P(x,D_x)f_\varepsilon(x)\|_\infty$

$$= \left\| \int_V ((P(x,D_x) - P^*(y,D_y))K_\varepsilon(x,y))f(y)dy \right\|_\infty \to 0 \quad \text{quand} \quad \varepsilon\to 0$$

par hypothèse, d'où le résultat.

Considérons un domaine de carte $U \subset V$ et $\Phi \begin{cases} U \to \Omega \subset \mathbb{R}^k \\ x \to \xi = (\xi_1,\ldots,\xi_k) \end{cases}$

l'application coordonnée associée.

Soit P un opérateur différentiel sur V alors il existe un opérateur différentiel P_Φ sur Ω tel que

$$\begin{aligned}(Pu)\circ\Phi^{-1} &= P_\Phi(u\circ\Phi^{-1}) \quad \forall u \in \mathcal{D}(U)\\ &= \sum_\alpha a_\alpha(\xi)\, D^\alpha_\xi(u\circ\Phi^{-1})\end{aligned}$$

où α est un multi-indice, les $a_\alpha(\xi)$ des fonctions $\mathcal{C}^\infty$ sur Ω à valeurs réelles et D^α_ξ représente la dérivation

$$\frac{\partial^{|\alpha|}}{\partial\xi_1^{\alpha_1}\ldots\partial\xi_k^{\alpha_k}} \quad \text{si } \alpha = (\alpha_1,\ldots,\alpha_k) \text{ et } |\alpha| = \alpha_1 + \ldots + \alpha_k \text{ (cf. [2])}$$

Dans la suite on notera D^α_x l'opérateur différentiel sur $V, P|_U$ tel que $P_\Phi = D^\alpha_\xi$.

PROPOSITION 3.1.4. <u>Soit</u> $(K_\varepsilon(x,y))_{\varepsilon\in\mathbb{R}^+}$ <u>une famille de noyaux sur</u> $V\times V$. <u>Pour que la famille</u> $(K_\varepsilon(x,y))_{\varepsilon\in\mathbb{R}^+}$ <u>vérifie pour tout opérateur différentiel</u> P

et toute fonction $f \in \mathcal{D}(V)$ la condition

$$(*) \qquad \lim_{\varepsilon\to 0} \left\| \int_V ((P(x,D_x) - P^*(y,D_y))K_\varepsilon(x,y))f(y)dy \right\|_\infty = 0$$

il faut et il suffit que pour toute $f \in \mathcal{D}(V)$ à support dans un domaine de carte et pour tout multi-indice α

$$(*') \qquad \lim_{\varepsilon\to 0} \left\| \int_V ((D_x^\alpha + (-1)^{|\alpha|+1} D_y^\alpha) K_\varepsilon(x,y)f(y)dy \right\|_\infty = 0.$$

Démonstration. 1/ Il est évident que (*) implique (*').

2/ Etudions la réciproque.

Soit $(U_i)_{i\in I}$ un recouvrement localement fini de V par des ouverts de carte et $(\chi_i)_{i\in I}$ une partition de l'unité subordonnée à ce recouvrement. Soit $f \in \mathcal{D}(V)$ et P un opérateur différentiel sur V ; supp f rencontrant seulement les ouverts (U_{i_r}), $r = 1,\dots,\ell$,

$$\int_V ((P(x,D_x) - P^*(y,D_y))K_\varepsilon(x,y))f(y)dy$$
$$= \sum_{r=1}^{\ell} \int_V ((P(x,D_x) - P^*(y,D_y))K_\varepsilon(x,y))\chi_{i_r}(y)f(y)dy$$

il suffit donc puisque $\chi_{i_r} f$ est à support dans le domaine de carte U_{i_r} de montrer (*) pour toute fonction $g \in \mathcal{D}(V)$ à support dans un domaine de carte.

De plus par linéarité il suffit de montrer que (*) est vérifiée pour tout opérateur P, pour toute carte (U,Φ), pour toute fonction g à support dans U, avec $P_\Phi = a(\xi)D_\xi^\alpha$.

Pour alléger l'écriture, dans la suite des calculs nous identifierons U et l'ouvert de $\mathbb{R}^k, \Omega = \Phi(U)$. P s'écrit alors $P(x,D_x) = a(x)D_x^\alpha$.

$$\int_V (P(x,D_x) - P^*(y,D_y))K_\varepsilon(x,y)g(y)dy$$

est la somme de trois termes :

$$(I) = a(x) \int_V [(D_x^\alpha + (-1)^{|\alpha|+1} D_y^\alpha)K_\varepsilon(x,y)]g(y)dy$$

$$(II) = (-1)^{|\alpha|} a(x) \int_V (D_y^\alpha K_\varepsilon(x,y))g(y)dy$$

$$(III) = -\int_V K_\varepsilon(x,y)P(y,D_y)g(y)dy \quad \text{(par définition de } P^*)$$

D'après (*'), l'intégrale de (I) tend vers 0 en $\| \ \|_\infty$ quand $\varepsilon \to 0$. En intégrant par parties, on obtient

(II) = $a(x)(D_y^\alpha g(y))_\varepsilon(x)$; d'après le lemme 2.4 , (II) converge en $\| \ \|_\infty$ vers $a(x)D_x^\alpha g(x)$ et (III) = $-(a(y)D_Y^\alpha g(y))_\varepsilon(x)$ converge en $\| \ \|_\infty$ vers $-a(x)D_x^\alpha(y)(x)$ quand $\varepsilon \to 0$, q.e.d.

COROLLAIRE 3.1.5. <u>Les conditions équivalentes données dans la proposition</u> 3.1.4 <u>sur la famille de noyaux</u> $(K_\varepsilon(x,y))_{\varepsilon\in\mathbb{R}^+}$ <u>sont des conditions nécessaires et suffisantes pour que, pour toute</u> $f \in \mathcal{D}(V)$, <u>si</u> $f_\varepsilon(x) = \int_V K_\varepsilon(x,y)f(y)dy$, <u>la famille</u> $(f_\varepsilon)_{\varepsilon\in\mathbb{R}^+}$ <u>converge vers</u> f <u>dans</u> $\mathcal{D}(V)$ <u>quand</u> ε <u>tend vers</u> 0 .

PROPOSITION 3.1.6. <u>La famille</u> $(T_\varepsilon)_{\varepsilon\in\mathbb{R}^+}$ <u>des régularisées de</u> T <u>converge faiblement vers</u> T <u>dans</u> $\mathcal{D}'(V)$ <u>quand</u> ε <u>tend vers</u> 0 <u>si et seulement si</u> $(K_\varepsilon)_{\varepsilon\in\mathbb{R}^+}$ <u>vérifie les conditions de la proposition</u> 3.1.4.

<u>D é m o n s t r a t i o n</u>. On a $\langle T - T_\varepsilon, \psi\rangle = \langle T, (\varphi - \varphi_\varepsilon)dx\rangle$, or d'après le corollaire 3.1.5 , φ_ε converge vers φ dans $\mathcal{D}(V)$ et donc par définition de la topologie faible

$$\lim_{\varepsilon\to 0} (T_\varepsilon - T) = 0 \text{ dans } \mathcal{D}'_{\text{faible}}(V) .$$

3.2. <u>Convergence forte des régularisées</u>.

Rappelons que la topologie forte de $\mathcal{D}'(V)$ est la topologie de la convergence uniforme sur les bornés de l'espace $\mathcal{D}^k(V)$ des formes différentielles $\mathcal{C}^\infty$ à support compact de degré maximal sur V (cf. [7]) .

PROPOSITION 3.2.1. <u>Pour que la famille des régularisées</u> $(T_\varepsilon)_{\varepsilon\in\mathbb{R}^+}$ <u>de</u> T <u>converge fortement vers</u> T <u>dans</u> $\mathcal{D}'(V)$ <u>quand</u> ε <u>tend vers</u> 0 <u>il faut et il suffit que pour tout opérateur différentiel</u> P <u>sur</u> V

$$(**) \qquad \lim_{\varepsilon\to 0} \left\| \int_V ((P(x,D_x) - P^*(y,D_y))K_\varepsilon(x,y))\varphi(y)dy \right\|_\infty = 0$$

<u>uniformément pour</u> φ <u>dans un borné de</u> $\mathcal{D}(V)$.

<u>D é m o n s t r a t i o n</u> . 1/ <u>Condition suffisante</u>.

$$\langle T - T_\varepsilon, \psi\rangle = \langle T , (\varphi - \varphi_\varepsilon)dx\rangle .$$

Il suffit donc de montrer qu'il y a équiconvergence de $(\varphi_\varepsilon)_{\varepsilon\in\mathbb{R}^+}$ vers φ dans $\mathcal{D}(V)$ sur tout borné de $\mathcal{D}(V)$, c'est-à-dire que pour tout opérateur différentiel P sur V

$$\|P(x),D_x)\varphi(x) - P(x,D_x)\varphi_\varepsilon(x)\|_\infty \to 0$$

uniformément pour φ dans un borné de $\mathcal{D}(V)$

$$\|P(x,D_x)\varphi(x) - P(x,D_x)\varphi_\varepsilon(x)\|_\infty \leqslant \|P(x,D_x)\varphi(x) - (P(y,D_y)\varphi(y))_\varepsilon(x)\|_\infty$$

$$+ \|\int_V ((P(x,D_x) - P^*(y,D_y))K_\varepsilon(x,y))\varphi(y)dy\|_\infty$$

en utilisant les calculs du paragraphe 3.1.

Le second terme tend vers 0 uniformément pour φ dans un borné de $\mathcal{D}(V)$ par hypothèse.

Il en est de même du premier terme. En effet il suffit de reprendre la démonstration du lemme 2.4 pour φ dans un borné de $\mathcal{D}(V)$.

2/ Condition nécessaire.

On suppose que $(T_\varepsilon)_{\varepsilon \in \mathbb{R}^+}$ converge vers T dans $\mathcal{D}'(V)$ muni de la topologie forte. Si P est un opérateur différentiel sur V et P^* son adjoint on a donc

$$\lim_{\varepsilon\to 0} P^*(x,D_x)T_\varepsilon - (P^*(y,D_y)T)_\varepsilon = 0 \text{ dans } \mathcal{D}'_{\text{fort}}(V) .$$

Or pour toute $\psi \in \mathcal{D}^k(V)$

$$<P^*(x,D_x)T_\varepsilon - (P^*(y,D_y)T)_\varepsilon , \psi > =$$
$$= <T,[(P(y,D_y)\varphi)_\varepsilon(x) - P(x,D_x)\,\varphi_\varepsilon(x)]\,dx>$$

si $\psi = \varphi\, dx$ avec $\varphi \in \mathcal{D}(V)$

donc $<P^*(x,D_x)T_\varepsilon - (P^*(y,D_y)T)_\varepsilon, \psi > = - <T , [\int_V ((P(x,D_x) - P^*(y,D_y))K_\varepsilon(x,y))\;\varphi(y)dy\;]dx >$,

d'où $(\int_V (P(x,D_x) - P^*(y,D_y))K_\varepsilon(x,y))\varphi(y)dy)_{\varepsilon\in\mathbb{R}^+}$ converge vers zéro dans $\mathcal{D}(V)$ uniformément pour φ dans un borné de $\mathcal{D}(V)$, donc en $\|\;.\;\|_\infty$. •

Nous allons donner maintenant, dans un domaine de carte de V une condition équivalente à celle de la proposition 3.2.1.

Nous reprenons les notations du paragraphe 3.1.

PROPOSITION 3.2.2. <u>Soit $(K_\varepsilon(x,y))_{\varepsilon\in\mathbb{R}^+}$ une famille de noyaux alors les conditions suivantes sont équivalentes</u> :

(i) <u>Pour tout opérateur P sur V</u>

(**) $\lim_{\varepsilon\to 0} \|\int_V ((P(x,D_x) - P^*(y,D_y))K_\varepsilon(x,y))\,\varphi(y)dy\|_\infty = 0$

<u>uniformément sur tout borné de $\mathcal{D}(V)$</u> .

(ii) <u>Pour tout domaine de carte et pour tout multi-indice</u> n

(**') $\lim_{\varepsilon\to 0} \|\int_V ((D_x^\alpha + (-1)^{|\alpha|+1} D_y^\alpha)K_\varepsilon(x,y))\varphi(y)dy\|_\infty = 0$

<u>uniformément sur tout borné de $\mathcal{D}(V)$ dont les fonctions sont à support dans un domaine de carte</u> .

LEMME 3.2.3. <u>Soit</u> $A \subset \mathcal{D}(V)$ <u>un borné de</u> $\mathcal{D}(V)$ <u>dont les éléments sont à support dans un même domaine de carte. Pour toute</u> $f \in A$, <u>si</u> (**') <u>est vérifiée sur</u> A <u>il y a équiconvergence de la famille</u> $(f_\varepsilon)_{\varepsilon \in \mathbb{R}^+}$ <u>vers</u> f <u>dans</u> $\mathcal{D}(V)$.

<u>Démonstration du lemme</u> . Les éléments de A étant à support dans un même domaine de carte nous devons donc montrer que pour tout multi-indice α

$$(\forall \theta > 0)(\exists \varepsilon_0)(\forall f \in A)(\varepsilon < \varepsilon_0 \Rightarrow \| D_x^\alpha (f - f_\varepsilon)\|_\infty < \theta) .$$

On a : $D_x^\alpha(f - f_\varepsilon)(x) = \int_V K_\varepsilon(x,y)\,(D_x^\alpha f(x) - D_y^\alpha f(y))dy$

$$- \int((D_x^\alpha + (-1)^{|\alpha|+1} D_y^\alpha)K_\varepsilon(x,y))f(y)dy$$

$$\|D_x^\alpha(f - f_\varepsilon)\|_\infty \leq \varepsilon' \sup_{\substack{\beta \in \mathbb{N}^k \\ |\beta| = |\alpha|+1}} \|D_x^\beta f\|_\infty + \|\int_V ((D_x^\alpha + (-1)^{|\alpha|+1} D_y^\alpha)K_\varepsilon(x,y))f(y)\|_\infty$$

ε' tendant vers 0 avec ε , d'après le théorème des accroissements finis ; or f est dans un borné A de $\mathcal{D}(V)$ donc il existe $M_{|\alpha|+1}$ indépendant de f telle que $\sup_{\substack{\beta \in \mathbb{N}^k \\ |\beta| = |\alpha|+1}} \|D_x^\beta f\|_\infty \leq M_{|\alpha|+1}$ et (**') est vérifié sur A ce qui démontre le lemme.

<u>Démonstration de la proposition 3.2.2.</u> 1/ Il est évident que (i) implique (ii) .

2/ Les fonctions d'un borné de $\mathcal{D}(V)$ étant à support dans un même compact K , par une partition de l'unité finie on se ramène à un borné de $\mathcal{D}(V)$ dont les fonctions sont à support dans un même domaine de carte. Comme dans le paragraphe 3.1 on peut considérer que $P(x,D_x) = a(x)D_x^\alpha$, alors

$$\| \int_V ((P(x,D_x) - P^*(y,D_y))K_\varepsilon(x,y))\varphi(y)dy\|_\infty$$

$$\leq \| \int a(x)[(D_x^\alpha + (-1)^{|\alpha|+1} D_y^\alpha)K_\varepsilon(x,y)]\varphi(y)dy \|_\infty \qquad \text{(I)}$$

$$+ \|a(x)(D_y^\alpha\varphi(y))_\varepsilon(x) - a(x)D_x^\alpha\varphi(x)\|_\infty \qquad \text{(II)}$$

$$+ \|(a(y)D_y^\alpha\varphi(y))_\varepsilon(x) - a(x)D_x^\alpha\varphi(x)\|_\infty \qquad \text{(III)}$$

$\lim_{\varepsilon \to 0} \text{(I)} = 0$ uniformément sur tout borné à support dans un domaine de carte d'après (**'), $\lim_{\varepsilon \to 0} \text{(II)} = 0$ et $\lim_{\varepsilon \to 0} \text{(III)} = 0$ uniformément sur tout borné à support dans un domaine de carte d'après le lemme 3.2.3.

D'où (ii) $\Rightarrow$ (i). q.e.d.

3.3. Convergence de la famille des régularisées dans $\mathcal{D}'_\Gamma(V)$.

Rappelons les définitions du front d'onde d'une distribution et de l'espace $\mathcal{D}'_\Gamma(V)$ (cf. [1], [3]) .

D é f i n i t i o n 3.3.1. Si T est une distribution définie sur un ouvert U de $\mathbb{R}^n$, le front d'onde $WF(T)$ est défini dans $V \times \mathbb{S}^{n-1}$ par :
$(x_0,\xi_0) \notin WF(T)$ si et seulement si il existe un voisinage U' de x_0 , un voisinage V de ξ_0 tels que pour toute $\Phi \in \mathcal{D}(U')$, pour tout $\xi \in V$, et pour tout $N \in \mathbb{N}$, $\mathcal{F}(\Phi T)(\tau\xi) = O(\tau^{-N})$ pour $\tau \in \mathbb{R}^+$ tendant vers l'infini uniformément en ξ sur V ($\mathcal{F}$ est la transformation de Fourier de $\mathcal{S}'(\mathbb{R}^n)$ dans lui-même).

D é f i n i t i o n 3.3.2. Si T est une distribution définie sur une variété $\mathcal{C}^\infty$, V , le front d'onde $WF(T)$ est le sous-ensemble du fibré cotangent en sphères $\mathbb{S}^*V$ défini dans chaque carte par la définition 3.3.1 .

D é f i n i t i o n 3.3.3. Si Γ est un fermé de $\mathbb{S}^*V$, on note $\mathcal{D}'_\Gamma(V)$ l'espace des distributions $T \in \mathcal{D}'(V)$ telles que $WF(T) \subset \Gamma$.

D'après DUISTERMAAT [1] on peut définir la topologie de $\mathcal{D}'_\Gamma(V)$ à l'aide des semi-normes de la topologie faible auxquelles on ajoute les semi-normes suivantes :

$$T \longmapsto \sup_{\substack{\tau \geq 1 \\ \xi \in A}} \tau^N |\mathcal{F}(\Phi T)(\tau\xi)|$$

où A est un compact de $\mathbb{S}^{n-1}$ ne rencontrant pas Γ et $\Phi \in \mathcal{D}(V)$ une fonction $\mathcal{C}^\infty$ à support dans un domaine de carte de V .

PROPOSITION 3.3.4. Soit $(K_\varepsilon)_{\varepsilon \in \mathbb{R}^+}$ une famille de noyaux sur $V \times V$ telle que pour tout compact V_i inclus dans un domaine de carte de V et tout multi-indice α

(***) $\lim\limits_{\varepsilon \to 0} \left\| \int_{V_i} \left|\left(\frac{\partial}{\partial x} + \frac{\partial}{\partial y}\right)^\alpha K_\varepsilon(y,x)\right| dy \right\|_\infty = 0$ et que la fonction $x \mapsto \int_V K_\varepsilon(y,x)dy$ converge vers 1 dans $\mathcal{C}^\infty(V)$ quand ε tend vers 0 , alors la famille $(T_\varepsilon)_{\varepsilon \in \mathbb{R}^+}$ des régularisées de $T \in \mathcal{D}'_\Gamma(V)$ converge vers T pour les semi-normes

$$T \mapsto \sup_{\substack{\tau \geq 1 \\ \xi \in A}} \tau^N \,|\mathcal{F}(\Phi T)(\tau\xi)| \; .$$

Démonstration. Soit $\Phi \in \mathcal{D}(V)$ une fonction $\mathcal{C}^\infty$ à support compact dans un domaine de carte de V et $\psi \in \mathcal{D}(V)$ telle que ψ soit égale à 1 au voisinage de $\operatorname{supp} \Phi$, alors si $T \in \mathcal{D}'_\Gamma(V)$

$$\Phi\,\psi\,T = \Phi T \quad \text{et} \quad \Phi(\psi T)_\varepsilon = \Phi T_\varepsilon \quad \text{pour } \varepsilon \text{ assez petit.}$$

Désignons par $\mathbb{F}$ ou par $\widehat{\cdot}$ la transformation de Fourier et considérons pour ε assez petit

$$F = \mathcal{F}(\Phi T_\varepsilon - \Phi T)(\xi) = \mathcal{F}(\Phi(\psi T)_\varepsilon - \Phi\psi T)(\xi) \; .$$

On remarque que

$$\psi T(y) = \int \widehat{\psi T}(\eta) e^{iy\eta}\, d\lambda(\eta) \quad \text{où } \lambda(\eta) \text{ désigne la mesure}$$

de Lebesgue sur $\mathbb{R}^k$.

Alors
$$F = \mathcal{F}\Big(\Phi(x)\int_V K_\varepsilon(y,x)(\psi T)(y)dy - \Phi(x)\int_V K_\varepsilon(y,x)(\psi T)(x)dy + (\Phi\psi T)(x)\Big(\int_V K_\varepsilon(y,x)dy - 1\Big)\Big)(\xi)$$

où $\int_V K_\varepsilon(y,x)(\psi T)(y)dy$ désigne $\langle(\psi T)(y), K_\varepsilon(y,x)dy\rangle$

$$F = \mathcal{F}\Big(\Phi(x)\int_V K_\varepsilon(y,x)dy\Big(\int \widehat{\psi T}(\eta)e^{iy\eta}\, d\lambda(\eta)\Big) - \Phi(x)\int_V K_\varepsilon(x,y)dy \Big(\int \widehat{\psi T}(\eta)\, e^{ix\eta}\, d\lambda(\eta)\Big)\Big)(\xi) + \underbrace{\mathcal{F}\Big((\Phi\psi T)(x)\Big(\int_V K_\varepsilon(y,x)dy - 1\Big)\Big)(\xi)}_{G}$$

$$F - G = \int \Phi(x) e^{-ix\xi}\, d\lambda(x)\Big(\int_V K_\varepsilon(y,x)dy\Big(\int \widehat{\psi T}(\eta)e^{iy\eta}\, d\lambda(\eta) - \int \widehat{\psi T}(\eta)e^{ix\eta} d\lambda(\eta)\Big)\Big)$$

$$= \int \widehat{\psi T}(\eta) d\lambda(\eta)\Big(\int\Big(\int \Phi(x)K_\varepsilon(y,x)e^{-i(x\xi - y\eta)}dy\Big)d\lambda(x) - \int\Big(\int_V \Phi(x)K_\varepsilon(y,x)e^{-i(x(\xi-\eta)+0.y)}dy\Big)d\lambda(x)\Big) \; .$$

Supposons les coordonnées choisies sur l'ouvert de carte contenant le support de Φ de telle sorte que dy coïncide avec $d\lambda(y)$ la mesure de Lebesgue sur $\mathbb{R}^n$.

Pour ε assez petit $\Phi(x)K_\varepsilon(x,y)$ est à support dans cet ouvert de carte et on a donc

$$F = \int \widehat{\psi T}(\eta) d\lambda(\eta)\Big(\iint \Phi(x)K_\varepsilon(y,x)e^{-i(x\xi - y\eta)} d\lambda(y)d\lambda(x) - \Phi(x)K_\varepsilon(y,x)e^{-i(x(\xi-\eta)+0.y)}d\lambda(y)d\lambda(x)\Big) + G$$

$$= \int \widehat{\psi T}(\eta)\big(\widehat{\Phi K_\varepsilon}(-\eta,\xi) - \widehat{\Phi K_\varepsilon}(0, \xi - \eta)\big)d\lambda(\eta) + \mathcal{F}\Big((\Phi\,\psi T)(x)\Big(\int_V K_\varepsilon(y,x)dy - 1\Big)\Big)(\xi) \; .$$

LEMME 3.3.5. Si les noyaux régularisants $K_\varepsilon(x,y)$ sont tels que les fonctions $x \mapsto \int_V K_\varepsilon(y,x)dy$ dans $\mathcal{E}^\infty(V)$ lorsque ε tend vers 0 alors

$$\lim_{\substack{\varepsilon\to 0}} \sup_{\substack{\tau\geq 1\\ \zeta\in A}} \tau^N \left|\mathcal{F}((\Phi\,\psi\,T)(x)\,(\int_V K_\varepsilon(y,x)dy - 1))(\tau\zeta)\right| = 0 .$$

D é m o n s t r a t i o n . Cela résulte immédiatement des propriétés de continuité de la transformation de Fourier relativement à la topologie de $\mathcal{E}^\infty(\mathbb{R}^k)$.

LEMME 3.3.6. Si A est un compact de $\mathbb{R}^k$ ne rencontrant pas $\mathbb{R}\Gamma$, soit λ tel que

$$\{\eta \,/\, \forall\xi \in \mathbb{R}A \ |\xi - \eta| < \lambda|\xi|\} \cap \mathbb{R}\Gamma = \emptyset$$

alors on a

$$\sup_{\substack{\tau\geq 1\\ \zeta\in A}} \tau^N \left|\int_{|\tau\zeta-\eta|<\lambda|\tau\zeta|} \widehat{\psi T}(\eta)(\widehat{\Phi K_\varepsilon}(-\eta,\tau\zeta) - \widehat{\Phi K_\varepsilon}(0,\tau\zeta-\eta))d\lambda(\eta)\right| \to 0$$

quand $\varepsilon \to 0$.

D é m o n s t r a t i o n . Grâce à l'hypothèse faite sur λ , $\widehat{\psi T}(\eta)$ est à décroissance rapide dans le cône

$$|\xi - \eta| < \lambda|\xi| \quad \text{où} \quad \xi \in \mathbb{R}A$$

car $WF(\psi T) \subset \Gamma$.

C'est-à-dire $(\forall m \in \mathbb{N})\ (\exists c_m)\ (\widehat{\psi T}(\eta) \leq c_m(1 + |\eta|)^{-m})$.

On a donc la majoration suivante pour tout N_1

$$\left|\int_{|\xi-\eta|<\lambda|\xi|} \widehat{\psi T}(\eta)(\widehat{\Phi K_\varepsilon}(-\eta,\xi) - \widehat{\Phi K_\varepsilon}(0,\xi-\eta))d\lambda(\eta)\right|$$

$$< c'_{N_1}(1 + |\xi|)^{-N_1}(1-\alpha)^{-N_1} \|\widehat{\Phi K_\varepsilon}(-\eta,\xi) - \widehat{\Phi K_\varepsilon}(0,\xi-\eta)\|_{\infty,\eta}$$

de plus

$$\|\widehat{\Phi K_\varepsilon}(-\eta,\xi) - \widehat{\Phi K_\varepsilon}(0,\xi-\eta)\|_{\infty,\eta} \leq \sup_\eta\left[\int |\Phi(x)|K_\varepsilon(y,x)|\,|e^{iy\eta}-e^{ix\eta}|\, d\lambda(x)d\lambda(y)\right]$$

$$\leq K(1 + \lambda)\ |\xi|\ \varepsilon\|\Phi\|_\infty$$

car $K_\varepsilon(x,y)$ est à support dans U_ε et $\int_V K_\varepsilon(x,y)dy$ converge vers 1 .

D'où le lemme en posant $\xi = \tau\zeta$ et en prenant $N_1 \geq N + 1$.

LEMME 3.3.7. Supposons que les noyaux $(K_\varepsilon)_{\varepsilon\in\mathbb{R}^+}$ vérifient pour tout multi-indice α et tout compact V_i contenu dans un domaine de carte et pour ε assez petit

$$\left\|\int_{V_i} \left|\left(\frac{\partial}{\partial x} + \frac{\partial}{\partial y}\right)^\alpha K_\varepsilon(y,x)\right|dy\right\|_\infty \leq M_{i,\alpha}$$

alors si $t \in \mathbb{R}^+$

$$\lim_{\varepsilon \to 0} \sup_{\substack{\tau \geq 1 \\ \zeta \in A}} \tau^N \left| \int_{|\tau\zeta - \eta| > t|\tau\zeta|} \widehat{\psi T}(\eta) \, (\widehat{\Phi K}_\varepsilon(-\eta, \tau\zeta) - \widehat{\Phi K}_\varepsilon(0, \tau\zeta - \eta)) \, d\lambda(\eta) \right| = 0$$

Démonstration. En intégrant par parties on obtient pour tout multi-indice $\alpha \in \mathbb{N}^k$

$$\widehat{\Phi K}_\varepsilon(-\eta, \xi) = \frac{1}{(i(\xi - \eta))^\alpha} \sum_{\beta \leq \alpha} \binom{\alpha}{\beta} \int_V e^{-iy\eta} d\lambda(y) \int_V e^{-ix(\xi-\eta)} \Phi^{(\alpha-\beta)}(x) D_x^\alpha K_\varepsilon(x-y, x) \, d\lambda(x)$$

d'où l'expression suivante :

$$\Phi K_\varepsilon(-\eta, \xi) - \Phi K_\varepsilon(0, \xi - \eta) = \frac{1}{(i(\xi-\eta))^\alpha} \sum_{\beta \leq \alpha} \binom{\alpha}{\beta} \int_{V \times V} e^{-ix\xi}(e^{iy\eta} - e^{ix\eta}) \Phi^{(\alpha-\beta)}_{(x)} \left(\frac{\partial}{\partial x} + \frac{\partial}{\partial y}\right)^\alpha K_\varepsilon(y, x) \, d\lambda(x) d\lambda(y)$$

$$|\widehat{\Phi K}_\varepsilon(-\eta, \xi) - \widehat{\Phi K}_\varepsilon(0, \xi - \eta)| \leq \frac{1}{|(\xi - \eta)^\alpha|} \sum_{\beta \leq \alpha} \binom{\alpha}{\beta} \varepsilon |\eta| \, \|\Phi^{(\alpha-\beta)}_{(x)}\|_1 \left\| \int_{\pi_2(\pi_1^{-1}(\mathrm{supp}\, \Phi) \cap U_\varepsilon)} \left|\left(\frac{\partial}{\partial x} + \frac{\partial}{\partial y}\right)^\alpha K_\varepsilon(y, x)\right| d\lambda(y) \right\|_\infty$$

où π_1 et π_2 sont les deux projections de $V \times V$ sur V.

D'autre part $\widehat{\psi T}(\eta)$ est à croissance lente

i.e. $\exists N_1$; $\widehat{\psi T}(\eta) \leq C_{N_1} (1 + |\eta|)^{N_1}$

en utilisant l'hypothèse du lemme on obtient donc la majoration suivante

$$\left| \int_{|\xi-\eta| > t|\xi|} \widehat{\psi T}(\eta) \, (\widehat{\Phi K}_\varepsilon(-\eta, \xi) - \widehat{\Phi K}_\varepsilon(0, \xi - \eta)) d\eta \right|$$

$$\leq \varepsilon \, C_{N_1} \left(\int_{|\xi-\eta| > t|\xi|} (1 + |\eta|)^{N_1} |\eta| \frac{1}{|(\xi - \eta)^\alpha|} M_{i,\alpha} \, d\eta \right) \sum_{\beta < \alpha} \binom{\alpha}{\beta} \|\Phi^{(\alpha-\beta)}\|_1$$

or $|\xi - \eta| \geq t|\xi| \Rightarrow |\eta| < (1 + \frac{1}{t}) \, |\xi - \eta|$.

Par un choix convenable de α on obtient alors

$$\leq \left(\sum_{\beta \leq \alpha} \binom{\alpha}{\beta} \|\Phi^{(\alpha-\beta)}\|_1 \right) M_{i,\alpha} \, C'_{N_1} (1 + \frac{1}{t})^{N_1} \times \frac{1}{t^N} |\xi|^{-N} \varepsilon$$

d'où le lemme en posant $\xi = \tau\zeta$.

Suite de la démonstration de la proposition 3.3.4. En utilisant les lemmes 3.3.5, 3.3.6 et 3.3.7 ainsi que l'hypothèse (***) on voit immédiatement que $(T_\varepsilon)_{\varepsilon \in \mathbb{R}^+}$ converge vers T quand ε tend vers 0 pour les semi-normes

$$T \mapsto \sup_{\substack{\tau \geq 1 \\ \xi \in A}} \tau^N |\mathcal{F}(\emptyset T)(\tau\xi)|$$

COROLLAIRE 3.3.8. *Si la famille des noyaux régularisants $(K_\varepsilon)_{\varepsilon \in \mathbb{R}^+}$ vérifie les hypothèses (*) (ou (*')), (***) et si $x \to \int_V K(y,x)dy$ converge vers 1 dans $\mathscr{C}^\infty(V)$, la famille $(T_\varepsilon)_{\varepsilon \in \mathbb{R}^+}$ des régularisées de $T \in \mathscr{D}'_\Gamma(V)$ converge vers T dans $\mathscr{D}'_\Gamma(V)$.*

D é m o n s t r a t i o n . Cela résulte immédiatement de la définition de la topologie de $\mathcal{D}'_\Gamma(V)$ et des proposition 3.1.6. et 3.3.4.

3.4. Lien entre les différentes conditions.

Dans ce paragraphe nous allons étudier les relations existantes entre les différentes conditions des paragraphes précédents.

Notons $(\widetilde{***})$ la condition suivante :

$(\widetilde{***})$ $\begin{cases} \text{Pour tout compact } V_i \text{ inclus dans un domaine de carte} \\ \text{de } V \text{ et tout multi-indice } \alpha . \\ \lim\limits_{\varepsilon\to 0} \| \int_{V_i} |(\frac{\partial}{\partial x} + \frac{\partial}{\partial y})^\alpha K_\varepsilon(x,y)|dy\|_\infty = 0 \end{cases}$

$(\widetilde{***})$ n'est autre que la propriété $(***)$ pour l'opérateur adjoint de l'opérateur associé au noyau $K_\varepsilon(x,y)$.

Considérons tout d'abord les conditions $(\widetilde{***})$ et $(*')$.
Il est facile de voir que la condition $(\widetilde{***})$ implique la condition suivante :

$(***')$ $\begin{cases} \text{Pour toute fonction } f \in \mathcal{D}(V) \text{ à support dans un domaine} \\ \text{de carte et pour tout multi-indice } \alpha \\ \lim\limits_{\varepsilon\to 0} \| \int_V ((\frac{\partial}{\partial x} + \frac{\partial}{\partial y})^\alpha K_\varepsilon(x,y))f(y)dy\|_\infty = 0 \end{cases}$

et on a la proposition suivante :

PROPOSITION 3.4.2. Soit $(K_\varepsilon)_{\varepsilon\in\mathbb{R}^+}$ une famille de noyaux sur $V\times V$. Les deux assertions suivantes sont équivalentes :

i/ Pour toute fonction $f \in \mathcal{D}(V)$ à support dans un domaine de carte et tout multi-indice α

$(*')$ $$\lim_{\varepsilon\to 0} \| \int_V ((\frac{\partial^{|\alpha|}}{\partial x^\alpha} + (-1)^{|\alpha|+1} \frac{\partial^{|\alpha|}}{\partial y^\alpha}) K_\varepsilon(x,y))f(y)dy\|_\infty = 0 .$$

ii/ Pour toute fonction $f \in \mathcal{D}(V)$ à support dans un domaine de carte et pour tout multi-indice α

$(***')$ $$\lim_{\varepsilon\to 0} \| \int_V ((\frac{\partial}{\partial x} + \frac{\partial}{\partial y})^\alpha K_\varepsilon(x,y))f(y)dy\|_\infty = 0 .$$

D é m o n s t r a t i o n . Notons $op(K_\varepsilon)$ l'opérateur sur $\mathcal{D}(V) : f\mapsto \int_V K_\varepsilon(x,y)f(y)dy$ et remarquons que si on pose pour deux opérateurs sur $\mathcal{D}(V)$, $[P,Q] = PQ - QP$ et $(ad(P))(Q) = [P,Q]$ on a

$$[D_x, op(K_\varepsilon)] = op [(\frac{\partial}{\partial x} + \frac{\partial}{\partial y})K_\varepsilon]$$
$$(ad(D_x))^\alpha(op(K_\varepsilon)) = op[(\frac{\partial}{\partial x} + \frac{\partial}{\partial y})^\alpha K_\varepsilon]$$
$$(ad(D_x^\alpha)) (op(K_\varepsilon)) = op[(\frac{\partial^{|\alpha|}}{\partial x^\alpha} + (-1)^{|\alpha|+1} \frac{\partial^{|\alpha|}}{\partial y^\alpha}) K_\varepsilon] .$$

LEMME 3.4.3. Soit A un anneau où l'on a défini

$$\forall (a,b) \in A^2 \qquad [a,b] = ab - ba$$

$$ad(a) = [a\ ,\ .\ ]$$

$$Ra = .\ a$$ (i.e. la multiplication à gauche par a),

alors on a la relation suivante :

$$\forall a = (a_1, \dots, a_p)\ ,\ \forall \alpha \in \mathbb{R}^p$$

$$(ad\ a)^\alpha = ad(a^\alpha) + \sum_{|i|<|\alpha|} K_\alpha^i\ ad(a^i)(Ra)^{\alpha-i}$$

si les $(a_j)_{1 \leqslant j \leqslant p}$ commutent entre eux.

Démonstration. On fait une récurrence sur la longueur du multi-indice α

. $|\alpha| = 2$

$$(ad\ a)^\alpha = (ad\ a_{i_1})(ad\ a_{i_2}) = ad(a_{i_1} a_{i_2}) - ad\ a_{i_1} Ra_2 - ad\ a_{i_2} Ra_1$$

. Supposons le résultat vrai pour $|\alpha| = k$.

. Montrons le pour : un multi-indice α' tel que $|\alpha'| = k+1$

$$(ad\ a)^{\alpha'} = (ad\ a_j)(ad\ a)^\alpha \qquad \alpha' = \alpha + (0,\dots,0,1,0,\dots,0)$$

$$= (ad\ a_j)(ad(a^\alpha)) + \sum_{|i|<|\alpha|} K_\alpha^i (ad\ a_j)(ad(a^i))\ (Ra)^{\alpha-i}$$

or $(ad\ a_j)(ad\ a^i) = ad(a_j\ a^i) - ad\ a_j\ R(a^i) - ad\ a^i\ Ra_j$ et

$R(a^i) = (Ra)^i$, donc $(ad\ a)^{\alpha'} = ad(a^{\alpha'}) + \sum_{|i|<|\alpha'|} K_\alpha^i\ ,\ ad(a^i)(Ra)^{\alpha'-i}$.

Fin de la démonstration de la proposition 3.4.2. On déduit donc du lemme 3.8. que

$op[(\frac{\partial}{\partial x} + \frac{\partial}{\partial y})^\alpha K_\varepsilon]$ s'exprime en fonction des opérateurs

$op[(\frac{\partial^{|\alpha-i|}}{\partial x^{\alpha-i}} + (-1)^{|\alpha-i|+1} \frac{\partial^{|\alpha-i|}}{\partial y^{\alpha-i}})\ K_\varepsilon] \circ \frac{\partial^i}{\partial y_i}$ où $i \in \mathbb{N}^k$

est un multi-indice de longueur strictement inférieure à $|\alpha|$

et de

$$op[(\frac{\partial^{|\alpha|}}{\partial x^\alpha} + (-1)^{|\alpha|+1} \frac{\partial^{|\alpha|}}{\partial y^\alpha})\ K_\varepsilon]$$

et réciproquement $op[(\frac{\partial^{|\alpha|}}{\partial x^\alpha} + (-1)^{|\alpha|+1} \frac{\partial^{|\alpha|}}{\partial y^\alpha})\ K_\varepsilon]$ s'exprime

en fonction des opérateurs $op[(\frac{\partial}{\partial x} + \frac{\partial}{\partial y})^{\alpha-i} K_\varepsilon] \circ \frac{\partial^i}{\partial y_i}$ où $i \in \mathbb{N}^k$

est un multi-indice de longueur strictement inférieure à $|\alpha|$

et de op $[(\frac{\partial}{\partial x}+\frac{\partial}{\partial y})^{\alpha} K_\varepsilon]$.

Cela démontre la proposition.

Par des méthodes analogues on obtient aussi le fait que (***) implique (**') (il suffit d'ajouter l'hypothèse d'uniformité par rapport aux bornés de $\mathcal{D}(V)$ dans la proposition 3.4.2) .

Par ailleurs on peut montrer sans difficultés que la condition (*') et le fait que la fonction $x \mapsto \int_V K_\varepsilon(x,y)\,dy$ converge vers 1 dans $\mathcal{E}^0(V)$ impliquent que cette même fonction converge vers 1 dans $\mathcal{E}^\infty(V)$.

4. Existence de noyaux.

Dans ([6] , § 15) de RHAM définit des opérateurs régularisants associés ([6], § 17) à des noyaux qui sont des fonctions $\mathcal{E}^\infty$ sur $V \times V$ à support dans un voisinage de la diagonale Δ de $V \times V$.

Rappelons la définition des opérateurs régularisants de de RHAM.

Considérons un recouvrement localement fini, dénombrable de V par des ouverts de coordonnées $V_i \subset\subset X$ homéomorphes à $\mathbb{R}^k$. Notons h_i l'homéomorphisme de V_i sur $\mathbb{R}^k$. On peut alors trouver un recouvrement de X par des ouverts $U_i \subset\subset V_i$ et des fonctions f_i , $\mathcal{E}^\infty$ à support compact dans V_i, telles que $f_i = 1$ sur $\overline{U_i}$.

Si T est une distribution sur V on pose $R_{i,\varepsilon_i} T = \overline{R}_{i,\varepsilon_i} f_i T + (1-f_i)T$ où $\overline{R}_{i,\varepsilon_i} = h_i^* r_{\varepsilon_i} h_{i*}$ et r_{ε_i} une convolution sur $\mathbb{R}^k$. Au voisinage de tout compact la suite des opérateurs $R^i(\varepsilon) = R_{i,\varepsilon_i} \dots R_{1,\varepsilon_1}$ est stationnaire et on pose $R_\varepsilon = \lim_{i\to\infty} R^i(\varepsilon)$.

Regardons par exemple le cas d'une variété recouverte par deux domaines de carte

$$RT = R_{1,\varepsilon_1} R_{2,\varepsilon_2} T = \overline{R}_{1,\varepsilon_1} f_1 \overline{R}_{2,\varepsilon_2} f_2 T + \overline{R}_1 f_1(1-f_2)T + (1-f_1)\overline{R}_2 f_2 T + \underbrace{(1-f_1)(1-f_2)T}_{0}$$

Le noyau associé s'écrit alors

$$K_{\varepsilon_1,\varepsilon_2}(x,z) = \int_V K_{1,\varepsilon_1}(x,y) f_1(y) K_{2,\varepsilon_2}(y,z) f_2(z)\,dy + K_{1,\varepsilon_1}(x,z) f_1(z)(1-f_2)(z) + (1-f_1)(x) K_{2,\varepsilon_2}(x,z) f_2(z).$$

Régularisation sur une variété

Soit R_ε un opérateur de de RHAM au voisinage d'un compact K de la variété V

$$R_\varepsilon = R_{n,\varepsilon_n} \ldots R_{1,\varepsilon_1} .$$

Notons $M_\varepsilon(x,y)$ le noyau associé à R_ε, M_ε se décompose de la manière suivante :

$M_\varepsilon(x,y) = \sum_{i\in I} G_{i,\varepsilon}(x,y) + \sum_{j\in J} \tilde{G}_{j,\varepsilon}(x,y)$, I et J étant des ensembles finis.

$G_{i,\varepsilon}$ est le noyau associé au composé de k_i opérateurs de la forme : $T \mapsto \bar{R}_{i,\varepsilon_\alpha} f_{i,\alpha} T$ où $f_{i,\alpha}$ est une fonction $\mathcal{C}^\infty$ à support compact dans un domaine de carte et $\bar{R}_{i,\varepsilon_\alpha}$ est l'image dans la variété d'un opérateur de convolution sur l'ouvert de carte contenant le support de $f_{i,\alpha}$.

On notera φ_i le produit des k_i fonctions $f_{i,\alpha}$ qui apparaissent dans l'opérateur auquel est associé $G_{i,\varepsilon}$.

$\tilde{G}_{j,\varepsilon}$ est le noyau associé au composé de k_j opérateurs de la forme : $T \to \bar{R}_{j,\varepsilon_\beta} f_{i,\beta} T$ et de l'opérateur $T \mapsto g_j T$ où g_j est une fonction $\mathcal{C}^\infty$.

On notera ψ_j le produit de g_j avec le produit des k_j fonctions $f_{j,\beta}$ qui apparaissent dans l'opérateur auquel est associé $\tilde{G}_{j,\varepsilon}$

Par définition de l'opérateur R_ε, les fonctions φ_i et ψ_j sont liées par la relation

$$\sum_{i\in I} \varphi_i + \sum_{j\in J} \psi_j = 1 .$$

Nous allons démontrer que les noyaux M_ε associés aux opérateurs R_ε sont des noyaux régularisants au sens de la définition 1.2 vérifiant les hypothèses (*'), (**'), (***) et $(\widetilde{***})$.

Notons $K_{i,\varepsilon_\alpha}(x,y)$ le noyau associé à $\bar{R}_{i,\varepsilon_\alpha}$ C'est un noyau régularisant et il vérifie les hypothèses (*'), (**') (***) et $(\widetilde{***})$ car c'est un noyau provenant d'une convolution.

Par ailleurs on peut remarquer que l'adjoint de l'opérateur R_ε est un opérateur de même type, les propriétés (***) et $(\widetilde{***})$ se correspondant par passage à l'adjoint on voit que pour que M_ε vérifie (***) et $(\widetilde{***})$ il suffit de montrer qu'il vérifie $(\widetilde{***})$.

On déduit alors du paragraphe 3.4 que pour obtenir le résultat cherché il suffit de montrer que M_ε est un noyau régularisant vérifiant la condition $(\widetilde{***})$.

PROPOSITION 4.1. <u>Les noyaux</u> M_ε <u>associés aux opérateurs régularisants de de RHAM</u> R_ε <u>vérifient les conditions suivantes</u>

(1) M_ε <u>est une fonction</u> $\mathcal{C}^\infty$ <u>sur</u> $V \times V$

(2) <u>les</u> M_ε <u>sont à support dans une famille décroissante d'entourages propres de la diagonale</u> Δ <u>de</u> $V \times V$

(3) $\lim\limits_{\varepsilon \to 0} \| \int_V M_\varepsilon(x,y)dy - 1 \|_{\infty,L} = 0$, <u>pour tout compact</u> L <u>de</u> V .

(4) <u>Pour tout compact</u> V_λ <u>inclus dans un domaine de carte de</u> V <u>et tout multi-indice</u> α <u>non nul</u>

$(\widetilde{***})$ $$\lim_{\varepsilon \to 0} \| \int_{V_\lambda} |(\frac{\partial}{\partial x} + \frac{\partial}{\partial y})^\alpha M_\varepsilon(x,y)|dy\|_\infty = 0 .$$

<u>Démonstration</u>: Les conditions (1) et (2) résultent de de RHAM ([6] , § 17).

Pour montrer que M_ε vérifie (3) considérons $(\varphi_k)_{k \in K}$ une partition de l'unité relative a des ouverts de carte de V . Soit L un compact de V , pour ε assez petit il existe un compact $L' \supset L$ tel que pour tout $x \in L$, $y \mapsto K_\varepsilon(x,y)$ soit à support dans L' . De plus il existe K' fini , $K' \subset K$ tel que pour tout $x \in L'$ $\sum\limits_{k \in K'} \varphi_k(x) = 1$.

On a donc pour tout $x \in L$

$$\int_V M_\varepsilon(x,y)dy = \sum_{k \in K'} \int_V M_\varepsilon(x,y)\varphi_k(y)dy .$$

Il suffit donc de montrer que $\lim\limits_{\varepsilon \to 0} \|R_\varepsilon \varphi_k - \varphi_k\|_\infty = 0$.

Cela se déduit facilement des propriétés des noyaux de convolution et de la multiplication par une fonction $\mathcal{C}^\infty$.

Considérons maintenant la condition (4) .

Reprenons la décomposition des noyaux M_ε

$$M_\varepsilon(x,y) = \sum_{i \in I} G_{i,\varepsilon}(x,y) + \sum_{j \in J} \widetilde{G}_{j,\varepsilon}(x,y) .$$

Notons $\overline{R}_{i,\alpha_1}$ et $\overline{R}_{j,\beta_1}$ les premiers opérateurs régularisants par convolution apparaissant dans les décompositions respectives de $G_{i,\varepsilon}$ et $\tilde{G}_{j,\varepsilon}$ et $K_{i,\varepsilon_{\alpha_1}}$ et $K_{j,\varepsilon_{\beta_1}}$ les noyaux associés .

On désigne par π^α l'opérateur $\left(\frac{\partial}{\partial x}+\frac{\partial}{\partial y}\right)^\alpha$.

On a alors

$$\int_{V_\lambda} |\pi^\alpha M_\varepsilon(x,y)|dy = \int_{V_\lambda} \Big| \sum_{i\in I} \pi^\alpha G_{i,\varepsilon}(x,y) + \sum_{j\in J} \pi^\alpha \tilde{G}_{j,\varepsilon}(x,y)\Big| dy$$

$$\text{(A)}\begin{cases} \leqslant \displaystyle\sum_{i\in I} \int_{V_\lambda} \Big|\pi^\alpha G_{i,\varepsilon}(x,y) - K_{i,\varepsilon_{\alpha_1}}(x,y)\frac{\partial^{|\alpha|}}{\partial y^\alpha}\varphi_i(y)\Big| dy \\ + \displaystyle\sum_{j\in J} \int_{V_\lambda} \Big|\pi^\alpha \tilde{G}_{j,\varepsilon}(x,y) - K_{j,\varepsilon_{\beta_1}}(x,y)\frac{\partial^{|\alpha|}}{\partial y^\alpha}\psi_j(y)\Big| dy \\ + \displaystyle\int_{V_\lambda} \Big| \sum_{i\in I} K_{i,\varepsilon_{\alpha_1}}(x,y)\frac{\partial^{|\alpha|}}{\partial y^\alpha}\varphi_i(y) + \sum_{j\in J} K_{j,\varepsilon_{\beta_1}}(x,y)\frac{\partial^{|\alpha|}}{\partial y^\alpha}\psi_j(y)\Big| dy \end{cases}$$

Nous allons étudier successivement chacun des termes de cette somme.

LEMME 4.2. <u>On a le résultat suivant</u> :

$$\lim_{\substack{\varepsilon\to 0\\ \varepsilon=(\varepsilon_1,\dots,\varepsilon_n)}} \Big\| \int_{V_\lambda} \Big| \sum_{i\in I} K_{i,\varepsilon_{\alpha_1}}(x,y)\frac{\partial^{|\alpha|}}{\partial y^\alpha}\varphi_i(y) + \sum_{j\in J} K_{j,\varepsilon_{\beta_1}}(x,y)\frac{\partial^{|\alpha|}}{\partial y^\alpha}\psi_j(y)\Big| dy \Big\|_\infty = 0$$

<u>D é m o n s t r a t i o n</u>. Comme les fonctions φ_i et ψ_j vérifient

$$\sum_{i\in I} \varphi_i + \sum_{j\in J} \psi_j = 1$$

on a pour tout multi-indice α non nul

$$\frac{\partial^{|\alpha|}}{\partial x^\alpha}\Big(\sum_{i\in I} \varphi_i(x) + \sum_{j\in J} \psi_j(x)\Big) = 0$$

et donc si K_ε est le noyau associé à une convolution sur l'ouvert de carte contenant V_λ

$$\begin{aligned} A_\varepsilon(x,y) &= \sum_{i\in I} K_{i,\varepsilon_{\alpha_1}}(x,y)\frac{\partial^{|\alpha|}}{\partial y^\alpha}\varphi_i(y) + \sum_{j\in J} K_{j,\varepsilon_{\beta_1}}(x,y)\frac{\partial^{|\alpha|}}{\partial y^\alpha}\psi_j(y) \\ &= \sum_{i\in I} (K_{i,\varepsilon_{\alpha_1}}(x,y) - K_\varepsilon(x,y))\frac{\partial^{|\alpha|}}{\partial y^\alpha}\varphi_i(y) \\ &\quad + \sum_{j\in J} (K_{j,\varepsilon_{\beta_1}}(x,y) - K_\varepsilon(x,y))\frac{\partial^{|\alpha|}}{\partial y^\alpha}\psi_j(y) \end{aligned}$$

et si on suppose de plus que K_ε est inférieur ou égal à

$K_{i,\varepsilon_{\alpha_1}}$ et $K_{j,\varepsilon_{\beta_1}}$ on a

$$\int_{V_\lambda} |A_\varepsilon(x,y)|dy \leq \sum_{i\in I} \int_{V_\lambda} (K_{i,\varepsilon_{\alpha_1}}(x,y)-K_\varepsilon(x,y))| \frac{\partial^{|\alpha|}}{\partial y^\alpha}\varphi_i(y)|dy$$

$$+ \sum_{j\in J} \int_{V_\lambda} (K_{j,\varepsilon_{\beta_1}}(x,y)-K_\varepsilon(x,y))| \frac{\partial^{|\alpha|}}{\partial y^\alpha}\psi_j(y)|dy$$

Les fonctions $|\frac{\partial^{|\alpha|}}{\partial y^\alpha}\varphi_i|$ et $|\frac{\partial^{|\alpha|}}{\partial y^\alpha}\psi_j|$ étant continues et les noyaux étant associés à des convolutions on a

$$\lim_{\varepsilon\to 0} \int_{V_\lambda} |A_\varepsilon(x,y)|dy = 0 .$$

LEMME 4.3. <u>Si</u> $G_{i,\varepsilon^i}(x,y)$ <u>est le noyau associé à l'opérateur</u> $\bar{R}_{\varepsilon_i} f_i \ldots \bar{R}_{\varepsilon_1} f_1$ <u>on a sur</u> V_λ

$$\lim_{\varepsilon^i=(\varepsilon_1,\ldots,\varepsilon_i)\to 0} \| \int_{V_\lambda} |\pi^\alpha G_{i,\varepsilon^i}(x,y)-\int_{V\times\ldots\times V} K_{\varepsilon_i}(x,x_{i-1})\ldots K_{\varepsilon_1}(x_1,y)\Phi_{i,\alpha}(f_1(y)f_2(x_1)\ldots f_i(x_{i-1})dx_1\ldots dx_i|dy\|_\infty = 0$$

<u>où</u> $\Phi_{i,\alpha}(f_1(x_1)\ldots f_i(x_i)) = (\frac{\partial}{\partial x_1}+\ldots+\frac{\partial}{\partial x_i})^\alpha (f_1(x_1)\ldots f_i(x_i))$

<u>D é m o n s t r a t i o n</u>. Elle se fait par récurrence sur i.

Considérons tout d'abord le cas $i = 1$, dans ce cas

$$G_{1,\varepsilon^1}(x,y) = K_{\varepsilon_1}(x,y)f_1(y)$$

$$\pi^\alpha G_{1,\varepsilon^1} = \sum_{|p|+|q|=|\alpha|} \frac{\alpha!}{p!\,q!} \pi^p K_{\varepsilon_1}(x,y) \frac{\partial^q}{\partial y^q} f_1(y)$$ où p et q sont des multi-indices

$$\int_{V_\lambda} |\pi^\alpha G_{1,\varepsilon^1}(x,y)-K_{\varepsilon_1}(x,y)\frac{\partial^\alpha}{\partial y^\alpha}f_1(y)|dy \leq \sum_{\substack{|p|+|q|=|\alpha| \\ p\neq 0}} \frac{\alpha!}{p!q!}\int_{V_\lambda}|\pi^p K_{\varepsilon_1}(x,y)\frac{\partial^q}{\partial y^q}f_1(y)|dy$$

or $\frac{\partial^q}{\partial y^q} f_1(y)$ est continue donc bornée sur V_λ et K_{ε_1} vérifie la condition $(\widetilde{***})$ et donc pour $x\in V_\lambda$

$$\lim_{\varepsilon^1\to 0} \| \int_{V_\lambda} |\pi^\alpha G_{1,\varepsilon^1}(x,y) - K_{\varepsilon_1}(x,y)\frac{\partial^\alpha}{\partial y^\alpha} f_1(y)|dy\|_\infty = 0$$

et de plus $\| \int_{V_\lambda} |\pi^\alpha G_{1,\varepsilon^1}(x,y)|dy\|_\infty$ est bornée quand ε^1 tend vers 0 .

On suppose que le résultat du lemme est vrai au rang i et que $\| \int_{V_\lambda} |\pi^\alpha G_{i,\varepsilon^i}(x,y)|dy\|_\infty$ est bornée quand ε^i tend vers 0 pour tout multi-indice α.

Etudions ce qui se passe au rang $i+1$.

On pose $L_{\varepsilon_{i+1}} = K_{\varepsilon_{i+1}}(x,y)\, f_{i+1}(y)$

alors $G_{i+1,\varepsilon^{i+1}}(x,z) = \int_V L_{\varepsilon_{i+1}}(x,y) G_{i,\varepsilon^i}(y,z)dy$ et

$$\pi^\alpha_{x,z}\, G_{i+1,\varepsilon^{i+1}}(x,z) = \sum_{|p|+|q|=|\alpha|} \frac{\alpha!}{p!q!} \int_V \pi^q_{x,y}\, L_{\varepsilon_{i+1}}(x,y)\pi^p_{y,z} G_{i,\varepsilon^i}(y,z)dy.$$

On note

$$A_{\varepsilon^{i+1}} = \pi^\alpha_{x,z}\, G_{i+1,\varepsilon^{i+1}}(x,z) - \int_{V\times\ldots\times V} K_{\varepsilon_{i+1}}(x,x_i)\ldots K_{\varepsilon_1}(x_1,z)\Phi_{i+1,\alpha}(f_1(z)\ldots f_{i+1}(x_i))dx_1\ldots dx_i$$

or $\Phi_{i,\alpha}$ vérifie la relation de récurrence suivante

$$\Phi_{i+1,\alpha}(f_1(x_1),\ldots,f_{i+1}(x_{i+1})) = \sum_{|p|+|q|=|\alpha|} \frac{\alpha!}{p!q!} \frac{\partial^{|p|}}{\partial x^p_{i+1}} f_{i+1}(x_{i+1})\, \Phi_{i,q}(f_1(x_1)\ldots f_i(x_i))$$

et donc

$$A_{\varepsilon^{i+1}} = \sum_{|p|+|q|=|\alpha|} \frac{\alpha!}{p!q!} \int_V [\pi^q_{x,y}\, L_{\varepsilon_{i+1}}(x,y)\pi^p_{y,z}\, G_{i,\varepsilon^i}(y,z) - K_{\varepsilon_{i+1}}(x,y)$$
$$\int_{V\times\ldots\times V} K_{\varepsilon_i}(y,x_{i-1})\ldots K_{\varepsilon_1}(x_1,z) \frac{\partial^q}{\partial y^q} f_{i+1}(y)\, \Phi_{i,p}(f_1(z)\ldots f_i(x_{i-1})dx_1\ldots dx_{i-1}]dy$$

Ajoutons et retranchons $\displaystyle\sum_{|p|+|q|=|\alpha|} \frac{\alpha!}{p!q!} \int_V K_{\varepsilon_{i+1}}(x,y) \frac{\partial^q}{\partial y^q} f_{i+1}(y)\, \pi^p_{y,z}\, G_{i,\varepsilon^i}(y,z)dy$

à l'expression de $A_{\varepsilon^{i+1}}$, on obtient alors

$$\int_{V_\lambda} |A_{\varepsilon^{i+1}}|dz \leq \sum_{|p|+|q|=|\alpha|} \frac{\alpha!}{p!q!}\; C_q\, D_p + B_{p,q}$$

avec $C_q = \int_V K_{\varepsilon_{i+1}}(x,y) \left| \frac{\partial^q}{\partial y^q} f_{i+1}(y)\right|dy$ qui est borné en $\|\ \|_\infty$ car il converge uniformément sur V vers $\left| \frac{\partial^q}{\partial x^q} f_{i+1}(x)\right|$

$$D_p = \| \int_V |\pi^p_{y,z} G_{i,\varepsilon^i}(y,z) - \int_{V\times\ldots\times V} K_{\varepsilon_i}(y,x_{i-1})\ldots K_{\varepsilon_1}(x,z)\Phi_{i,p}(f_1(z)\ldots f_i(x_{i-1}))\, dx_1\ldots dx_{i-1}|dz\,\|_{\infty,y}$$

qui tend vers 0 avec ε^i d'après l'hypothèse de récurrence

$$B_{p,q} = \int_{V_\lambda} |\int_V [\pi^q_{x,y} L_{\varepsilon_{i+1}}(x,y)\pi^p_{y,z} G_{i,\varepsilon^i}(y,z) - K_{\varepsilon_{i+1}}(x,y)\frac{\partial^q}{\partial y^q} f_{i+1}(y)\, \pi^p_{y,z} G_{i,\varepsilon^i}(y,z)]dy|dz$$

$$\leqslant \int_V \pi^q_{x,y} L_{\varepsilon_{i+1}}(x,y) - K_{\varepsilon_{i+1}}(x,y)\frac{\partial^q}{\partial y^q} f_{i+1}(y)\,|\,(\int_{V_\lambda} |\pi^p_{y,z} G_{i,\varepsilon^i}(y,z)|dz)dy$$

$$\leqslant \|\int_{V_\lambda} |\pi^p_{y,z} G_{i,\varepsilon^i}(y,z)|dz\,\|_{\infty,y} \int_V |\pi^q_{x,y} L_{\varepsilon_{i+1}}(x,y) - K_{\varepsilon_{i+1}}(x,y)\frac{\partial^q}{\partial y^q} f_{i+1}(y)|dy\,.$$

Le premier terme du majorant de $B_{p,q}$ est borné par hypothèse de récurrence et le second tend vers 0 en norme $\|\ \|_\infty$ d'après l'étude du cas $i = 1$.

Et finalement on a $\lim_{\varepsilon^{i+1}\to 0} \|\int_{V_\lambda} A_{\varepsilon^{i+1}}(x,z)\, dz\|_\infty = 0$.

Il reste finalement à vérifier que $\|\int_V |\pi^\alpha_{x,z} G_{i+1,\varepsilon^{i+1}}(x,z)|dz\|_\infty$ reste borné quand ε^{i+1} tend vers 0

$$\|\int_{V_\lambda} |\pi^\alpha G_{i+1,\varepsilon^{i+1}}(x,z)|dz\|_\infty \leqslant \sum_{|p|+|q|=|\alpha|} \|\int_{V\times V_\lambda} |\pi^q L_{\varepsilon_{i+1}}(x,y)|\, |\pi^p G_{i,\varepsilon^i}(y,z)|dy\,dz\|_\infty$$

$$\leqslant \sum_{|p|+|q|=|\alpha|} \frac{\alpha!}{p!q!} \|\int_V |\pi^q L_{\varepsilon_{i+1}}(x,y)|dy\|_\infty \|\int_{V_\lambda} |\pi^p G_{i,\varepsilon^i}(y,z)\, dz\|_\infty\,.$$

Le premier terme de l'expression est borné d'après l'étude du cas $i = 1$ et le second par hypothèse de récurrence.

LEMME 4.4. <u>On a le résultat suivant</u> :

$$\lim_{(\varepsilon_2\ldots\varepsilon_i)\to 0} \int_{V\times\ldots\times V} K_{\varepsilon_i}(x,x_{i-1})\ldots K_{\varepsilon_1}(x_1,y)\Phi_{i,\alpha}(f_1(y)f_2(x_1)\ldots f_i(x_{i-1}))\, dx_{i-1}\ldots dx_1$$

$$= \Phi_{i,\alpha}(f_1(x_1)f_2(x_2)\ldots f_i(x_i))|_{(x_1,\ldots,x_i)=(y,x\ldots,x)} \times K_{\varepsilon_1}(x,y)$$

<u>uniformément par rapport à x et y sur</u> V_λ <u>et</u> ε_1 <u>au voisinage de</u> 0 .

D é m o n s t r a t i o n . Montrons le résultat par récurrence sur i .

Pour $i = 1$ il y a égalité.

Supposons le résultat vrai au rang i .

Considérons le cas $i+1$

$$\int_{V\times\dots\times V} K_{\varepsilon_{i+1}}(x,x_i)\dots K_{\varepsilon_1}(x_1,y)\Phi_{i+1,\alpha}(f_1(y)f_2(x_1)\dots f_{i+1}(x_i))dx_i\dots dx_1$$

$$= \sum_{|p|+|q|=|\alpha|} \frac{\alpha!}{p!q!} \int_V K_{\varepsilon_{i+1}}(x,x_i) \frac{\partial^p}{\partial x_i^p} f_{i+1}(x_i)\Big(\int_{V\times\dots\times V} K_{\varepsilon_i}(x_i,x_{i-1})\dots K_{\varepsilon_1}(x_1,z)$$

$$\Phi_{i,q}(f_1(y)\dots f_i(x_{i-1}))dx_{i-1}\dots dx_1\Big)dx_i$$

$$\xrightarrow[(\varepsilon_i,\dots,\varepsilon_i)\to 0]{} \sum_{|p|+|q|=|\alpha|} \frac{\alpha!}{p!q!} \int_V K_{\varepsilon_{i+1}}(x,x_i) \frac{\partial^p}{\partial x_i^p} f_{i+1}(x_i)$$

$$\Phi_{i,q}(f_1(x_1)\dots f_i(x_i))\Big|_{(x_1,\dots,x_i)=(y,x,\dots,x)} K_{\varepsilon_1}(x,y)\, dx_i$$

grâce à l'hypothèse de récurrence uniformément par rapport à x , ε_1 et z .

De plus $\lim_{\varepsilon_{i+1}\to 0} K_{\varepsilon_{i+1}}(x,x_i) \frac{\partial^p}{\partial x_i^p} f_{i+1}(x_i)dx_i = \frac{\partial^p}{\partial x^p} f_{i+1}(x)$

uniformément par rapport à x sur V_λ et donc l'expression considérée converge vers le produit de $K_{\varepsilon_1}(x,y)$ et de

$$\sum_{|p|+|q|=|\alpha|} \frac{\alpha!}{p!q!} \frac{\partial^p}{\partial x^p} f_{i+1}(x)\Phi_{i,q}(f_1(x_1)\dots f_i(x_i))\Big|_{(x_1,\dots,x_i)=(y,x,\dots,x)}$$

$$= \Phi_{i+1,\alpha}(f_1(x_1)\dots f_i(x_i))\Big|_{(x_1,\dots,x_i)=(y,x,\dots,x)}$$

uniformément par rapport à x et y dans V_λ et ε_1 au voisinage de 0 .

LEMME 4.5. <u>On a le résultat suivant</u>

$$\lim_{\varepsilon_1\to 0} \int_{V_\lambda} |K_{i,\varepsilon_1}(x,y) \; (\Phi_{i,\alpha}(f_1(x_1)\dots f_i(x_i))\Big|_{(x_1,\dots,x_i)=(y,x,\dots,x)}$$

$$- \frac{\partial^{|\alpha|}}{\partial y^\alpha}(f_1(y)\dots f_i(y)))|dy = 0$$ <u>uniformément en</u> x <u>sur</u> V_λ .

D é m o n s t r a t i o n . Puisque K_{i,ε_1} est positif (noyau associé à une convolution) l'expression considérée est égale à

$$\int_{V_\lambda} K_{i,\varepsilon_1}(x,y) \left| \Phi_{i,\alpha}(f_1(x_1)\dots f_i(x_i)) \right|_{(x_1,\dots,x_i)=(y,x,\dots,x)} - \frac{\partial^{|\alpha|}}{\partial y^\alpha}(f_1(y)\dots f_i(y)) \,|\, dy$$

et converge uniformément sur V_1 grâce aux propriétés de la convolution vers

$$\left| \Phi_{i,\alpha}(f_1(x_1)\dots f_i(x_i)) \right|_{(x_1,\dots,x_i)=(x,\dots,x)} - \frac{\partial^{|\alpha|}}{\partial x^\alpha}(f_1(x)\dots f_i(x)) \Big| = 0$$

par définition de $\Phi_{i,\alpha}$.

Fin de la démonstration de la proposition 4.1. Reprenons la décomposition (A) de $\int_V |\pi^\alpha M_\varepsilon(x,y)| dy$.

On déduit des lemmes 4.3, 4.4 et 4.5 que le premier terme tend vers 0 en $\| \ \|_\infty$ avec ε .

Par ailleurs $\tilde{G}_{j,\varepsilon_j}(x,y) = g_j(x) H_{j,\varepsilon_j}(x,y)$ où $H_{j,\varepsilon_j}(x,y)$ est un noyau de la même forme que les G_{i,ε^i} et on montre donc facilement un résultat identique pour le second terme.

Le comportement du troisième terme résulte du lemme 4.2

q.e.d.

On déduit alors immédiatement des résultats des paragraphes 3 et 4 le corollaire suivant :

COROLLAIRE 4.6. Les régularisées T_ε d'une distribution $T \in \mathcal{D}'(V)$ par les opérateurs de de RHAM convergent, quand ε tend vers 0 vers T pour les topologies faibles et fortes de $\mathcal{D}'(V)$ et pour la topologie de $\mathcal{D}'_\Gamma(V)$ si $T \in \mathcal{D}'_\Gamma(V)$.

Remarque. La convergence pour les topologies faibles et fortes de $\mathcal{D}'(V)$ est démontrée par de RHAM dans ([6] , § 15).

5 . Régularisation des courants et une de ses applications.

5.1. Régularisation des courants.

Dans [6] de RHAM définit les formes doubles sur le produit $V \times W$ de deux variétés V et W comme les formes différentielles sur V à coefficients dans l'espace vectoriel des formes différentielles sur W à coefficients dans l'espace vectoriel des formes différentielles sur V .

Pour régulariser les courants sur V , il suffit de remplacer les noyaux des paragraphes précédents par des formes différentielles doubles sur $V \times V$, $\mathcal{C}^\infty$, à support dans un système fondamental $(U_\varepsilon)_{\varepsilon \in \mathbb{R}^+}$ d'entourages de la diagonale Δ de $V \times V$: $(\Psi_\varepsilon)_{\varepsilon \in \mathbb{R}^+}$.

Si $\varphi \in \mathcal{D}^r(V)$ est une forme $\mathcal{C}^\infty$ de degré n , à support compact dans un domaine de carte U tel que $\pi^{-1}(U) \cap U_\varepsilon$ soit contenu dans un domaine de carte de $V \times V$ ($\pi : V \times V \to V$) , on a

$$\varphi(x) = \sum_{|I| = n} \varphi_I(x) dx_I \quad \text{sur } U$$

$$\Psi_\varepsilon(x,y) = \sum_{\substack{I \cup J = \{1,\ldots,p\} \\ I \cap J = \emptyset}} K_{\varepsilon,I,J}(x,y) dx_I dy_J \quad \text{sur } \pi^{-1}(U) \cap U_\varepsilon$$

on pose $\varphi_\varepsilon(x) = \int_V \emptyset_\varepsilon(x,y)\varphi(y)$

$$= \sum_{|I| = r} (-1)^{\sigma(\tau)} \left(\int_U K_{\varepsilon,I,J}(x,y)\varphi_I(y)dy\right)dx_I$$

où $\sigma(\tau)$ est la signature de la permutation $(J,I) \mapsto (1,\ldots,p)$.

On peut faire une étude analogue à celle des paragraphes précédents et l'on obtient dans les cartes de $V \times V$ des conditions sur les coefficients $K_{\varepsilon,I,J}$ de Ψ_ε identiques à celles des noyaux K_ε .

Dans ([6], § 15) de RHAM définit les opérateurs régularisants qui sont associés ([6], § 17) à des formes noyaux vérifiant les hypothèses des paragraphes 1,2 et 3 ainsi que nous l'avons vu au paragraphe 4.

5.2. Application.

On peut utiliser les résultats précédents pour simplifier la démonstration d'un des résultats de [5].

Si T et S sont deux courants sur V de degrés complémentaires sur V et dont l'intersection des supports est compacte, on dit que l'indice de Kronecker de T et S , $\mathcal{K}(T,S)$ est défini au sens de de RHAM si pour toutes familles régularisantes $(R_\varepsilon)_{\varepsilon > 0}$, $(R'_{\varepsilon'})_{\varepsilon' > 0}$, $\langle R_\varepsilon T \wedge R'_{\varepsilon'} S, 1\rangle$ tend vers une limite indépendante des familles choisies lorsque ε et ε' tendent vers 0 , limite que l'on note $\mathcal{K}(T,S)$ ([6], § 20).

Si de plus $WF(T) \cap \overset{\vee}{WF}(S) = \emptyset$ on peut définir le produit extérieur $T \wedge S$ de T et S (cf. [5]).

PROPOSITION 2.1. Soient T et S deux courants sur une variété V telle que $d^\circ T + d^\circ S = \dim V$ et $\operatorname{supp} T \cap \operatorname{supp} S$ soit compact. Si $WF(T) \cap \overset{\vee}{WF}(S) = \emptyset$ alors $\mathcal{K}(T,S)$ existe au sens de de RHAM et on a :

$$\langle T \wedge S\ ,\ 1\rangle = \mathcal{K}(T,S)\ .$$

Démonstration. On considère des régularisants vérifiant les hypothèses de la proposition 3.3.4.

Soient Γ_1 et Γ_2 deux fermés de $S^* V$ tels que $WF(T) \subset \Gamma_1$, $WF(\overset{\vee}{S}) \subset \Gamma_2$ et $\Gamma_1 \cap \Gamma_2 = \emptyset$, la continuité de $(T,S) \to \langle T \wedge S, 1\rangle$ sur $\mathcal{D}'^{\bullet}_{\Gamma_1}(V) \times \mathcal{D}'^{\bullet}_{\Gamma_2}(V)$ et la proposition 3.3.4. impliquent que

$$\langle T \wedge S\ ,\ 1\rangle = \lim_{\substack{\varepsilon \to 0 \\ \varepsilon' \to 0}} \langle T_\varepsilon \wedge S_{\varepsilon'},\ 1\rangle$$ où T_ε et $S_{\varepsilon'}$ sont les régularisées de T et S

$$= \mathcal{K}(T,S) \text{ par définition.}$$

6. Régularisation dans les espaces de Sobolev.

Définition 6.1. On note $\mathcal{H}^s_{loc}(V)$, $s \in \mathbb{R}$, l'espace des distributions T sur V telles que pour tout opérateur pseudo-différentiel π d'ordre inférieur ou égal à s on ait $\pi T \in \mathcal{L}^2_{loc}(V)$.

Remarque. Cette définition est équivalente à la définition traditionnelle des espaces de Sobolev par transformation de Fourier (cela se déduit facilement des propriétés des opérateurs pseudo-différentiels données dans ([4],th.3.6)).

PROPOSITION 6.2. Soit $T \in \mathcal{H}^s_{loc}(V)$, la famille $(T_\varepsilon)_{\varepsilon \in \mathbb{R}^+}$ des régularisées de T converge vers T dans $\mathcal{H}^s_{loc}(V)$ quand ε tend vers 0 si les noyaux $K_\varepsilon(x,y)$ vérifient :

1/ $\int_V K_\varepsilon(y,x)dy \to 1$ dans $\mathcal{C}^\circ(V)$

2/ $\int_V (\pi_x - \pi^*_y)K_\varepsilon(y,x)\psi(y)T(y)dy \to 0$ dans $\mathcal{L}^2_{loc}(V)$

pour toute $\psi \in (V)$ et $T \in \mathcal{H}^s_{loc}(V)$ et tout opérateur pseudo-différentiel π d'ordre $< s$.

D é m o n s t r a t i o n . Il suffit de démontrer que pour tout opérateur pseudo-différentiel π d'ordre inférieur ou égal à s et toute fonction φ, $\mathcal{C}^\infty$ à support compact dans V on a

$$\lim_{\varepsilon\to 0} \| \varphi\,\pi\, T - \varphi\,\pi\, T_\varepsilon \|_2 = 0$$

$$\|\varphi\,\pi\,T - \varphi\,\pi\,T_\varepsilon\|_2 = \|\varphi\,\pi\,T - \varphi(\pi\,T)_\varepsilon\|_2 + \|\varphi(\pi\,T)_\varepsilon - \varphi\,\pi\,T_\varepsilon\|_2 .$$

Etudions successivement chaque terme du second membre .
Puisque $T \in \mathcal{H}^s_{loc}(V)$, πT est dans $\mathcal{L}^2_{loc}(V)$ et d'après l'étude du paragraphe 2 , $(\pi T)_\varepsilon$ converge vers πT dans $\mathcal{L}^2_{loc}(V)$ et par conséquent $\lim_{\varepsilon\to 0} \| \varphi\pi T - \varphi(\pi T)_\varepsilon \|_2 = 0$.

Considérons une fonction ψ, $\mathcal{C}^\infty$ à support compact dans V égale à 1 sur $\pi_2(\pi_1^{-1}(\operatorname{supp}\varphi) \cap U_1)$ alors

$$\varphi(\pi\,T)_\varepsilon = \varphi(\pi\,\psi T)_\varepsilon \quad \text{et} \quad \varphi\,\pi\,T_\varepsilon = \varphi\,\pi(\psi T)_\varepsilon$$

$$(\pi\,\psi\,T)_\varepsilon - \pi(\psi\,T)_\varepsilon(x) = \int_V (\pi^*_y - \pi_x) K_\varepsilon(y,x)\psi(y)T(y)\,dy$$

et cela converge vers 0 dans $\mathcal{L}^2_{loc}(V)$ par hypothèse.

PROPOSITION 6.3. Dans le cas où $s > 0$ la condition 2/ de la proposition précédente peut être remplacée par

2'/ $(\pi_x - \pi^*_y)K_\varepsilon(x,y) \to 0$ dans $\mathcal{L}^2_{loc}(V \times V)$ pour tout opérateur pseudo-différentiel π d'ordre inférieur ou égal à s .

D é m o n s t r a t i o n . Soit $\varphi \in \mathcal{D}(V)$, la condition 2/ s'écrit :

$$A_\varepsilon = \|\varphi(x) \int_V (\pi_x - \pi^*_y) K_\varepsilon(y,x)\psi(y)T(y)\,dy\|_2 \xrightarrow[\varepsilon\to 0]{} 0$$

or $A_\varepsilon \leqslant \|\varphi(x)\, |(\pi_x - \pi^*_y)\, K_\varepsilon(y,x)\, |\|_{2,V\times V}\ \|\psi T\|_{2,V}$

car si $s > 0$, $\mathcal{H}^s_{loc}(V) \subset \mathcal{L}^2_{loc}(V)$.

PROPOSITION 6.4. Dans le cas où $s \in \mathbb{N}$, les conditions peuvent être remplacées par :

2"/ Pour tout multi-indice $\alpha, |\alpha| \leqslant s$ et tout domaine de carte V_i

$$\left(\left(\frac{\partial}{\partial x}\right)^\alpha + (-1)^{|\alpha|+1}\left(\frac{\partial}{\partial y}\right)^\alpha\right) K_\varepsilon(x,y) \xrightarrow[\varepsilon\to 0]{} 0 \ \text{dans} \ L^2_{loc}(V_i \times V_i)$$

ou $\left(\frac{\partial}{\partial x} + \frac{\partial}{\partial y}\right)^\alpha K_\varepsilon(x,y) \longrightarrow 0$ dans $L^2_{loc}(V_i \times V_i)$ quand $\varepsilon \to 0$.

D é m o n s t r a t i o n . Si s est entier, il suffit alors de considérer les opérateurs différentiels d'ordre $|\alpha| \leq s$.

Par une démonstration analogue à celle de la proposition 3.1.4 on montre que 2/ est équivalent à

$$\int_V \left(\left(\frac{\partial}{\partial x} \right)^\alpha + (-1)^{|\alpha|+1} \left(\frac{\partial}{\partial y} \right)^\alpha \right) K_\varepsilon(x,y)\psi(y)T(y)dy \to 0 \text{ dans } \mathcal{L}^2_{loc}(V)$$

pour tout $\psi \in \mathcal{D}(V_i)$ et $T \in \mathcal{H}^s_{loc}(V)$ et grâce à la proposition 3.4.2. on voit que l'on peut remplacer

$$\left(\frac{\partial}{\partial x} \right)^\alpha + (-1)^{|\alpha|+1} \left(\frac{\partial}{\partial y} \right)^\alpha \quad \text{par} \quad \left(\frac{\partial}{\partial x} + \frac{\partial}{\partial y} \right)^\alpha .$$

On obtient enfin 2"/ grâce à la proposition 6.3.

Puisque $\mathcal{E}^\infty(V) = \bigcap_{s \in \mathbb{N}} \mathcal{H}^s_{loc}(V)$, cette intersection étant topologique on en déduit la condition suffisante suivante pour que la suite des régularisées $(T_\varepsilon)_{\varepsilon \in \mathbb{R}^+}$ d'une distribution T converge faiblement vers T .

COROLLAIRE 6.5. <u>Soit</u> $T \in \mathcal{D}'(V)$, <u>la famille</u> $(T_\varepsilon)_{\varepsilon \in \mathbb{R}^+}$ <u>des régularisées de</u> T <u>converge faiblement vers</u> T <u>si pour tout multi-indice</u> α <u>et tout ouvert de carte</u> V_i

$$\left(\frac{\partial}{\partial x} + \frac{\partial}{\partial y} \right)^\alpha K_\varepsilon(x,y) \xrightarrow[\varepsilon \to 0]{} 0 \quad \underline{\text{dans}} \quad L^2_{loc}(V_i \times V_i) .$$

B I B L I O G R A P H I E

[1] DUISTERMAAT , J.J. : Fourier integral operators, Courant Institute, Lecture Notes, New York, 1973.

[2] HÖRMANDER , L. : Linear partial differential operators, Springer, Berlin, 1963.

[3] —— Fourier Integral operators, Acta Math., 127, 1971, 79-183.

[4] —— Pseudo-differential operators, Com.on pure and applied Math., 18, 1965, 501-517.

[5] LAURENT-THIEBAUT , Ch.: Produits de courants et formules des résidus, Bull. Sc. Math., 2e série, 105, 1981, 113-158.

[6] RHAM , G. de : Variétés différentiables, Hermann, Paris, 1960.

[7] SCHWARTZ , L. : Théorie des distributions, Hermann, Paris, 1966.

Mathématiques, L.A. 213 du C.N.R.S.
Université Pierre et Marie Curie
4, Place Jussieu
F-75230 Paris Cédex 05, France

HURWITZ PAIRS EQUIPPED WITH COMPLEX STRUCTURES

Julian Ławrynowicz and Jakub Rembieliński (Łódź)

Contents | page

Summary. In 1923 there appeared a paper of A. Hurwitz, where he solved the problem of finding all the pairs of positive integers (n,p) and all systems of $c^k_{j\alpha} \in \mathbb{R}$, $j, k = 1,\dots,n$; $\alpha = 1,\dots,p$, $p \leq n$, such that the set of bilinear forms $\eta_j = x_\alpha c^k_{j\alpha} y_k$ satisfied the condition $\Sigma_j \eta_j^2 = \Sigma_\alpha x_\alpha^2 \Sigma_k y_k^2$. The present authors reformulate the Hurwitz problem and its solution in the language of Clifford algebras, introducing the notion of the Hurwitz pair of two real unitary vector spaces V and S. Then they study the dependence of these pairs on induced symplectic decomposition and arrive in a natural way at a complex geometry induced by a given symplectic decomposition in connection with distinguishing a fixed direction in S, what introduces a kind of anisotropy.

Introduction

We are dealing with the problem of A. Hurwitz [3] formulated in the Summary in the sense of constructing a suitable complex vector space and studying its properties. The main results are formulated as Theorems 1, 2, and 3 below. They show a deep connection of the problem with the theory of Clifford algebras and explain how the variety of Hurwitz pairs could be put together by a suitable "calibration" of the natural complex structure or, in the terminology precised in the paper, by introducing a suitable supercomplex structure and a suitable

Hurwitz-type vector space of endomorphisms of the higher-dimensional member of the Hurwitz pair, connected with distinguishing a fixed direction in the lower-dimensional member of the Hurwitz pair, what introduces a kind of anisotropy.

The present results have an almost immediate application to studying deformations of physical spaces describing systems of elementary particles, what is a key to the selection rules and classification schemes for the particles; cf. earlier papers of the first author [6], [7], [5]. In particular the method arising from the present paper generalizes the approach introduced by the second author for studying broken symmetries and the algebraic confinement of coloured states of elementary particles [8]. These questions will be treated in detail in subsequent papers.

Our interest in the topic goes back to the unforgetable Mathematician and Friend, Professor Aldo Andreotti, whose illuminating influence on the first author concerning the expected role of the Hurwitz result in physics attains now a form of concrete theorems.

1. Hurwitz pairs and Clifford algebras

Following Hurwitz [3] we are dealing with pairs of positive integers (n,p) and all systems of $c^k_{j\alpha} \in \mathbb{R}$, $j, k = 1,\dots,n$; $\alpha = 1,\dots,p$, $p \leq n$, such that the set of bilinear forms $\eta_j = x_\alpha c^k_{j\alpha} y_k$ satisfies the condition

$$\Sigma_j \eta_j^2 = \Sigma_\alpha x_\alpha^2 \, \Sigma_k y_k^2 \, .$$

In a special case $p = n$ the solution of this problem (Hurwitz [2]) leads to the well known algebras of real numbers, complex numbers, quaternions, and octonions. We are going to reformulate the original general Hurwitz solution in the language of Clifford algebras.

Consider two real unitary vector spaces V and S of dimension n and p, respectively. In both cases we denote the usual norms by $\| \;\|$. Let (e_j) be an orthonormal basis in V and (ε_α) in S. Define a mapping $V \times S \longrightarrow V$ (<u>multiplication of</u> V <u>by</u> S) with the properties:

(i) $f(a+b) = fa + fb$ and $(f+g)a = fa + ga$ for $f, g \in V$ and $a, b \in S$;

(ii) $\|f\| \, \|a\| = \|fa\|$ (<u>the Hurwitz condition</u>);

(iii) there exists the unit element ε_0 in S with respect to the multiplication $\cdot$: $f \cdot \varepsilon_0 \equiv f \varepsilon_0 = f$.

If $\cdot$ does not leave invariant proper subspaces of V, the corresponding pair (V,S) is said to be <u>irreducible</u>. In such a case we call (V,S) the <u>Hurwitz pair</u>.

Clearly, the product fa is uniquely determined by the multiplication scheme

(1) $\quad e_j \varepsilon_\alpha = c^k_{j\alpha} e_k, \; \alpha = 1,\dots,p; \; j,k = 1,\dots,n,$

and the Hurwitz condition (ii) is equivalent to the matrix condition [3]:

(2) $\quad C^T_\alpha C_\beta + C^T_\beta C_\alpha = 2I_n \delta_{\alpha\beta}, \; \alpha, \beta = 1,\dots,p,$

where $C_\alpha = [c^k_{j\alpha}]$, I_n is the identity $n \times n$-matrix, and $\delta_{\alpha\beta}$ denotes the Kronecker symbol. Note that

(3) $\quad C^T_\alpha C_\alpha = I_n, \; \alpha = 1,\dots,p,$

and thus C_α are orthogonal. Setting

(4) $\quad C_\alpha = i\, C_p \gamma_\alpha, \; \alpha = 1,\dots, p-1,$

where i denotes the imaginary unit, we observe that the system of equations (2) is equivalent to the system of (3) for $\alpha = p$, (4), and

(5) $\quad \gamma_\alpha \gamma_\beta + \gamma_\beta \gamma_\alpha = 2I_n \delta_{\alpha\beta}, \; \alpha, \beta = 1,\dots, p-1,$

(6) $\quad \gamma^T_\alpha = -\gamma_\alpha, \; \mathrm{re}\, \gamma_\alpha = 0, \; \alpha = 1,\dots, p-1.$

In particular we conclude that the matrices γ_α are hermitian: $\gamma^+_\alpha = \gamma_\alpha$.

Therefore γ_α are the generating elements of a real Clifford algebra. The bases (e_j) and (ε_α) are chosen with the exactness to a pair of orthogonal transformations R and $\mathcal{O}$, respectively. Hence the general formula for an admissible matrix C'_α reads as follows:

(7) $\quad C'_\alpha = \Sigma_\beta\, \mathcal{O}_{\alpha\beta} R C_\beta R^T, \; \alpha = 1,\dots,p,$

where $RR^T = I_n$, $R \subset O(n)$ and $\mathcal{O}\mathcal{O}^T = I_p$, $\mathcal{O} \subset O(p)$. By (iii) and (7), without any loss of generality we may set $C_p = I_n$. Consequently we have

(8) $\quad C_p = I_n, \; C_\alpha = i\gamma_\alpha, \; \alpha = 1,\dots, p-1,$

so now $\mathcal{O} \subset O(p-1)$. Thus we have proved the following

<u>LEMMA 1</u>. <u>The problem of classifying the Hurwitz pairs</u> (V,S) <u>is equivalent to the classification problem for real Clifford algebras with</u> $i\gamma_\alpha$ <u>real and antisymmetric</u>. <u>The relationship is given by formulae</u> (5), (6), <u>and</u> (8).

2. Dependence of Hurwitz pairs on induced symplectic decompositions

The property mentioned in the title of the section will arise in a natural way after proving the following analogue of Hurwitz's conclusion of Section IV in [3]:

LEMMA 2. *The Hurwitz pairs* (V,S) *are of bidimension* (n,p), $n = \dim V$, $p = \dim S$, *with*

$$(9)\qquad n = \begin{cases} 2^{[\frac{1}{2}p-\frac{1}{2}]} & \text{for } p \equiv 7, 0, 1 \pmod 8, \\ 2^{[\frac{1}{2}p+\frac{1}{2}]} & \text{for } p \equiv 2, 3, 4, 5, 6 \pmod 8, \end{cases}$$

where [ ] *stands for the function "entier".*

Proof. Let $\mathbb{F} = \mathbb{R}, \mathbb{C}$, and $\mathbb{H}$ denote the real, complex, and quaternion number fields, and let $\mathbb{M}(r,\mathbb{F})$ be the algebra of r-dimensional matrices over $\mathbb{F}$. It is well known (cf. e.g. [1], p. 11, table 1, the column for C_k', where $p-1$ has to be substituted for k) that the classification scheme for irreducible real Clifford algebras satisfying (5) reads as follows (in the case $p \equiv 5, 6, 7 \pmod 8$ the factor 2 in the formula for r comes from the fact that $\mathbb{H}$ is considered as a subalgebra of $\mathbb{M}(2,\mathbb{C})$:

p	Clifford algebra isomorphic to:	r
1	$\mathbb{M}(2^{\frac{1}{2}p-\frac{1}{2}}, \mathbb{R})$	$2^{[\frac{1}{2}p-\frac{1}{2}]}$
2	$\mathbb{M}(2^{[\frac{1}{2}p-\frac{1}{2}]}, \mathbb{R}) \oplus \mathbb{M}(2^{[\frac{1}{2}p-\frac{1}{2}]}, \mathbb{R})$	$2\cdot 2^{[\frac{1}{2}p-\frac{1}{2}]}$
3	$\mathbb{M}(2^{\frac{1}{2}p-\frac{1}{2}}, \mathbb{R})$	$2^{[\frac{1}{2}p-\frac{1}{2}]}$
4	$\mathbb{M}(2^{[\frac{1}{2}p-\frac{1}{2}]}, \mathbb{C})$	$2^{[\frac{1}{2}p-\frac{1}{2}]}$
5	$\mathbb{M}(\frac{1}{2}\cdot 2^{\frac{1}{2}p-\frac{1}{2}}, \mathbb{H})$	$2\cdot\frac{1}{2}\cdot 2^{[\frac{1}{2}p-\frac{1}{2}]}$
6	$\mathbb{M}(\frac{1}{2}\cdot 2^{[\frac{1}{2}p-\frac{1}{2}]}, \mathbb{H}) \oplus \mathbb{M}(\frac{1}{2}\cdot 2^{[\frac{1}{2}p-\frac{1}{2}]}, \mathbb{H})$	$2\cdot 2^{[\frac{1}{2}p-\frac{1}{2}]}$
7	$\mathbb{M}(\frac{1}{2}\cdot 2^{\frac{1}{2}p-\frac{1}{2}}, \mathbb{H})$	$2\cdot\frac{1}{2}\cdot 2^{[\frac{1}{2}p-\frac{1}{2}]}$
8	$\mathbb{M}(2^{[\frac{1}{2}p-\frac{1}{2}]}, \mathbb{C})$	$2^{[\frac{1}{2}p-\frac{1}{2}]}$

Next we observe that in the case $p \equiv 6, 7, 0, 1, 2 \pmod 8$ the conditions (6) are satisfied, so $n = r$, whereas for $p \equiv 3, 4, 5 \pmod 8$ in order to arrive at (6) we have to double the dimension r: $n = 2r$ by taking the direct sum of two irreducible copies of the corresponding Clifford algebras. In such a way, by Lemma 1, the conclusion of Lemma 2 is proved for all n, as desired.

Now we are going to make a symplectic decomposition of V, chosen as described below, where we will also precise the notion of a symplectic decomposition of V. To this end we need (cf. e.g. [4], vol. II, p. 115):

LEMMA 3. *Let* n *and* p *be positive integers satisfying* (9) *and* $n > 1$. *Then for any system of real numbers*

$$(10) \quad n^\alpha,\ \alpha = 1, \ldots, p-1,\ \Sigma_\alpha (n^\alpha)^2 = 1,$$

there exists a system of imaginary $n \times n$ *matrices*

$$(11) \quad \gamma_\alpha,\ \alpha = 1, \ldots, p-1, \text{ satisfying (5) and (6)},$$

with the property

$$(12) \quad in^\alpha \gamma_\alpha = J_0 = \begin{bmatrix} 0 & I_{\frac{1}{2}n} \\ -I_{\frac{1}{2}n} & 0 \end{bmatrix}, \quad \text{where } I_{\frac{1}{2}n} \text{ is the identity } \tfrac{1}{2}n \times \tfrac{1}{2}n\text{-matrix.}$$

Proof. The orbit $\mathcal{O}(n)$ in the family of all antisymmetric real $n \times n$-matrices A, i.e. in the adjoint representation, is determined by $[\frac{1}{2}n]$ invariants which are the even moments $\operatorname{Tr} A^{2k}$, $k = 1, \ldots, [\frac{1}{2}n]$; the odd moments being equal to zero. By (10), the even moments of $in^\alpha \gamma_\alpha$ are equal to

$$\operatorname{Tr}[(-1)^k (n^\alpha \gamma_\alpha)^{2k}] = (-1)^k \operatorname{Tr} I_{\frac{1}{2}n} = (-1)^k n,$$

and the analogous moments J_0 are equal to $\operatorname{Tr}(-I_{\frac{1}{2}n})^k = (-1)^k n$, so both matrices belong to the same orbit. Consequently, by the transitivity of the action of $\mathcal{O}(n)$ in this orbit, for each system (n^α) in question there is an orthogonal transformation of one matrix to the other, and thus the proof is completed.

Since the matrix J_0 is invariant with respect to the transformations of the group $Sp(\frac{1}{2}n, \mathbb{R})$, from Lemma 3 we deduce

PROPOSITION 1. *The little (stability) group of* J_0 *in* $\mathcal{O}(n)$ *is* $\mathcal{O}(n) \cap Sp(\frac{1}{2}n, \mathbb{R}) \sim U(\frac{1}{2}n)$, *so the space of all matrices* $J = in^\alpha \gamma_\alpha$, *not necessarily satisfying* (12), *is isomorphic to the coset space* $\mathcal{O}(n)/U(\frac{1}{2}n)$.

We are in a position to prove

THEOREM 1. *Consider a Hurwitz pair* (V,S) *of bidimension* (n,p), $n > 1$, *and some orthonormal bases* (e_j) *in* V *and* (ε_α) *in* S. *Let* (n^α) *be an arbitrary system of real numbers* (10) *and* (γ_α) *a system of imaginary* $n \times n$-*matrices* (11) *with the property* (12). *Suppose that* f *is an arbitrary vector of* V *and let* $e_j f^j_{\mathbb{R}}$ *be its decomposition (in* V).

Then this decomposition can be rearranged into the form

$$(13)\qquad f=\Sigma_{j\le\frac{1}{2}n} e_j f^j, \text{ where } f^j=\varepsilon_o f^j_{\mathbb{R}}+\tilde{n} f^{j+\frac{1}{2}n}_{\mathbb{R}},\ \tilde{n}=\varepsilon_\alpha n^\alpha\in S.$$

P r o o f. The problem whose solution is formulated as Theorem 1 is well posed by Lemma 2. Taking into account (3) and (8), we have, as a consequence of Lemma 1, the following form of the multiplication scheme (1):

$$(14)\qquad e_j\varepsilon_o=e_j,\ e_j\varepsilon_\alpha=e_k i\gamma^k_j,\ \alpha=1,\dots,p-1;\ j,k=1,\dots,n,$$

ε_p being identified with ε_o. Hence, with the choice of (n^α) and (γ_α) as in Lemma 3, by this lemma we have

$$(15)\qquad e_j\tilde{n}=e_{j+\frac{1}{2}n},\ e_{j+\frac{1}{2}n}\tilde{n}=-e_j,\ j=1,\dots,\tfrac{1}{2}n.$$

Thus, for every $f=e_j f^j_{\mathbb{R}}$ we get

$$f=\Sigma_{j\le\frac{1}{2}n}(e_j f^j_{\mathbb{R}}+e_{j+\frac{1}{2}n} f^{j+\frac{1}{2}n}_{\mathbb{R}})=\Sigma_{j\le\frac{1}{2}n} e_j(\varepsilon_o f^j_{\mathbb{R}}+\tilde{n} f^{j+\frac{1}{2}n}_{\mathbb{R}}),$$

and hence (13) follows. The uniqueness of this decomposition is, clearly, a consequence of the uniqueness of $f=e_j f^j_{\mathbb{R}}$.

D e f i n i t i o n s. The decomposition (13) of $f\in V$ will be called the *canonical symplectic decomposition* of f in V. This decomposition generates in a natural way the *canonical symplectic decomposition* of V:

$$(16)\qquad V=\oplus_{j\le\frac{1}{2}n}\mathbb{C}_j(\varepsilon_o,\tilde{n}).$$

Indeed, each f^j belongs to the real two-dimensional subspace $\mathbb{R}^2(\varepsilon_o,\tilde{n})$ of S, spanned by ε_o and $\tilde{n}$, while $\mathbb{C}_j(\varepsilon_o,\tilde{n})$ are complex one-dimensional subspaces of V generated by e_j, $j=1,\dots,\frac{1}{2}n$, in dependence on ε_o and $\tilde{n}$. We remark that, clearly, for each j the expression $\tilde{n}e_j$ is n o t d e f i n e d although the expression $e_j\tilde{n}$ is well defined.

The adjective "symplectic" is motivated by Proposition 1 and by the following consequence of Theorem 1 (cf. e.g. [4], pp. 119, 147, 149):

COROLLARY 1. *The linear endomorphism* $J_o: V\to V$, *determined by*

$$(17)\qquad Je_j=e_k J^k_j,\ j=1,\dots,n, \text{ where } [J^k_j]=J,$$

with $J=J_o$ *given in* (12) *has the following property: the bilinear form* $\Phi_o(f,g)=(J_of,g)_{\mathbb{R}}$ *for* $f,f\in V$ *is skew-symmetric, where* $(\ ,\ )_{\mathbb{R}}$ *denotes the usual (real) scalar product in* V.

P r o o f. The conclusion follows immediately from the relation

$(J_o f, J_o g)_{\mathbb{R}} = (f,g)_{\mathbb{R}}$.

Definitions. The bilinear form Φ_o given in Corollary 1 will be called the canonical symplectic structure on V. We may also introduce an arbitrary symplectic decomposition of f in V, symplectic decomposition of V, and symplectic structure Φ on V, considering matrices J obtained from the canonical form (12) by transformations R belonging to the subspace $O(n)/U(\frac{1}{2}n)$ of $O(n)$ (cf. (7) and Corollary 1):

$$\Phi(f,g) = (f, Jg)_{\mathbb{R}} \quad \text{for } f, g \in V,$$

$$J = R\, J_o\, R^T = in^{\alpha}\gamma_{\alpha}. \tag{18}$$

From Theorem 1 and Proposition 1 we easily deduce

COROLLARY 2. The choice of a symplectic decomposition of all vectors of V is uniquely determined by a given point (18) of the homogeneous space $O(n)/U(\frac{1}{2}n)$. With this choice the formula (13) holds true with e_j and $f_{\mathbb{R}}^k$ replaced by $(R^T e)_j$ and $(R^T f_{\mathbb{R}})^k$, respectively:

$$f = \Sigma_{j \le \frac{1}{2}n} (R^T e)_j\, f^j, \ \text{where} \ f^j = \varepsilon_o (R^T f_{\mathbb{R}})^j + \tilde{n}(R^T f_{\mathbb{R}})^{j+\frac{1}{2}n}. \tag{19}$$

3. The complex geometry induced by a given symplectic decomposition

Any symplectic decomposition of V and the corresponding symplectic structure Φ of V are determined, according to Proposition 1, by fixing a point in the orbit $O(n)/U(\frac{1}{2}n)$, i.e. by fixing any matrix (18), where (n^{α}) is an arbitrary system of real numbers (10) and (γ_{α}) is an arbitrary system of imaginary $n \times n$-matrices (11). In particular we may take $J = J_o$; cf. formula (12).

It is clear that the same matrix J can be expressed by different pairs of systems (10) and (11) related by rotations belonging to the group $O(p-1)$. Thus the triples (V,S,J) are not satisfactory for our purposes: they do not reflect the variety of possible directions of vectors

$$\tilde{n} = \varepsilon_{\alpha} n^{\alpha} \in S, \ \Sigma_{\alpha}(n^{\alpha})^2 = 1; \tag{20}$$

cf. formula (13). Consequently we are led to replacing the triples (V, S,J) by the quadruples $(V,S,J,\tilde{n})$. Following this way we have immediately (cf. e.g. [4], vol. II, p. 6):

PROPOSITION 2. The set of all vectors (20) forms a $(p-2)$-dimensional sphere in S, which is isomorphic to the Stiefel manifold $O(p-1)/O(p-2)$ of 1-frames in $\mathbb{R}^{p-1}$.

The quadruple $(V,S,J,\tilde{n})$ can be turned into a $\frac{1}{2}n$-dimensional complex vector space $(V,J,\cdot)$ by checking that J is a complex structure of V and $\cdot$ is a scalar multiplication by complex numbers. In addition the space constructed will have an extra structure $(\tilde{n},\mathcal{E})$. We may formulate our results as follows (the structure $\mathcal{E}$ will be precised in Theorem 3 below):

THEOREM 2. *Consider a Hurwitz pair of bidimension* (n,p), $n>1$, *with a fixed direction* $(n^{\alpha})=(n^1,\dots,n^{p-1})$ *in* S, *i.e. a vector* (20), *where* $(\varepsilon_o,\varepsilon_\alpha)$ *is an orthonormal basis in* S. *We further equip the triple* $(V,S,\tilde{n})$ *with a matrix* (18) *with* R *belonging to* $O(n)/U(\frac{1}{2}n)$, *where* (γ_α) *is a system of imaginary* $n\times n$-*matrices* (11). *Next we derive from it the linear endomorphism* $f\mapsto f\tilde{n}$ *of* V, *also denoted by* J:

$$(21)\qquad Jf=f\tilde{n}\quad\text{for}\quad f\in V.$$

Then we have the following assertions.

A1. *The basis* $(\varepsilon_o,\varepsilon_\alpha)$ *satisfies the formal rules*

$$(22)\qquad \varepsilon_o^2=\varepsilon_o,\ \varepsilon_o\tilde{n}=\tilde{n}\varepsilon_o=\tilde{n},\ \tilde{n}^2=-\varepsilon_o.$$

A2. *An orthonormal basis* (e_j) *of* V *has to satisfy the property* (17); *such a basis always exists.*

A3. *The symplectic decomposition* (16) *of* V *holds and, by* (17), *it depends on* J. *It is generated by the symplectic decomposition* (19) *of all vectors* $f=e_j f_{\mathbb{R}}^j\in V$.

A4. *The Hurwitz pair* (V,S) *or, more exactly, the quadruple* $(V,S,J,\tilde{n})$ *can be turned into a* $\frac{1}{2}n$-*dimensional complex vector space* $(V,J,\cdot)$, *whose complex structure is defined as the linear endomorphism* J, *and the scalar multiplication* $\cdot$ *by complex numbers is defined by*

$$(23)\qquad (q+is)\cdot f=fq+Jfs\quad\text{for}\quad f\in V\ \text{and}\ q,s\in\mathbb{R}.$$

A5. *The relation* $\cdot$ *has the property*

$$(24)\qquad (q+is)\cdot f=f(\varepsilon_o q+\tilde{n}s)\quad\text{for}\quad f\in V\ \text{and}\ q,s\in\mathbb{R}.$$

In particular, for complex components $[(q+is)\cdot f]^j=f^j(\varepsilon_o q+\tilde{n}s)$ *this gives*

$$(25)\qquad [(q+is)\cdot f]^j=\varepsilon_o[(R^T f_{\mathbb{R}})^j q-(R^T f_{\mathbb{R}})^{j+\frac{1}{2}n}s] + \tilde{n}\,[(R^T f_{\mathbb{R}})^j q-(R^T f_{\mathbb{R}})^{j+\frac{1}{2}n}s],\quad j=1,\dots,\tfrac{1}{2}n.$$

Proof. The formal rules (22) are a consequence of the equalities

(26) $(f\varepsilon_0)\varepsilon_0 = f\varepsilon_0 = f, \; (f\tilde{n})\varepsilon_0 = (f\varepsilon_0)\tilde{n} = f\tilde{n}, \; (f\tilde{n})\tilde{n} = f(-\varepsilon_0)$

which follow from the definition (20), so we have A1. In order to prove A2 let us start with (18), where J is meant as a matrix. We have

$$e_k f^j_{\mathbb{R}} J^k_j = e_k f^j_{\mathbb{R}} in^\alpha \gamma^k_{j\alpha} \quad \text{for} \quad f = e_j f^j_{\mathbb{R}} \in V.$$

Hence, by (14), the expression equals $e_j f^j_{\mathbb{R}} \varepsilon_\alpha n^\alpha$ and, consequently, by (20), $e_j f^j_{\mathbb{R}} \tilde{n}$. Thus, by the arbitrariness of $f^j_{\mathbb{R}}$, $j = 1,\dots,n$, we arrive at (17), so such a choice of (e_j) appears in a natural way when considering an arbitrary Hurwitz pair (V,S). This justifies also the notation J for the linear endomorphism $V \ni f \longmapsto f\tilde{n}$.

The assertion A3 is a straightforward consequence of Theorem 1 and Proposition 1 or, equivalently, of Corollary 2. In order to conclude A4 it suffices to observe that in our context the equalities (15) have to be replaced by

$$(R^T e)_k J^k_j = (R^T e)_{j+\frac{1}{2}n}, \quad (R^T e)_k J^k_{j+\frac{1}{2}n} = -(R^T e)_j, \quad j = 1,\dots,\tfrac{1}{2}n,$$

so $(R^T e)_{j+\frac{1}{2}n} = J(R^T e)_j$, $j = 1,\dots,\frac{1}{2}n$. Finally, (24) follows from (23) by (21) and by $f\varepsilon_0 = f$ for $f \in V$ which is included in (26). If we take into account Theorem 1 and Proposition 1 or, equivalently, Corollary 2, especially formula (19), we arrive at (25), so we have also A5. Therefore the proof is completed.

The formula (24) can be interpreted in a way that $\tilde{n}$ is a counterpart of the imaginary unit, which replaces $i \in \mathbb{C}$ in the field of "numbers" $\varepsilon_0 q + \tilde{n}s$ (more exactly that $(\varepsilon_0, \tilde{n})$ replaces $(1,i)$) or, alternatively, that we equip the Hurwitz pair (V,S) with the "complex" structure $J_{\tilde{n}} = (J,\tilde{n})$. Therefore we are led to introducing the <u>supercomplex structure</u> $(J,\tilde{n})$ of V, determined by the relations (21), (18), and (20), which in fact precises the anisotropy caused by distinguishing the direction (n^α) in S.

It seems still convenient to get rid of the vector space S as an element of our construction since the basis $(\varepsilon_0, \varepsilon_\alpha)$ of S is somehow already involved in the chosen direction $\tilde{n}$. Thus we are led to introducing the <u>Hurwitz-type vector space</u> $\mathcal{E}$, defined as the p-dimensional subspace of the space $\operatorname{End} V$ of endomorphisms of V, which consists of all endomorphisms E, not leaving invariant proper subspaces of V, determined by the following analogue of the Hurwitz condition for V and S:

(27) $\|Ef\| = \|E\|\,\|f\|$ for $f \in V$, $E \in \mathcal{E}$, where $\|E\| = (\mathrm{Tr}\, E^T E)^{\frac{1}{2}}$,

$E^T E$ being considered in an arbitrary matrix representation. All the corresponding basic matrices E_0, E_α can be generated by the formulae

(28) $E_0 e_j = e_j$, $E_\alpha e_j = i\gamma^k_{j\alpha} e_k$, $\alpha = 1, \ldots, p-1$; $j, k = 1, \ldots, n$.

Definition. A supercomplex vector space $(V, J, \tilde{n}, \cdot, \mathcal{E})$ is a complex vector space $(V, J, \cdot)$ equipped with a supercomplex structure $(J, \tilde{n})$ and a Hurwitz-type vector space $\mathcal{E}$ of endomorphisms $E: V \to V$, satisfying the relation (23). (By definition, it has to satisfy also the relations (21), (18), (20), and (27) with $\varepsilon_0, \varepsilon_\alpha$ replaced by E_0, E_α, respectively, where $\alpha = 1, \ldots, p-1$. The matrices E_0, E_α are determined by (28), (5), and (6), where $\gamma_\alpha = [\gamma^k_{j\alpha}]$.)

We remark that if we equip $(V, J, \cdot)$ with $(J, \tilde{n})$ and $\mathcal{E}$ at the same time, we eliminate $\varepsilon_0, \varepsilon_\alpha$ also from the definition of $(J, \tilde{n})$: as above, we can replace (20) and (22) by

(29) $\tilde{n} = E_\alpha n^\alpha$, $\Sigma_\alpha (n^\alpha)^2 = 1$,

and

(30) $E_0^2 = E_0$, $E_0 \tilde{n} = \tilde{n} E_0 = \tilde{n}$, $\tilde{n}^2 = -E_0$,

respectively. Moreover, by Theorems 1 and 2, we may summarize our considerations as follows:

THEOREM 3. The construction described in Theorem 2 determines the supercomplex vector space $(V, J, \tilde{n}, \cdot, \mathcal{E})$ with $\mathcal{E}$ generated by the formulae (28).

Now, with help of the complex structure J we can introduce the complex scalar product $(\ ,\ ): V \times V \to \mathbb{C}$ as follows:

(31) $(f,g) = (f,g)_{\mathbb{R}} + i(Jf, g)_{\mathbb{R}}$ for $f, g \in V$ (cf. Corollary 1).

By (21) and (23), the definitions of $(\ ,\ )$ and $(\ ,\ )_{\mathbb{R}}$ yield easily the standard properties of $(\ ,\ )$:

PROPOSITION 3. For $f, g \in V$ and $g, s \in \mathbb{R}$ we have $(f,g) = (g,f)^*$, $(f, g+h) = (f,g) + (f,h)$, $(f, (q+is)g) = (q+is)(f,g)$, and $(f,f) = \|f\|^2$, where $*$ denotes the complex conjugation.

By the orthonormality of (e_j) and the symplectic decomposition (19) of f in Corollary 2 we obtain for $(\ ,\ )$ a suggestive formula:

PROPOSITION 4. We have

(32) $(f,g) = \Sigma_{j\leq\frac{1}{2}n} (f^j_{\mathbb{C}})^* g^j_{\mathbb{C}}$ for $f, g \in V$,

where $f^j_{\mathbb{C}} = f^j_{\mathbb{R}} + if^{j+\frac{1}{2}n}_{\mathbb{R}}$, $g^j_{\mathbb{C}} = g^j_{\mathbb{R}} + ig^{j+\frac{1}{2}n}_{\mathbb{R}}$, $j = 1,\ldots,\frac{1}{2}n$. The product (f,g) is invariant under action of the unitary group $U(\frac{1}{2}n)$.

Finally we introduce the $\mathbb{C}$-linear operators on V in a standard way: a $\mathbb{C}(E_0,\tilde{n})$-linear endomorphism L of V, a $\mathbb{C}(E_0,\tilde{n})$-hermitian operator H of V, and a $\mathbb{C}(E_0,\tilde{n})$-unitary operator U of V have to satisfy the equalities

$$L(zf + ug) = z(Lf) + u(Lg) \quad \text{for } f \in V;\ z, u \in \mathbb{C}(E_0,\tilde{n}),$$

$$(Hf, g) = (f, Hg) \text{ for } f, g \in V, \quad \text{and} \quad (Uf, g) = (f, U^{-1}g) \text{ for } f, g \in V,$$

respectively, where $\mathbb{C}(E_0,\tilde{n}) \sim \mathbb{R}^2(E_0,\tilde{n})$; $\mathbb{C}(E_0,\tilde{n})$ being a complex one-dimensional subspace of $\mathcal{E}$, generated by E_0 and $\tilde{n}$, determined by (29) and (30), in dependence on ε_0 and $\tilde{n}$, determined by (20) and (22). The above definitions depend on a fixed symplectic decomposition of V, which, by (16) and the assertion A3 in Theorem 2, has the form $V = \bigoplus_{j\leq\frac{1}{2}n} \mathbb{C}_j(E_0,\tilde{n})$, all the addends being dependent on J.

References

[1] ATIYAH, M.F., R. BOTT, and A. SHAPIRO: Clifford modules, Topology 3, Suppl. 1 (1964), 3-38.

[2] HURWITZ, A.: Über die Komposition der quadratischen Formen von beliebig vielen Variablen, Nachrichten von der königlichen Gesellschaft der Wissenschaften zu Göttingen, Math.-phys. Kl. 1898, 309-316; reprinted in: A. HURWITZ, Mathematische Werke II, Birkhäuser Verl., Basel 1933, pp. 565-571.

[3] ——: Über die Komposition der quadratischen Formen, Math. Ann. 88 (1923), 1-25; reprinted in: A. HURWITZ, Mathematische Werke II, Birkhäuser Verl., Basel 1933, pp. 641-666.

[4] KOBAYASHI, S. and K. NOMIZU: Foundations of differential geometry I-II (Interscience Tracts in Pure and Applied Mathematics 15), Interscience Publishers, New York - London - Sydney 1963-1969.

[5] ŁAWRYNOWICZ, J.: Application à une théorie non lineaire de particules élémentaires. Appendice pour le travail de B. GAVEAU et J. ŁAWRYNOWICZ: Intégrale de Dirichlet sur une variété complexe I, in: Séminaire Pierre Lelong - Henri Skoda (Analyse), Années 1980/81 (Lecture Notes in Math. 919), Springer-Verlag, Berlin-Heidelberg - New York 1982, pp. 152-166.

[6] —— and L. WOJTCZAK: A concept of explaining the properties of elementary particles in terms of manifolds, Z. Naturforsch. 29a (1974), 1407-1417.

[7] —— and ——: On an almost complex manifold approach to elementary particles, ibid. 32a (1977), 1215-1221.

[8] REMBIELIŃSKI, J.: Algebraic confinement of coloured states, J. Phys. A: Math. Gen. 14 (1981), 2609-2624.

Institute of Mathematics of the
Polish Academy of Sciences, Łódź Branch
Narutowicza 56, PL-90-136 Łódź, Poland
(J. Ławrynowicz)

Institute of Physics
University of Łódź
Nowotki 149/153
PL-90-236 Łódź, Poland
(J. Rembieliński)

AN OPEN PROBLEM ON BOUNDARY BEHAVIOUR OF HOLOMORPHIC MAPPINGS

Ewa Ligocka (Warszawa)

In the study of the boundary behaviour of proper holomorphic mappings between weakly pseudoconvex domains with C^∞-boundaries the following property plays the fundamental role:

Property R. A domain D has the property R iff the Bergman projection $P: L^2(D) \longrightarrow L^2H(D)$ is a continuous operator from $C^\infty(\overline{D})$ into itself with the usual Fréchet topology.

A sufficient (but not necessary) condition for the property R is the following: A smooth pseudoconvex domain has the property R if for each $s > 0$ there exists an operator solving the $\bar\partial$-problem, which is a compact operator T_s from the space $W^s_{\langle 0,1\rangle}$ of $\bar\partial$-closed $\langle 0,1\rangle$-forms with the s-th Sobolev norm into the Sobolev space $W^s(D)$.

The PROBLEM posed is to characterize the domains D for which such compact operators exist. In particular, does there exist a domain D for which the subelliptic estimates for the $\bar\partial$-Neumann problem are not valid and for which there exist compact operators which solve the $\bar\partial$-problem?

Institute of Mathematics of the
Polish Academy of Sciences (Śniadeckich 8)
PL-00-950 Warszawa, P.O. Box 137, Poland

THE REGULARITY OF THE WEIGHTED BERGMAN PROJECTIONS

Ewa Ligocka (Warszawa)

Contents page

Summary. In the paper the following fact is proved: If D is a smooth pseudoconvex bounded domain such that for some $s > 0$ there exists a compact operator $T_s : W^s_{\langle 0;1\rangle}(D) \to W^s(D)$ solving the $\bar\partial$-problem ($\bar\partial T_s w = w$), then for each $w \in C^\infty(\bar D)$, the weighted Bergman projection with weight e^w is a continuous operator from $W^s(D)$ into $W^s(D)$.

We also study some other weighted Bergman projections related to the defining function ρ of the domain D.

1. Introduction and statement of results

Let D be a bounded domain in $\mathbb{C}^n$ and let w be a real function on $\bar D$, $w \in C^\infty(\bar D)$. We define a scalar product $\langle\ ;\ \rangle_w$ on the space $L^2(D)$ by

$$\langle f,g\rangle_w = \int_D f\,\bar g\, e^w\, dV.$$

Let $L^2H(D)$ denote the subspace of $L^2(D)$ consisting of holomorphic functions.

Definition. A weighted Bergman projection with weight w is a projection from $L^2(D)$ onto $L^2H(D)$ which is orthogonal with respect to the scalar product $\langle\ ;\ \rangle_w$. We shall denote it by P_w.

It is clear that if we put $w \equiv 0$ then we get the usual Bergman projection.

The projection P_w can be expressed as follows:

$$P_w(f)(z) = \int_D K_w(z,t)f(t)dV_t,$$

where $K_w(z,t)$ is a reproducing kernel for a Hilbert space $L^2H(D)$ with scalar product $\langle \ , \ \rangle_w$.

Now, let $W^s(D)$ denote the s-th Sobolev space, let $H^s(D)$ denote its subspace consisting of holomorphic functions and $A^s_{\langle 0,1\rangle}$ - the space of $\bar\partial$-closed $\langle 0,1\rangle$ forms on D with coefficients from $W^s(D)$ with Hilbert norm

$$\|w\|^2 = \sum_{i=1}^n \|w_i\|^2, \quad w = \sum_{i=1}^n w_i \, d\bar z_i .$$

The aim of this note is to prove the following

THEOREM 1. Let D be a bounded pseudoconvex domain with C^∞-boundary. Suppose that for some $s > 0$ there exists a compact operator $T_s : A^s_{\langle 0,1\rangle}(D) \to W^s(D)$ such that $\bar\partial T_s w = w$. Then for every real function $w \in C^\infty(\bar D)$, the weighted Bergman projection P_w maps continuously $W^s(D)$ onto $H^s(D)$.

The important role in the study of boundary behaviour of biholomorphic and proper holomorphic mappings is played by the property R: A domain D has this property iff the Bergman projection P maps continuously $C^\infty(\bar D)$ onto $A^\infty(\bar D)$ (see [2], [3], and [4]).

Our theorem yields the following

COROLLARY. If D is a bounded smooth pseudoconvex domain such that for each $s > 0$ there exists a compact operator $T_s : A^s_{\langle 0,1\rangle}(D) \to W^s(D)$ satisfying the condition $\bar\partial T_s w = w$, then D has property R.

It should be mentioned that the existence of compact operators T_s is not a necessary condition for property R.

We shall give an example of smooth bounded circular pseudoconvex domain D, which for each $s \geq 0$ does not admit any compact operator T_s solving $\bar\partial$-problem.

Bell and Boas proved in [8] that every smooth bounded circular domain has property R, so D has it as well.

This example shows also that Kohn's estimates for $\bar\partial$-problem, given in [6], are sharp.

The Regularity of the Weighted Bergmann Projections

The last part of this note will be devoted to study weighted projections with weights, which can be expressed by the defining functions for domain D.

Definition. Let D be a bounded smooth domain in $\mathbb{C}^n$. A real function $\rho \in \mathbb{C}^\infty(\mathbb{C}^n)$ is called a defining function for D iff $D = \{z \in \mathbb{C}^n : \rho(z) > 0\}$ and $\operatorname{grad} \rho \neq 0$ on ∂D.

Note that if ρ_1 and ρ_2 are two defining functions for the same domain D, then $\rho_1 = e^h \cdot \rho_2$, $h \in C^\infty(\mathbb{C}^n)$. For a defining function ρ for D we shall consider the space

$$L^2_{\rho,\kappa}(D) = \{f : \int_D |f|^2 \rho^\kappa < \infty\}$$

and $L^2_{\rho,\kappa}H(D)$ - its subspace consisting of holomorphic functions. These spaces are larger than the spaces $L^2(D)$ and $L^2H(D)$.

We shall denote by $P_{\rho,\kappa}$ the projection from $L^2_{\rho,\kappa}(D)$ onto $L^2_{\rho,\kappa}H(D)$ orthogonal with respect to the scalar product $\langle f,g\rangle_{\rho,\kappa} = \int_D f\bar{g}\rho^\kappa$ and by $K_{\rho,\kappa}(z,t)$ - the reproducing kernel for $L^2_{\rho,\kappa}H(D)$. We have $P_{\rho,\kappa}f = \int_D K_{\rho,\kappa}(z,t)f(t)\rho^\kappa(t)dV_t$.

THEOREM 2. Let D be a smooth bounded domain in $\mathbb{C}^n$ and ρ a defining function for D such that $\tilde{D} = \{(z_0,z) \in \mathbb{C}^{n+1} : |z_0|^2 < \rho(z)\}$ has the property R (that means that the Bergman projection P is continuous from $C^\infty(\bar{\tilde{D}})$ onto $A^\infty(\bar{\tilde{D}})$). Then for each $k \geq 1$ the projection $P_{\rho,\kappa}$ maps continuously $C^\infty(\bar{D})$ onto $A^\infty(\bar{D})$.

COROLLARY. If

1) D is a bounded strictly pseudoconvex domain and ρ is a defining function for D, ρ strictly plurisuperharmonic in the neighbourhood of $\bar{D}$

or

2) D has a real analytic defining function ρ, plurisuperharmonic on D,

then for each s and $k \geq 1$, $P_{\rho,\kappa}$ maps continuously $W^{s+k-1}(D)$ into $H^s(D)$.

If $k = 1$, then Corollary holds for each defining function for D.

In the case when D is a smooth bounded strictly pseudoconvex domain and $k = 1$ we get the following more precise estimates.

THEOREM 3. If D is a strictly pseudoconvex domain with smooth boundary then, for each defining function ρ for D and for every $\alpha > 0$ $P_\rho = P_{\rho,1}$ maps continuously the Hölder space Λ_α into Λ_α.

2. Proof of Theorem 1

Let s be a fixed number greater than zero. Since D is a pseudoconvex domain with smooth boundary, then from Kohn's results [6] it follows that there exists $N > 0$ such that the projection P_u with weight $u = -N \sum_{i=1}^{n} |z_i|^2$ is a continuous operator from W^s onto H^s. Let $w \in C^\infty(\bar{D})$. Since for every $g \in W^s$ $g - P_w g \perp L^2H(D)$ with respect to the scalar product $\langle \, , \rangle_w$, then $e^{w-u} g - e^{w-u} P_w(g) \perp L^2H(D)$ with respect to the scalar product $\langle \, , \rangle_w$. Thus $P_u(e^{w-u} g) = P_u(e^{w-u} P_w(g))$.

Let A denote the linear operator $Af = P_u(e^{w-u} f)$ which maps continuously H^s into H^s. To prove our theorem it suffices to show that A has a continuous inverse operator $A^{-1} : H^s \to H^s$, because then $P_w(g) = A^{-1}(P_u(e^{w-u} g))$ will be a continuous operator from W^s onto H^s. Let T_s be a compact operator from $W^s_{\langle 0,1 \rangle}$ into W^s, which solves $\bar{\partial}$-problem. Then the operator $T_u = (I - P_u)T_s$ is also compact. We have

$$Af = P_u(e^{w-u} f) = e^{w-u} f - T_u(f \wedge \bar{\partial}(e^{w-u})).$$

Now we shall extend A to an operator $\tilde{A}$ from W_s into W_s by

$$\tilde{A}(g) = e^{w-u} g - T_u(P_u(g) \wedge \bar{\partial} e^{w-u} = e^{w-u}[g - {}^{u-w} T_u(P_u(g) \wedge \bar{\partial} e^{w-u}].$$

The operator in square brackets is a Fredholm operator. To prove that $\tilde{A}$ is invertible it suffices to show that $\ker \tilde{A} = \{0\}$. To prove that A is invertible (as an operator from H^s into H^s) we have to show that if $\tilde{A}g \in H^s$ then $g \in H^s$.

Let $\tilde{A}g \in H^s$. We have

$$\bar{\partial}(e^{w-u} g - T_u(P_u(g) \wedge \bar{\partial} e^{w-u})) = 0$$

and

$$\bar{\partial}(e^{w-u} g - T_u(P_u(g) \wedge \bar{\partial} e^{w-u})) = \bar{\partial}(e^{w-u} g) - P_u(g) \wedge \bar{\partial} e^{w-u}$$

$$= \bar{\partial}(e^{w-u}(g - P_u(g)) = 0.$$

It means that $e^{w-u}(g - P_u(g)) \in H^s \subset L^2H(D)$. On the other hand, $g - P_u(g) \perp L^2H(D)$ with respect to the scalar product $\langle \, , \rangle_u$. It implies that $e^{w-u}(g - P_u(g))$ is orthogonal to $L^2H(D)$ with respect to the scalar product $\langle \, , \rangle_{2u-w}$. Thus, $e^{w-u}(g - P_u(g)) = 0$ and $g = P_u(g) \in H^s$.

Finally, let $\tilde{A}g = 0$. We have $g = P_u(g)$ and $e^{w-u} g = Tu(g \wedge \bar\partial(e^{w-u})) \perp L^2H(D)$ with respect to the scalar product $\langle\ ,\ \rangle_u$. Thus, g is orthogonal to $L^2H(D)$ with respect to the scalar product $\langle\ ,\ \rangle_w$ and $g \equiv 0$. Our theorem has been proved.

3. Example

Let $\varphi: \langle 0,1\rangle \to \langle 0,1\rangle$ be a C^∞, concave function such that $\varphi \equiv 1$ on $\langle 0,\frac{1}{4}\rangle$ and $\varphi(r) = 1 - r^2$ on $\langle \frac{3}{4},1\rangle$. Define $D \subset \mathbb{C}^2$ as follows:

$$D = \{(z_1 z_2) : |z_2|^2 \le \varphi(|z_1|)\}.$$

The domain D is a smooth 2-circular convex domain.

We shall prove that there cannot exist a compact operator $T_s : W^s_{\langle 0,1\rangle}(D) \to W^s(D)$ solving $\bar\partial$-problem. Suppose that for some $s \ge 0$ there exists such an operator. Since D is convex then as in the proof of Theorem 1 we can use Kohn's estimates [6] and get the continuity in the g-th Sobolev norm projection P_u, $u = -M|z|^2$ for some sufficiently large M. Consider an integral operator

$$Af = \int_D (\bar z_1 - \bar t_1) K_u(z,t) f(t) e^{u(t)} dV_t .$$

Then we have

$$\bar\partial Af = P_u(f) \quad \text{and} \quad Af \perp L^2H(D)$$

with respect to the scalar product $\langle\ ,\ \rangle_u$. Thus $Af = (I - P_u)(T_s(P_u f))$. Since T_s is compact, then Af is also a compact operator from $W^s_u(D)$ into $W^s(D)$.

Now, let $f \in H^s(\Delta)$, Δ - the unit disc in $\mathbb{C}$. Extend f to D putting $\tilde f(z_1,z_2) = f(z_2)$. Then we have

$$A\tilde f = \int_D (\bar z_1 - \bar t_1) K_u(z,t) f(t_2) e^{-M(|t_1|^2 + |t_2|^2)} dV_t$$

$$= z_1 f(z_2) - \int_{|t_2|<1} f(t_2) \Big[\int_{D_{t_2}} \bar t_1 K_u(z,t) e^{-M|t_1|^2} dt_1 \Big] e^{-M|t_2|^2} dt_2 .$$

Since D_{t_2} is a disc, this last integral vanishes. Thus $A\tilde f = z_1 f(z_2)$.

Since D contains a polidisc $\{|z_2| < 1 \ \ |z_1| < \frac{1}{4}\}$, then the following norms are equivalent:

$$\|f\|_{H^s(\Delta)} \quad \|\tilde f\|_{H^s(D)} \quad \text{and} \quad \|z_1 f(z_2)\|_{H^s(D)} .$$

Thus the compactness of A yields that the unit ball in $H^s(\Delta)$ is compact, so we get a contradiction.

Remark. The Bergman kernel of the domain D of our Example has the following interesting property: The singularity of $K_D(z,t)$ is not of the type $1/|z-t|^M$ for any $M>0$. If it could be possible to estimate $|K_D(z,t)| < C/|z-t|^M$, then for sufficiently large n the operator

$$A_n f = \int_D (\overline{z}_1 - \overline{t}_1)^n K_D(z,t) f(t) dV_t$$

must be compact from $L^2(D)$ into $L^2(D)$. However, we have as before

$$A_n(f(z_2)) = \overline{z}_1^{\,n} f(z_2).$$

It implies that A_n is not compact.

4. Proofs of Theorems 2 and 3

It is easy to check that functions of the form $f(z)z_0^k$ and $f(z) \in L^2_{\rho,k+1}H(D)$ form a linearly dense subset of $L^2H(\tilde{D})$. Moreover, if $k \neq j$ then

$$f(z)z_0^k \perp g(z)z_0^j$$

and

$$\langle f(z)z_0^k, g(z)z_0^k \rangle_{L^2H(D)} = \int_D f(z)\cdot g(z)\cdot \rho^{k+1}(z)\cdot \frac{\Pi}{k+1}.$$

Thus the Bergman function of D can be expressed in the form

$$K_{\tilde{D}}(z_0,z,t_0,t) = \sum_{k=0}^{\infty} c_k z_0^k K_{\rho,k+1}(z\,t)\,\overline{t}_0^{\,k}.$$

It implies that for every $f \in L^2_{\rho,k+1}(D)$

$$P(z_0^k f(z)) = \frac{c_k\Pi}{k+1} P_{\rho,k+1}(f)\cdot z_0^k.$$

Now, fix $s>0$. If D has the property R that there exist $m \geq 0$ such that P maps $W^{s+m+k}(\tilde{D})$ continuously into W^{s+k}, then $P_{\rho,k+1}(f)$ maps continuously W^{s+m+k} into W^s. This concludes the proof of Theorem 2 and the Corollary.

It remains to prove Theorem 3. If ρ is a strictly plurisuperharmonic on $\overline{D}$ defining function, then $\tilde{D}$ is also strictly pseudoconvex. Thus, P maps continuously $\Lambda_\alpha(D)$ into $\Lambda_\alpha(\tilde{D})$ (see [1] and

[7] . $P_\rho(f) = c\,P(f)$ and that is why P_ρ is continuous from $\Lambda_\alpha(D)$ into $\Lambda_\alpha(D)$.

If ρ_1 is any other defining function, then $\rho_1 = e^h \cdot \rho$. Similarly as in the proof of Theorem 1, we must prove that $P_\rho(e^h \cdot f)$ is an automorphism of $\Lambda_\alpha H(D)$ ($\Lambda_\alpha H$ - the subspace of Λ_α consisting of holomorphic functions). Since D is strictly pseudoconvex, then there exists an operator T solving $\bar\partial$-problem which for every α maps continuously $A_{\alpha(0,1)}(D)$ into $\Lambda_{\alpha+\frac{1}{2}}(D)$ (see [5]) ($A_{\alpha(0,1)}$ - the space of ∂-closed (0,1) forms with coefficients from Λ_α). As in the proof of Theorem 1 we have

$$P_\rho(e^h \cdot f) = e^h f - (I - P_\rho)(T(\bar\partial e^h \wedge f)) = e^h f - Af,$$

where A is compact from $\Lambda_\alpha H$ into Λ_α. It implies that $P_\rho(e^h \cdot f)$ has a closed range.

Since P_ρ is continuous from W^s onto H^s for each s, then we can prove as in the proof of Lemma 2 that $P_\rho(e^h H^s)$ is dense in H^s for each s. Moreover, since A^∞ is dense in H^s then $P_\rho(e^h A^\infty)$ is dense in H^s for each s. For sufficiently large s, H^s imbeds in $\Lambda_\alpha H$ as its dense subspace and that is why $P_\rho(e^h \cdot A^\infty)$ is dense in Λ_α. Thus, $P_\rho(e^h f)$ maps $\Lambda_\alpha H$ onto $\Lambda_\alpha H$ and is an automorphism of $\Lambda_\alpha H$.

Institute of Mathematics of the
Polish Academy of Sciences
Śniadeckich 8, P.O.Box 137
PL-00-950 Warszawa, Poland

References

[1] AHERN, P. and R. SCHNEIDER, Holomorphic Lipschitz functions in pseudoconvex domains, Amer. J. Math. 101 (1979), 543-565.

[2] BELL, S., Proper holomorphic mappings and the Bergman projection, Duke Math. J. 48 (1981), 167-175.

[3] ——, Biholomorphic mappings and the $\bar\partial$-problem, Ann. of Math. 114 (1981), 103-113.

[4] —— and E. LIGOCKA, A simplification and extension of Fefferman's theorem on biholomorphic mappings, Invent. Math. 57 (1980), 283-289.

[5] HENKIN, G. and . ROMANOV, Exact Hölder estimates for solutions of $\bar\partial$-problem (in Russian), Izv. Akad. IV Ser. Mat. 35 (1971), 1171-1183.

[6] KOHN, J.J., Global regularity for $\bar\partial$ on weakly pseudoconvex manifolds, Trans. Amer. Math. Soc. 181 (1973), 273-291.

[7] LIGOCKA, E., The Hölder continuity of the Bergman projection and proper holomorphic mappings, Studia Math., to appear.

[8] BELL, S. and H. BOAS, Regularity of the Bergman projections in weakly pseudoconvex domains, Ann. of Math. 257 (1981), 23-30.

TRANSCENDENTAL BÉZOUT ESTIMATE BY THE LOGARITHMIC FUNCTION IN $\mathbb{C}^n$

Masami Okada (Sendai)

Contents page

Summary and Introduction

In [4] we gave a transcendental Bézout estimate for equi-dimensional entire mappings as an application of an inequality for their Jacobians. In this note we study the same problem for holomorphic mappings defined in the unit ball in $\mathbb{C}^n$. For simplicity we consider only the case $n=2$. The idea is that we perform our calculations in the unit ball with respect to the Bergman metric using an approximative analogue of "$\Delta \log |z| = \delta_0$" for the case $n=1$ (cf. [4] for the background material).

1. Approximative analogue of log

In the following we use a family of convex functions $\{q_\lambda(t)\}_{\lambda>0}$, which satisfies the following properties:

(i) q_λ is convex on R and derivable,

(ii) $q_\lambda(0) = q'_\lambda(-\infty) = 0$, where $q'_\lambda(t) = \frac{d}{dt} q_\lambda(t)$,

(iii) $q'_\lambda(t) \to 1$ $(\lambda \to 0)$ $\forall t \in R$.

We remark that $q_\lambda(\log\|z\|)$ is then a plurisubharmonic function and while λ tends to zero, $dd^c q_\lambda(\log\|z\|) \wedge dd^c q_\lambda(\log\|z\|)$ tends to the Dirac's delta.

LEMMA 1. *Let* $D(0,s)\subset \mathbb{C}$ *be the disk of centre* 0 *and of radius* s *and let* $\phi_\lambda = \Delta q_\lambda(\log|z|)\,dV_1$, *where* $dV_1 = i\,dz\wedge d\bar{z}$. *Then there is a positive constant, independent of* λ *and* s *and such that*

$$\int_{D(0,s)} \phi_\lambda = c q'_\lambda(\log s).$$

Proof. Since $\Delta = \frac{\partial^2}{\partial r^2} + \frac{1}{r}\frac{\partial}{\partial r} + \frac{1}{r^2}\cdot\frac{\partial^2}{\partial\theta^2}$, then

$$\int_{D(0,s)} \phi_\lambda = c\int_0^s r\left(\frac{\partial^2}{\partial r^2} + \frac{1}{r}\frac{\partial}{\partial r}\right) q_\lambda(\log r)\,dr = c\int_0^s q''_\lambda(\log r)\,\frac{1}{r}\,dr$$
$$= c q'_\lambda(\log s)\,, \quad \text{q.e.d.}$$

For $\Lambda = (\lambda,\mu)\in \mathbb{R}^2_+$ we define w_Λ by

$$w_\Lambda = dd^c q_\lambda(\log|z_1|)\ dd^c q_\mu(\log|z_2|) = \Delta_{z_1} q_\lambda(\log|z_1|)\Delta_{z_2} q_\mu(\log|z_2|)dV,$$

where dV is the canonical volume form of $\mathbb{C}^2 = \mathbb{R}^4$.

Now, we suppose we are given a holomorphic mapping $F = (f,g)$, where f and g are defined in B. For convenience we may suppose that $|f(0)| = |g(0)| = 1$.

2. Majoration of F^*w_Λ

Let $M_r(f) = \sup\{|f(z)|;\ \|z\| < (r+1)/2\}$, $M_r(g) = \sup\{|g(z)|;\ \|z\| < (r+1)/2\}$, and $M_r(F) = (M_r(f)^2 + M_r(g)^2)^{\frac{1}{2}}$. Then we have the following proposition.

PROPOSITION 1. There exists a constant c independent of μ, λ, $\frac{1}{2} < r < 1$ and F such that

$$\int_{B(0,r_1)} F^*w_\Lambda \le c(1-r)^{-2} + q_\lambda(\log M_r(f))q_\mu(\log M_r(g)),$$

where $r_1 = (3r+1)/4$.

Proof. We give here a somewhat brief proof and we hope the readers will see [4] and [3] for detailed arguments.

1-st step. Let $r_2 = (2r+1)/3$ and $r_3 = (r+1)/2$. We define a non-positive plurisubharmonic function which is -1 on $B(0,r_1)$ by

$$U(z) = \min\{\max(\log\|z\|,\ \log r_1) - \log r_2,\ 0\}/(\log r_2 - \log r_1).$$

Then, we take into account tnat F^*w_Λ is positive and

$$\int_{B(0,r_1)} F^*w_\Lambda = \int_B -UF^*w_\Lambda = -\int U\, dd^c q_\lambda(\log|f|) \wedge dd^c q_\mu(\log|g|).$$

By Stokes' theorem this is equal to $-\int q_\lambda(\log|f|)\, dd^cU \wedge dd^c q_\mu(\log|g|) + \int_{|z|=r_2} q_\lambda(\log|f|) d^cU \wedge dd^c q_\mu(\log|g|)$. Now, dd^cU is found to be equal to

$$(2\|z\|^2 \log r_2/r_1)^{-1}\, 1\!\!1_{B(r_1,r_2)}\, dz_t \wedge d\bar z_t$$

$$+ (r_1 \log r_2/r_1)^{-1}\, \delta_{\|z\|=r_1}\, dz_n \wedge d\bar z_n,$$

where $B(r_1,r_2) = \{z \in B;\ r_1 < \|z\| < r_2\}$ and z_t is the complex tangential direction and so on. We have

$$\int_{B(0,r_1)} F^*w_\Lambda = -q_\lambda(-\infty)(2r_1^2 \log r_2/r_1)^{-1} \int_{B(r_1,r_2)} \Delta_N q_\mu(\log|g|)\, dV$$

$$+ q_\lambda(\log M_r(f))\left(r_2 \log \frac{r_2}{r_1}\right)^{-1} \int_{|z|=r_2} \Delta_T q_\mu(\log|g|)\, d\sigma$$

where $\Delta_N = \partial^2/\partial z_n \partial \bar z_n$ and $\Delta_T = \partial^2/\partial z_t \partial \bar z_t$.

2-nd step. Let I denote the first integral in question and J the second one. For I we remark that $\Delta_N q_\mu(.) = \Delta_E q_\mu(.)$, where Δ_E is the usual Laplacian in $\mathbb{R}^4$. In order to estimate J we have to remember that $\Delta_T = r_1^{-2}(\Delta_K + \partial/\partial n)$, where Δ_K is the Laplacian of Kohn and $\partial/\partial n$ is the normal derivative. Since the integral of Δ_K over a sphere is equal to zero, we have

$$J \le r_1^{-2} \int_{\|z\|=r_1} \frac{\partial q_\mu}{\partial n}(\log|g|)\, d\sigma,$$

the R.H.S. being equal to $r_1^{-2} \int_{\|z\|<r_1} \Delta_E q_\mu(\log|g|)\, dV$ by Green's theorem.

To estimate

$$K = \int_{\|z\|<r_2} \Delta_E q_\mu(\log|g|)\, dV$$

we introduce the Green kernel of $B(0,r_3)$. Then, $\Delta_E q_\mu$ is nonnegative and we get

$$K \le \frac{cr_3}{r_3 - r_2} \int \Delta_E q_\mu(\log|g|)(z) G(0,z)\, dV(z)$$

$$= \frac{cr_3}{r_3 - r_2}\left[\int_{\partial B(0,r_3)} q_\mu(\log|g|)\, d\sigma - q_\mu(\log|g|(0))\right]$$

$$\le \frac{c}{r_3 - r_2} q_\mu(\log M_r(g))$$

$$\leq \frac{c}{1-r} q_\mu (\log M_r(g)).$$

Finally, we remark that $\log(r_2/r_1) \sim (1-r)/12$, when r tends to 1, q.e.d.

3. Minoration

Let $E_r = F^{-1}(0) \cap B(0,r)$. Then, Lemma 1 and Proposition 1 give the following

PROPOSITION 2. Let $\{a_1, a_2, \dots\}$ be a subset of E_r whose each point a_j has an open neighbourhood $V(a_j)$ disjoint with each other and such that $V(a_j) \subset B(0,r_1)$ and $F(V(a_j))$ contains $D(0,s_j) \times D(0,t_j)$, where s_j and t_j are positive. Then there is a constant c, independent of λ, μ, r, j, and g, such that

$$\Sigma_j q'_\lambda(\log s_j)\, q'_\mu(\log t_j) \leq c + q_\lambda(\log M_r(f))(1-r)^{-2} q_\mu(\log M_r(g)).$$

Remark. If $q_\mu(t) = t$ $(t \geq 0)$ or $q_\mu(t) \to t$ $(\mu \to 0)$, we have, letting μ to infinity,

$$\Sigma_j q'_\lambda(\log s_j) \leq c + q_\lambda(\log M_r(f))(1-r)^{-2} \log M_r(g).$$

To get an estimate we need the next lemma of Ono.

LEMMA 2. Let $\Phi : B(0,1) \to B(0,M)$ be a holomorphic mapping such that $\Phi(0) = 0$ and $|\Phi'(0)| > 0$. Define s_0 and s by $s_0 = M^{-3}|\Phi'(0)|^2/12$ and $s = M^{-2}|\Phi'(0)|^2/6$. Then Φ is univalent in $B(0,s)$ and $\Phi(B(0,s))$ contains $B(0,s_0)$.

Now, let $a = (a^1, 0) \in E_r$. We apply this lemma to

$$\Phi(z) = F((1-r)z_1/4 + a^1, (1-r)^{1/2} z_2/4).$$

Thus,

$$M_a = \sup_{z \in B(0,1)} \|\Phi(z)\| = \sup_{z \in B(0,r_1)} \|F(z)\| = M_r(F),$$

$$|\Phi'(0)|^2 = 4^{-4}(1-r)^3 |F'(a)|^2,$$

and $s_a = c\, M_r(F)^{-3}(1-r)^3 |F'(a)|^2$. Hence, we have the following Proposition and Theorem.

PROPOSITION 3. There exists c independent of λ, μ, r, and q, such that

(i) $$\sum_{a\in E_r} q_\lambda'(\log s_a/2)\, q_\mu'(\log t_a/2) \le c + q_\lambda(\log M_r(f))(1-r)^{-2} q_\mu(\log M_r(g)).$$

Furthermore, if $q_\mu(t) \to t$ $(\mu \to 0)$, then

(ii) $$\sum_{a\in E_r} q_\lambda'(\log s_a/2) \le c + q_\lambda(\log M_r(f))(1-r)^{-2} \log M_r(g).$$

THEOREM 1. *There exists* c *independent of* λ, μ, r, *and* F, *such that*

(i) $$\sum_{a\in E_r} |F'(a)|^{2\lambda} \le c + M_r(f)^\lambda (1-r)^{-2-3\lambda} M_r(F)^{3\lambda} \log M_r(g)$$

and

(ii) $$\sum_{a\in E_r} [x_a(1+\log x_a)^{1+\gamma}]^{-1} \le c_\gamma + \log M_r(f)(1-r)^{-2} \log M_r(g),$$

where $\gamma > 0$, c_γ *are constants, and* x_a *is defined by*

$$e^{x_a} = \max[(1-r)^{-1} M_r(F)|F'(a)|^{-2/3}, c_1]$$

with a fixed sufficiently large constant c_1.

Proof of the theorem. We apply Proposition 2, where $q_\lambda(t) = (e^{\lambda t} - 1)/\lambda$ and

$$q_\lambda(t) = \begin{cases} t & \text{for } t \ge 0 \\ \lambda^{-1}[1+\log(1+\log(1-\lambda t))]^{-1} - \lambda^{-1} & \text{for } t < 0. \end{cases}$$

COROLLARY 1. *If* $M_r(F) = \mathcal{O}(e^{(1-r)^{-\beta}})$, $\beta > 0$, *then*

$$\sum_{a\in E_r} |F'(a)|^{(1-r)^\beta} \le c_\beta (1-r)^{-2\beta-2}.$$

COROLLARY 2. *In the case when* $M_1(F) = 10$, *we have*

(i) $$\sum_{a\in E_r} |F'(a)|^{2\lambda} \le c + 3\,(1-r)^{-2-3\lambda}(100)^\lambda.$$

In particular, for $\lambda = (-\log(1-r))^{-1}$ *we obtain*

$$\sum_{a\in E_r} |F'(a)|^{(-\log(1-r))^{-1}} \le c(1-r)^{-2}(-\log(1-r)).$$

We also have

(ii) $$\sum_{a\in E_r} [y_a(\log y_a)^{1+\gamma}]^{-1} \le c_\gamma(1-r)^{-2},$$

where $y_a = \max(-\log[(1-r)^{3/2}|F'(a)|], 2)$.

In particular, for $E'_r = \{a\in E_r;\ |F'(a)| \le 1-r\}$, we obtain

$$\sum_{a\in E'_r} (-\log|F'(a)|)^{-1}[\log(-\log|F'(a)|)]^{-1-\gamma} \le c_\gamma(1-r)^{-2}.$$

Proof of Cor. 1. We put $\lambda = (2\log M_r(F))^{-1}$.

Proof of Cor. 2. (i) Easy from Th. 1 (i). (ii) Put $\lambda = 1$ in Th. 1 (ii).

4. Application to modified Bézout estimates in transcendental case

THEOREM 2. (i) If $M_r(F) = \mathcal{O}(e^{(1-r)^{-\beta}})$, $\beta > 0$, then

$$\#\{a\in E_r;\ |F'(a)| > e^{-k(1-r)^{-\beta}}\} \le c_{k,\beta}(1-r)^{-2\beta-2}.$$

(ii) If $f, g\in H^\infty(B)$, then $\#\{a\in E_r; |F'(a)| > (1-r)^k\} \le c'_k(1-r)^{-2}(-\log(1-r))$, where c'_k depends only on the positive integer k and $M_1(F)$.

References

[1] BEDFORD, E. and TAYLOR, B.A., A new capacity for plurisubharmonic functions, Acta Math. 149 (1982), 1-44.

[2] GAVEAU, B., Intégrales de courbure et potentiels sur les hypersurfaces de $\mathbb{C}^n$ ou de la boule, C.R. Acad. Sci. Paris 293 (1981), 253-255.

[3] MALLIAVIN, P., Équation de la chaleur associée à une fonction plurisousharmonique d'exhaustion et comportement frontière, Ann. Inst. Fourier 25 (1975), 447-464.

[4] OKADA, M., Un théorème de Bézout transcendant sur $\mathbb{C}^n$, J. of Functional Anal. 45 (1982), 236-244.

[5] ONO, I., Analytic vector functions of several complex variables, J. Math. Soc. Japan 8 (1956), 216-246.

[6] STOLL, W., A Bézout estimate for complete intersections, Ann. of Math. 96 (1972), 361-401.

Institute of Mathematics
Tohoku University
980, Sendai, Japan

DAS SPEKTRUM TORSIONSFREIER GARBEN II

Christian Okonek und Heinz Spindler (Göttingen)

Inhaltverzeichnis

Abstract. Die Arbeit stellt eine Fortsetzung einer früheren Arbeit [12] über torsionsfreie Garben auf $\mathbb{P}^3$ dar. Es werden unter anderem auch einige globale Modulräume stabiler torsionsfreier Garben auf $\mathbb{P}^3$ untersucht. Die Arbeit berührt also Fragen der Deformationstheorie algebraischer Strukturen. Wir wenden die Methoden und Techniken aus [12] auf die folgenden Probleme an: 1) Schranken und Lücken für die dritten Chernklassen torsionsfreier Garben auf $\mathbb{P}^3$, 2) Schranken für das arithmetische Geschlecht 2-codimensionaler Unterschemata von $\mathbb{P}^3$, 3) Klassifikation von Garben mit extremen Chernklassen, 4) Kompaktifizierung einiger Modulräume stabiler lokalfreier und reflexiver Garben auf $\mathbb{P}^3$.

0. Einleitung

Sei k ein algebraisch abgeschlossener Körper, $\mathbb{P}^3 = \mathbb{P}^3_k$ der 3-dimensionale projektive Raum über k. In [12] haben wir Techniken zur Untersuchung kohärenter torsionsfreier Garben von beliebigem Rang auf $\mathbb{P}^3$ bereitgestellt. Wir wollen hier diese Techniken anwenden. Dabei beschränken wir uns auf die Untersuchung torsionsfreier Garben mit generischem Spaltungstyp.

Wir erhalten Abschätzungen für die dritte Chernklassen c_3 solcher Garben, die frühere Abschätzungen [5], [8], [11] verallgemeinern. Falls die betrachtete Garbe keine instabilen Ebenen besitzt, ergeben sich Abschätzungen, die analog zur Ungleichung von Castelnuovo [4] für Raumkurven sind. Anschließend zeigen wir, daß unterhalb der angegebenen Schranken für c_3 weitere Lücken auftreten.

Im zweiten Abschnitt wenden wir die erhaltenen Ergebnisse auf die Untersuchung abgeschlossener Unterschemata $Y \subset \mathbb{P}^3$ der Dimension 1 an. Wir erhalten Ungleichungen für das aritmetische Geschlecht solcher Kurven, die für glatte, zusammenhängende Kurven klassisch sind. Als eine einfache Folgerung ergibt sich die Ungleichung von Castelnuovo

$$p_a \leq \frac{1}{4}d^2 - d + 1 - \frac{1}{4}\varepsilon, \quad \varepsilon = \begin{cases} 1, & d \equiv 1\ (2), \\ 0, & d \equiv 0\ (2) \end{cases}$$

für das arithmetische Geschlecht einer integren, nicht entarteten Kurve $Y \subset \mathbb{P}^3$. Wir beschreiben dann die geometrischen Punkte der Hilbertschemata abgeschlossener 1-dimensionaler Unterschemata $Y \subset \mathbb{P}^3$ vom Grad d mit

$$p_a = \frac{1}{2}(d-1)(d-2) \quad \text{und} \quad p_a = \frac{1}{2}(d-2)(d-3).$$

Zum Schluß dieses Abschnittes klassifizieren wir die Macauley Kurven mit $p_a = \frac{1}{4}d^2 - d + 1 - \frac{1}{4}\varepsilon$. Sie liegen alle auf Quadriken.

Im dritten Teil untersuchen wir torsionsfreie Garben, für die die am Anfang gewonnenen Ungleichungen Gleichungen sind. Ihre Beschreibung läßt sich im wesentlichen auf die Klassifikation der entsprechenden Idealgarben reduzieren, die wir schon durchgeführt haben. Mit ähnlichen Methoden kann man noch Garben in einem Bereich nahe den extremen Fällen behandeln. Wir tun dies soweit, wie wir es in Abschnitt 4 benötigen.

In dem folgenden vierten Abschnitt bestimmen wir einige vollständige Modulschemata semi-stabiler torsionsfreier Garben. Sei ${}^r\overline{M}_{\mathbb{P}^3}(c_1, c_2, c_3)$ das Modulschema, dessen abgeschlossene Punkte den S-Äquivalenzklassen torsionsfreier Garben F vom Rang r auf $\mathbb{P}^3$ mit den Chernklassen $c_i(F) = c_i$ entsprechen. Wir untersuchen 3 Fälle: ${}^2\overline{M}_{\mathbb{P}^3}(0, c_2, c_2^2 - c_2 + 2)$ ist – zumindest für $c_2 \geq 6$ – ein projektives Bündel über einem Produkt $\mathbb{P}^3 \times \mathbb{P}^3$. Eine universelle Familie läßt sich über diesem Modulschema auch dann konstruieren, wenn die hinreichende Bedingung $\delta(H) = 1$ [7] nicht erfüllt ist. Das beantwortet eine Frage von Maruyama. Ähnlich einfach ist der Modulraum ${}^3\overline{M}_{\mathbb{P}^3}(0, c_2, c_2^2 - c_2)$ zu beschreiben. Er ist isomorph zu einem projektiven Bündel über $\mathbb{P}^3$ und besitzt ebenfalls eine universelle Familie. Schließlich untersuchen wir noch den offenen Teil

$$\overline{M}_0 \subset {}^r\overline{M}_{\mathbb{P}^3}(1-r, c_2, c_2^2 - (r^2 - 2r)c_2 + \binom{r}{3} + \binom{r-1}{2}^2)$$

der torsionsfreien Garben mit generischem Spaltungstyp. Für $r=2,3$ ist dies in $\operatorname{char} k=0$ der ganze Modulraum. Es stellt sich heraus, daß $\overline{M}_0$ eine Zusammenhangskomponente ist und mit einem Grassmannbündel über $\mathbb{P}^3$ identifiziert werden kann.

1. Abschätzungen und Lücken für c_3

Sei F eine torsionsfreie Garbe vom Rang r auf $\mathbb{P}^3$ mit:
Spaltungstyp [12] $a_F=(a_1,\dots,a_s;\ r_1,\dots,r_s)$,
Chernklassen [12] $c_i=c_i(F)$, $i=0,1,2,3$ und
Spektrum [12] $k_F=(k_1,\dots,k_m)$.
Nach Normieren können wir uns auf den Fall $-r<c_1\le 0$ beschränken. Weiter setzen wir voraus, daß für den Spaltungstyp von F gilt $d(F)=a_s-a_1$. Dies ist der generische Fall. Es gibt also ein $a=-c_1$ mit $0\le a<r$ und

$$(1)\qquad a_F=\begin{cases}(0;\ r) & \text{für } a=0,\\ (-1,0;\ a,r-a) & \text{für } a\neq 0.\end{cases}$$

Sei schließlich $\bar{s}=\bar{s}_F=h^0(\underline{\mathrm{Ext}}^2(F,\mathcal{O}))$, $s=s_F=h^0(\underline{\mathrm{Ext}}^1(F_H,\mathcal{O}_H))$ für generische Ebenen $H\subset\mathbb{P}^3$. Mit $|k_F|$ bezeichnen wir die Quersumme $|k_F|=\Sigma_{i=1}^m k_i$. Nach [12] gilt dann ($H\subset\mathbb{P}^3$ ist eine generische Ebene):

$$(2)\qquad m=-\chi(F_H(-1))=c_2-\binom{a}{2},$$

$$(3)\qquad -|k_F|=\chi(F(-2))+\bar{s}=\frac{1}{2}c_3+\frac{1}{2}ac_2-\binom{a+1}{3}+\bar{s},$$

$$(4)\qquad s\le m.$$

Das Spektrum besitzt folgende Eigenschaften [12]:

(5) Ist $k_1\le a_1-2$, so kommen alle Zahlen k mit $k_1\le k\le a_1-1$ im Spektrum vor.

(6_s) Ist $s<m$ und $k_{m-s}>1$, so kommen alle Zahlen k mit $1\le k\le k_{m-s}$ im Spektrum vor.

(7) Ist $k_1\le a_1-3$ und kommt k_{i_0} mit $k_1<k_{i_0}\le a_1-2$ genau einmal im Spektrum vor, so kommen auch alle Zahlen k mit $k_1\le k\le k_{i_0}$ genau einmal im Spektrum vor.

Nach (5) ist — bei festen a_1, m — $\overline{k}=(a_1-m,\ a_1-m+1,\dots,\ a_1-1)$ das Spektrum mit maximaler negativer Quersumme $-|\overline{k}|=\binom{m-a_1+1}{2}+a_1$.
Aus (3) folgt

PROPOSITION 1.1. Für eine torsionsfreie Garbe F auf $\mathbb{P}^3$ mit Spaltungstyp (1) gilt

$$(8) \qquad \chi(F(-2)) + \bar{s} \leq \binom{m-a_1+1}{2} + a_1 .$$

KOROLLAR 1.2. Für eine torsionsfreie Garbe F auf $\mathbb{P}^3$ mit Spaltungstyp (1) gilt

$$(9) \qquad c_3 \leq \begin{cases} c_2^2 + c_2 - 2\bar{s} & \text{für } c_1 = 0, \\ c_2^2 + 2c_2 - 2\bar{s} & \text{für } c_1 = -1, \\ c_2^2 - c_2 - 2\bar{s} & \text{für } c_1 = -2. \end{cases}$$

Aus den Eigenschaften (5) und (7) des Spektrums erhält man eine erste Aussage über Lücken für die dritte Chernklasse.

PROPOSITION 1.3. Sei F eine torsionsfreie Garbe auf $\mathbb{P}^3$ mit Spaltungstyp (1) und $m \geq 3$. Entweder ist

$$(10) \qquad \chi(F(-2)) + \bar{s} = \binom{m-a_1+1}{2} + a_1 ,$$

oder es gilt

$$(11) \qquad \chi(F(-2)) + \bar{s} \leq \binom{m-a_1}{2} + 2 .$$

Beweis. Nach (5) ist $k_{m-2} \geq a_1 - 3$. Ist $k_{m-2} = a_1 - 3$, so folgt $k_{m-1} = a_1 - 2$, $k_m = a_1 - 1$, also mit (7) $k_F = \bar{k}$ und es gilt (10). Sei jetzt $k_{m-2} \geq a_1 - 2$. Aus (5) folgt dann $k_{m-i} \geq a_1 - i$ für $i \geq 2$, $k_{m-1} \geq a_1 - 2$ und $k_m \geq a_1 - 1$. Man bekommt daraus

$$\chi(F(-2)) + \bar{s} = -|k_F| \leq \binom{m-a_1}{2} + 2 .$$

KOROLLAR 1.4. Für eine torsionsfreie Garbe F auf $\mathbb{P}^3$ mit Spaltungstyp (1) gilt

$$(12) \qquad c_3 \leq \begin{cases} c_2^2 - c_2 + 4 - 2\bar{s} & \text{für } c_1 = 0, \\ c_2^2 + 4 - 2\bar{s} & \text{für } c_1 = -1, \\ c_2^2 - 3c_2 + 6 - 2\bar{s} & \text{für } c_1 = -2, \end{cases}$$

oder es gilt Gleichheit in (9).

Bei festen Invarianten $c_1, c_2, \bar{s}$ nimmt $\chi(F(-2)) + \bar{s}$ nicht die $m-3$ Werte $\binom{m-a_1}{2} + 2 + l$, $l = 1, \ldots, m-3$, an. Dies ergibt die ersten Lücken für c_3. Aus [12] bekommt man

PROPOSITION 1.5. Sei F eine torsionsfreie Garbe auf $\mathbb{P}^3$ mit

Spaltungstyp (1).

i) *Ist* $k_1 \leq a_1 - 3$ *und* $\mathrm{Hom}(F, \mathcal{O}_H(k_1)) = 0$ *für alle Ebenen* $H \subset \mathbb{P}^3$, *so gilt* $m \geq 4$ *und* $\chi(F(-2)) + \bar{s} \leq \frac{1}{4}m^2 + m - \frac{1}{4}\varepsilon - a_1 m$, *wobei* $\varepsilon = 1$ *für* $m \equiv 1\ (2)$ *und* $\varepsilon = 0$ *für* $m \equiv 0\ (2)$ *ist*.

ii) *Ist* $k_1 \geq a_1 - 2$, *so gilt* $\chi(F(-2)) + \bar{s} \leq 2m - 1 - a_1 m$.

B e w e i s. Die zweite Aussage folgt leicht aus (3) und (5). Ist $k_1 \leq a_1 - 3$ und $\mathrm{Hom}(F, \mathcal{O}_H(k_1)) = 0$ für alle Ebenen $H \subset \mathbb{P}^3$, so kommt nach [12] jede Zahl k mit

$$(13) \qquad k_1 < k \leq a_1 - 2$$

mindestens zweimal im Spektrum k_F von F vor. Daraus folgt schon $m \leq 4$. Wir betrachten nun die möglichen Spektren für F. Das Spektrum k^* mit der maximalen negativen Quersumme $-|k^*|$ ist

$$k^* = \begin{cases} (a_1-t, a_1-t, \ldots, a_1-2, a_1-2, a_1-1) & \text{für } m = 2t-1, t \geq 3, \\ (a_1-t-1, a_1-t, a_1-t, \ldots, a_1-2, a_1-2, a_1-1) & \text{für } m = 2t, t \geq 2. \end{cases}$$

Es folgt: $\chi(F(-2)) + \bar{s} = -|k_F| \leq |k^*| = \frac{1}{4}m^2 + m - \frac{1}{4}\varepsilon - a_1 m$.

KOROLLAR 1.6. *Sei* F *eine torsionsfreie Garbe auf* $\mathbb{P}^3$ *mit Spaltungstyp* (1) *und Spektrum* $k_F = (k_1, \ldots, k_m)$. *Gilt* $k_1 \leq a_1 - 3$ *und* $\mathrm{Hom}(F, \mathcal{O}_H(k_1)) = 0$ *für alle Ebenen* $H \subset \mathbb{P}^3$, *so folgt*

$$(14) \qquad c_3 \leq \begin{cases} \frac{1}{2}c_2^2 + 2c_2 - \frac{1}{2}\varepsilon - 2\bar{s} & \text{für } c_1 = 0, \\ \frac{1}{2}c_2^2 + 3c_2 - \frac{1}{2}\varepsilon - 2\bar{s} & \text{für } c_1 = -1, \\ \frac{1}{2}c_2^2 + c_2 - 1 + \frac{1}{2}\varepsilon - 2\bar{s} & \text{für } c_1 = -2. \end{cases}$$

mit $\varepsilon = 1$ *für* $c_2 \equiv 1\ (2)$ *und* $\varepsilon = 0$ *für* $c_2 \equiv 0\ (2)$.

Der folgende Satz liefert weitere L ü c k e n für c_3 unter der zusätzlichen Voraussetzung $s = 0$.

THEOREM 1.7. *Sei* F *eine torsionsfreie Garbe auf* $\mathbb{P}^3$ *mit Spaltungstyp* (1). *Es gelte* $s_F = 0$ *und* $\chi(F(-2)) + \bar{s} > \frac{1}{4}m^2 + m - \frac{1}{4}\varepsilon - a_1 m$. *Sei* $q := k_1 - a_1 + m$. *Dann gilt*:

$$(15) \qquad 0 \leq q \leq m - 2 - [\tfrac{1}{2}m]$$

und

$$(16) \qquad \binom{m-a_1+1}{2} + a_1 - q(m-a_1+1) \leq \chi(F(-2)) + \bar{s} \leq \binom{m-a_1+1}{2} + a_1 - q(m-q-1).$$

B e w e i s. Aus der Voraussetzung und aus Proposition 1.5 ii) folgt

$m \geq 3$ und $k_1 \leq a_1 - 3$; nach Proposition 1.5 i) gibt es also eine Ebene H mit $\mathrm{Hom}(F, \mathcal{O}_H(k_1)) \neq 0$. Aus [12] folgt, daß es einen Index i_0 gibt mit $k_1 < k_{i_0} \leq a_1 - 2$ und $\# \{i \mid k_i = k_{i_0}\} = 1$.

Wir beweisen zunächst (15), oder äquivalent

$$(17) \quad a_1 - m \leq k_1 \leq a_1 - 2 - [\tfrac{1}{2} m].$$

Die Ungleichung $a_1 - m \leq k_1$ ist klar. Wäre $k_1 \geq a_1 - 1 - [\frac{1}{2}m]$, so erhielte man – für $m \geq 4$ –

$$\chi(F(-2)) + \bar{s} = -|k_F| \leq -|\tilde{k}| = -|k^*| - 1 - 2\varepsilon < -|k^*|.$$

Hierbei ist k^* das in Proposition 1.5 erwähnte Spektrum $k^* = (k_1^*, k_2^*, \ldots, k_m^*)$, $\tilde{k}$ die folgende Abänderung von k^* $(k_1^* = a_1 - 1 - [\frac{1}{2}m])$:

$$\tilde{k} = \begin{cases} (k_1^*, k_1^* - 1, k_1^* - 2, k_1^* - 2, k_5^*, \ldots, k_m^*) & \text{für } m \equiv 0\ (2), \\ (k_1^*, k_1^* - 1, k_1^* - 2, k_4^*, \ldots, k_m^*) & \text{für } m \equiv 1\ (2). \end{cases}$$

Für $m = 3$ ist notwendig $k_F = (a_1 - 3, a_1 - 2, a_1 - 1)$, die Ungleichung (17) ist bewiesen.

Um (16) zu beweisen, betrachten wir die Mengen

$$K^s = K^s(m, a_1) = \{(k_1, \ldots, k_m) \in \mathbb{Z} \mid k_1 \leq k_2 \leq \ldots \leq k_m, \text{ es gelten } (5), (6_s), (7), (17)\ \exists\, i_0 \text{ mit } k_1 < k_{i_0} \leq a_1 - 2 \text{ mit } \#\{i \mid k_i = k_{i_0}\} = 1\}.$$

Nach Voraussetzung gilt also $k_F \in K^0$. Für jedes s ist K^s die disjunkte Vereinigung $K^s = \dot\bigcup_{q=0, m-2-[\frac{1}{2}m]} K_q^s$ der Mengen

$$K_q^s = \{(k_1, \ldots, k_m) \in K^s \mid k_1 = a_1 - m + q\}.$$

Wir bestimmen die Unter- bzw. Obergrenzen $\min\{-|k| \mid k \in K_q^0\}$ bzw. $\max\{-|k| \mid k \in K_q^0\}$ der negativen Quersummen für Elemente $k \in K_q^0$. Dies liefert dann sofort die behaupteten Ungleichungen (16) für $\chi(F(-2)) + \bar{s}$.

Aus (5) und (6_0) folgt sofort, daß $\underline{k}^q = (a_1 - m + q, \ldots, a_1 - 1, 1, \ldots, q)$ das Minimum liefert, also

$$\min\{-|k| \mid k \in K_q^0\} = -|\underline{k}^q| = \binom{m - a_1 + 1}{2} + a_1 - q(m - a_1 + 1).$$

Der maximale Wert $\max\{-|k| \mid k \in K_q^0\}$ wird durch

$$\bar{k}^q = (a_1 - m + q, \ldots, a_1 - q - 2, \underbrace{a_1 - q - 1, a_1 - q - 1, \ldots, a_1 - 2a_1 - 2}_{\text{jeweils doppelt}}, a_1 - 1)$$

gegeben, es folgt $\max\{-|k| \mid k \in K_q^0\} = -|\bar{k}^q| = \binom{m - a_1 + 1}{2} + a_1 - q(m - q - 1)$. Damit ist Theorem 1.7 bewiesen.

B e m e r k u n g 1.8. Für $s>0$ gilt $\min\{-|k| \mid k\in K_q^s\} = -\infty$, da keine Zusammenhangseigenschaften im "positiven" Bereich $1,\dots,k_m$ bestehen. Für die Bestimmung des Maximums spielt dies keine Rolle, es gilt also $\max\{-|k| \mid k\in K_q^s\} = -|\overline{k}^q|$ für alle $s\geq 0$.

KOROLLAR 1.9. *Sei $q\geq 0$ eine ganze Zahl mit $m\geq (q+2)^2-a_1q$. Es gibt keine torsionsfreie Garbe F auf $\mathbb{P}^3$ vom Spaltungstyp (1), für die gilt $s_F=0$ und $-|\overline{k}^q|<\chi(F(-2))+\bar{s}<-|\underline{k}^q|$.*

B e w e i s. Ist $q\geq 0$, $m\geq (q+2)^2-a_1q$, so gilt auch $q\leq m-2-[\frac{1}{2}m]$. Lücken für mögliche Werte $\chi(F(-2))+\bar{s}$ treten genau dann auf, wenn $0\leq |\overline{k}^{q+1}|-|\underline{k}^q|-2$ gilt. Es gilt aber $|\overline{k}^{q+1}|-|\underline{k}^q|-2=-(q+2)^2+a_1q+m$.

B e i s p i e l 1.10. Wir wollen das Ergebnis am Beispiel $c_1=0$ illustrieren. Sei F also torsionsfrei auf $\mathbb{P}^3$ mit trivialem Spaltungstyp und $s_F=0$. Ist $\frac{1}{2}c_3+\bar{s}>\frac{1}{4}c_2^2+c_2-\frac{1}{4}\varepsilon$, so gibt es ein q mit $0\leq q\leq c_2-2-[\frac{1}{2}c_2]$ und

$$\binom{c_2+1}{2}-q(c_2+1)\leq \frac{1}{2}c_3+\bar{s}\leq \binom{c_2+1}{2}-q(c_2-q-1).$$

Ist $q\geq 0$ und $c_2\geq (q+2)^2$, so kommen die Werte

$$\binom{c_2+1}{2}-q(c_2+1)-l,\quad l=1,\dots,c_2-(q+2)^2+1,$$

nicht als $\frac{1}{2}c_3+\bar{s}$ vor. Für $q=0$ ergeben sich die Lücken aus Proposition 1.2. Wir stellen die ersten auftretenden Lücken für $\frac{1}{2}c_3+\bar{s}$ ($q=0,1$) in einer Tabelle dar:

c_2	$\binom{c_2+1}{2}$	Lücken für $\frac{1}{2}c_3$	$\frac{1}{4}c_2^2+c_2-\frac{1}{4}\varepsilon$
3	6		5
4	10	9	8
5	15	14, 13	11
6	21	20, 19, 18	15
7	28	27, 26, 25, 24	19
8	36	35, 34, 33, 32, 31	24
9	45	44, 43, 42, 41, 40, 39; 34	29
10	55	54, 53, 52, 51, 50, 49, 48; 43, 42	35

2. Anwendungen auf Kurven im $\mathbb{P}^3$

In diesem Abschnitt spezialisieren wir die Ergebnisse aus 1. auf den Fall torsionsfreier Garben vom Rang 1. Wir untersuchen also Idealgarben I_Y abgeschlossener Unterschemata $Y\subset\mathbb{P}^3$ der Dimension ≤ 1.

Ist $Y \subset \mathbb{P}^3$ 1-dimensional, so gilt für das Hilbertpolynom:

(18) $\chi(\mathcal{O}_Y(l)) = dl + 1 - p_a$.

Hier ist d der Grad, p_a das arithmetische Geschlecht von Y. Für die Idealgarbe I_Y erhält man

(19) $\chi(I_Y(l)) = \binom{l+3}{3} - dl + p_a - 1$.

Wie in 1. setzen wir $\bar{s}_Y = \bar{s}_{I_Y} = h^0(\underline{Ext}^2(I_Y, \mathcal{O}))$. Y ist genau dann eine Macauley Kurve, wenn $\bar{s} = 0$ ist. Man sieht leicht, daß $s_Y = s_{I_Y} = d$ gilt. Sei $k_Y = k_{I_Y} = (k_1, \ldots, k_m)$ das Spektrum von I_Y. Es ist [12]: $m = d$ und

(20) $\chi(I_Y(-2)) = p_a - 1 + 2d = -|k_Y| - \bar{s}$.

Bemerkung 2.1. Ist $Y \subset \mathbb{P}^3$ ein 1-dimensionales abgeschlossenes Unterschema, A die Menge der 0-dimensionalen assoziierten Punkte von $\mathcal{O}_Y$, so definiert $\mathcal{H}^0_A\, \mathcal{O}_Y \subset \mathcal{O}_Y$ eine Macauley Kurve $Y' \subset Y$ mit $\deg(Y') = \deg(Y)$, $k_{Y'} = k_Y$ und $p_a(Y') = p_a(Y) + \bar{s}_Y$. Wir nennen Y' die 1-dimensionale Komponente von Y und $\bar{s}_Y = h^0(\mathcal{H}_A\, \mathcal{O}_Y)$ die Anzahl der 0-dimensionalen assoziierten Punkte von Y.

Aus den Sätzen 1.1 und 1.3 bekommt man unmittelbar folgende Anwendung:

THEOREM 2.2. *Für ein 1-dimensionales abgeschlossenes Unterschema $Y \subset \mathbb{P}^3$ mit Hilbertpolynom (18) gilt*

(21) $p_a + \bar{s} \leq \frac{1}{2}(d-1)(d-2)$.

Ist $d \geq 3$, *so gilt*

(22) $p_a + \bar{s} = \frac{1}{2}(d-1)(d-2)$

oder

(23) $p_a + \bar{s} \leq \frac{1}{2}(d-2)(d-3)$.

Für Macauley Kurven, die generisch lokal vollständige Durchschnitte sind, wurde dieses Resultat von Sauer [13] mit Hilfe reflexiver Garben bewiesen.

Spezialisiert man Proposition 1.5 auf Idealgarben, so erhält man

THEOREM 2.3. *Sei $Y \subset \mathbb{P}^3$ ein abgeschlossenes 1-dimensionales Unterschema mit Hilbertpolynom (18) und* $e = e_Y = \max\{l \mid h^1(\mathcal{O}_Y(l)) \neq 0\}$. *Dann gilt:*

i) *Für* $e < 0$ *ist* $p_a + \bar{s} \leq 0$.

ii) Ist $e \geq 0$ und gilt für alle Ebenen $H \subset \mathbb{P}^3$: $h^1(\mathcal{O}_{Y\cap H}(e)) = 0$, so folgt

$$(24) \qquad p_a + \bar{s} \leq \frac{1}{4}d^2 - d + 1 - \frac{1}{4}\varepsilon .$$

Hier ist ε definiert durch $\varepsilon = 1$ für $d \equiv 1\,(2)$ und $\varepsilon = 0$ für $d \equiv 0\,(2)$.

Beweis. Sei $k = (k_1, \ldots, k_d)$ das Spektrum von I_Y. Nach Definition ist genau dann $k_1 > 0$, wenn für alle $l \geq -3$ $h^2(I_Y(l)) = 0$ ist. Wegen $h^1(\mathcal{O}_Y(l)) = h^2(I_Y(l))$ ist dies genau dann der Fall, wenn $e < -3$ ist. Ist $k_1 \leq 0$, so gilt $e = -k_1 - 3$. Es gilt $H^1(\mathcal{O}_{Y\cap H}(e)) = \mathrm{Hom}(I_Y, \mathcal{O}_H(-e-3))$. Die Voraussetzungen von Proposition 1.5 sind daher erfüllt.

Aus $e < 0$ folgt also $k_1 \geq -2$ und die Aussage i) folgt mit Proposition 1.5 ii). Ist $e \geq 0$, also $k_1 \leq -3$, so folgt ii) aus Proposition 1.5 i).

KOROLLAR 2.4 (CASTELNUOVO). Sei $Y \subset \mathbb{P}^3$ integer und nicht entartet mit $p_a > 0$. Dann gilt $d \geq 4$ und

$$(25) \qquad p_a \leq \frac{1}{4}d^2 - d + 1 - \frac{1}{4}\varepsilon .$$

Beweis. Y ist Macauleysch, also $\bar{s}_Y = 0$. Wegen $0 < p_a = h^1(\mathcal{O}_Y)$ ist $e \geq 0$. Da Y nicht eben ist, gilt für jede Ebene $H \subset \mathbb{P}^3$

$$\mathrm{Hom}(I_Y, \mathcal{O}_H(-e-3)) = H^0(\mathcal{O}_H(-e-3)) = 0 .$$

Bemerkung 2.5. Die Aussage von 2.4 bleibt richtig für reduzierte Kurven ohne ebene Komponenten vom Grad $\geq e+3$.

Wir beschreiben nun die Kurven, für die Gleichheit in (21), (23) oder (24) gilt.

PROPOSITION 2.6. Sei $Y \subset \mathbb{P}^3$ ein abgeschlossenes Unterschema des $\mathbb{P}^3$ mit Hilbertpolynom (18), $d \geq 1$ und $p_a = \frac{1}{2}(d-1)(d-2)$. Dann ist Y ein vollständiger Durchschnitt einer Ebene und einer Fläche vom Grad d.

Beweis. Es ist $\bar{s}_Y = 0$, Y also Macauleysch. Das Spektrum von Y ist notwendig $k_Y = (-d, -d+1, \ldots, -1)$. Es folgt:

$$h^0(I_Y(1)) \geq \chi(I_Y(1)) - h^2(I_Y(1)) = 1 ,$$

$$h^0(I_Y(d)) \geq \chi(I_Y(d)) = \binom{d+2}{3} + 2d + 1 .$$

Sei $h \in H^0(I_Y(1)) \setminus \{0\}$. Aus Dimensionsgründen gibt es ein $f \in H^0(I_Y(d)) \setminus \{0\}$, das nicht von der Form $f = gh$ ist. Die ebene Kurve $Y' = (f,h)_0$ enthält Y. Aus $\chi(\mathcal{O}_{Y'}(l)) = \chi(\mathcal{O}_Y(l))$ folgt daher $Y = Y'$.

PROPOSITION 2.7. Sei $Y \subset \mathbb{P}^3$ ein abgeschlossenes Unterschema mit

Hilbertpolynom (18), $d \geq 3$ und $p_a = \frac{1}{2}(d-2)(d-3)$. Dann gilt entweder

i) Die 1-dimensionale Komponente Y' von Y ist eben vom Grad d und Y besitzt $\bar{s}_Y = d-2$ 0-dimensionale assoziierte Punkte oder

ii) Y ist Macauleysch. Ist $d \geq 5$, so enthält Y eine ebene Kurve C vom Grad $d-1$, eine Gerade L, und es besteht eine exakte Sequenz

$$(26) \quad 0 \to \mathcal{O}_L(-1) \to \mathcal{O}_Y \to \mathcal{O}_C \to 0.$$

Ist $d = 4$, so besitzt I_Y eine Auflösung

$$(27a) \quad 0 \to \mathcal{O}(-4) \oplus \mathcal{O}(-3) \to \mathcal{O}(-3) \oplus \mathcal{O}(-2)^{\oplus 2} \to I_Y \to 0$$

und ist im allgemeinen vollständiger Durchschnitt zweier Quadriken. Für $d = 3$ ist Y eine Determinantenvarietät und es existiert eine exakte Sequenz

$$(27) \quad 0 \to \mathcal{O}^{\oplus 2} \to \mathcal{O}(1)^{\oplus 3} \to I_Y(3) \to 0.$$

Beweis. Sei $Y' \subset Y$ die 1-dimensionale Komponente. Es gilt $\deg(Y') = \deg(Y) = d$ und $p_a(Y') = p_a + \bar{s}$. Ist also $\bar{s} \neq 0$, so muß nach Theorem 2.2 gelten $p_a(Y') = \frac{1}{2}(d-1)(d-2)$ und $\bar{s}_Y = d-2$. Aus Proposition 2.6 folgt dann i). Ist $\bar{s}_Y = 0$, Y also eine Macauley Kurve, so hat I_Y notwendig das Spektrum $k_Y = (-d+1, \ldots, -4, -3, -2, -2, -1)$. Sei zunächst $d > 4$. Nach [12] gibt es dann eine instabile Ebene H_o und einen Reduktionsschritt

$$(28) \quad 0 \to I_{\tilde{Y}}(-1) \to I_Y \to I_{Z,H_o}(-d+1) \to 0.$$

Hier ist $\tilde{Y}$ eine Macauley Kurve. $Z \subset H_o$ 0-dimensional. Es gilt $\deg(\tilde{Y}) = 1$, $p_a(\tilde{Y}) = l(\mathcal{O}_Z)$, also — wegen $p_a \leq \frac{1}{2}(d-1)(d-2) - l(\mathcal{O}_Z) = 0$, d.h. $Z = \emptyset$. Y' ist also eine Gerade, $Y' = L$ und die Sequenz (26) folgt aus (28). Für $d = 3,4$ ergibt sich das Ergebnis aus der Spektralsequenz von Beilinson [10].

Bemerkung 2.8. Die in 2.7 ii) erwähnte Gerade L schneidet C im generischen Fall in genau einem Punkt. Durch spezialisieren erhält man eine ebene Kurve vom Grad d mit einer Geraden als Komponente und $d-2$ 0-dimensionalen assoziierten Punkten. Auf diese Weise hängen die beiden Komponenten des Hilbertschemas der 1-dimensionalen abgeschlossenen Unterschemata $Y \subset \mathbb{P}^3$ vom Grad $d \geq 5$ mit $p_a = \frac{1}{2}(d-2) \times (d-3)$ zusammen.

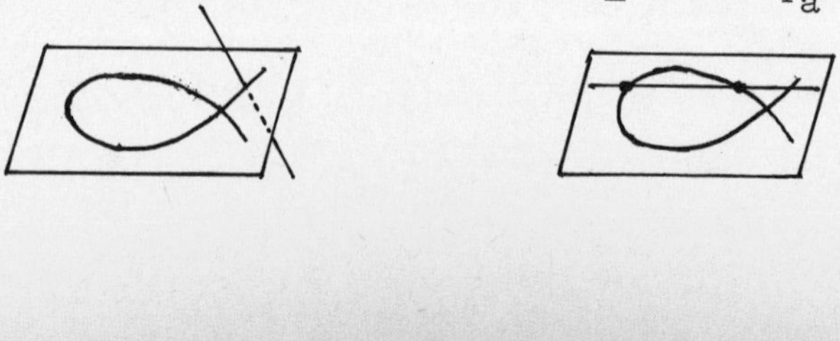

Bemerkung 2.9. Theorem 1.7 läßt sich nicht auf den Fall von 1-dimensionalen abgeschlossenen Unterschemata $Y \subset \mathbb{P}^3$ anwenden. Man kann also keine Lücken für das arithmetische Geschlecht von Macauley Kurven vom Grad d unterhalb von $\frac{1}{2}(d-2)(d-3)$ erwarten. Tatsächlich hat Sauer [13] gezeigt, daß es zu jedem Paar (d,g) ganzer Zahlen mit $d \geq 3$, $g \geq 0$, $g \leq \frac{1}{2}(d-2)(d-3)$ eine Macauley Kurve Y vom Grad d mit $p_a = g$ gibt. Man kann sogar erreichen, daß Y generisch ein lokal vollständiger Durchschnitt ist. Diese Kurven wurden in [13] mit Hilfe instabiler, reflexiver Garben F konstruiert: $0 \to \mathcal{O} \xrightarrow{s} F \to I_Y(c_1) \to 0$. Sauer definiert für solche Garben F ein "Spektrum". Diesem "Spektrum" entspricht bei uns das Spektrum von I_Y (Y ist hier durch F festgelegt).

Natürlich ist die Klassifikation der Macauley Kurven mit $p_a \leq \frac{1}{2}(d-2)(d-3)$ viel schweriger. Man kann allerdings leicht Kurven mit bestimmten Spektren beschreiben. Sei etwa $0 < q < d-1-[\frac{1}{2}d]$, I_Y die Idealgarbe einer Macauley Kurve vom Grad d mit Spektrum (vgl. 1.7):

$$\bar{k}^q = (-d+q, \ldots, -q-2, \underbrace{-q-1, -q-1, \ldots, -2, -2}_{\text{jeweils doppelt}}, -1).$$

Es gilt dann $p_a = \binom{d-q-1}{2} + \binom{q}{2}$, man kann einen Reduktionsschritt durchführen und erhält $0 \to I_{\tilde{Y}}(-1) \to I_Y \to \mathcal{O}_{H_0}(-d+q) \to 0$. Hier ist $\tilde{Y}$ eben vom Grad q, und man bekommt die Extension (vgl. [13]):

$$(29) \qquad 0 \to \mathcal{O}_{\tilde{Y}}(-1) \to \mathcal{O}_Y \to \mathcal{O}_C \to 0$$

mit einer weiteren ebenen Kurve C vom Grad $d-q$. Für die Idealgarbe I_Y erhält man die folgende Auflösung:

$$(30) \qquad 0 \to \mathcal{O}(-d+q-1) \oplus \mathcal{O}(-q-2) \to \mathcal{O}(-d+q) \oplus \mathcal{O}(-q-1) \oplus \mathcal{O}(-2) \to I_Y \to 0.$$

Als Spezialfall $(q=1)$ bekommt man 2.7 ii).

Als letztes beschreiben wir diejenigen Macauley Kurven, für die in (24) Gleichheit gilt.

THEOREM 2.10. Sei $Y \subset \mathbb{P}^3$ eine Macauley Kurve vom Grad $d \geq 4$ mit $p_a = \frac{1}{4}d^2 - d + 1 - \frac{1}{4}\varepsilon$ und $0 \leq e := \max\{l \mid h^1(\mathcal{O}_Y(l)) \neq 0\}$. Es gelte $h^1(\mathcal{O}_{Y \cap H}(e)) = 0$ für alle Ebenen $H \subset \mathbb{P}^3$. Dann folgt:

i) Ist $d \equiv 0\ (2)$, so ist Y ein vollständiger Durchschnitt einer Quadrik mit einer Fläche vom Grad $\frac{1}{2}d$.

ii) Ist $d \equiv 1\ (2)$, so existiert eine Gerade L, für die $Y \cup L$ ein ein vollständiger Durchschnitt einer Quadrik und einer Fläche vom Grad $\frac{1}{2}d + \frac{1}{2}$ ist.

Beweis. Es gilt $\mathrm{Hom}(I_Y, \mathcal{O}_H(-e-3)) \cong H^1(\mathcal{O}_{Y\cap H}(e)) = 0$ für alle Ebenen H. Daher hat Y folgendes Spektrum:

$$k_Y = \begin{cases} (-t,-t,\dots,-2,-2,-1) & \text{für } d = 2t-1, \\ (-t-1,-t,-t,\dots,-2,-2,-1) & \text{für } d = 2t. \end{cases}$$

Damit bekommt man nach einer kleinen Rechnung

$$h^0(I_Y(2)) \geq \chi(I_Y(2)) - h^2(I_Y(2)) > 0,$$

$$h^0(I_Y(t)) \geq \chi(I_Y(t)) \geq h^0(\mathcal{O}_{\mathbb{P}^3}(t-2)) + 1 + \varepsilon.$$

Es gibt also eine Quadrik $Q = (q)_0$, die Y enthält sowie eine Fläche $X = (f)_0$ vom Grad t die Y enthält, aber Q nicht als Komponente hat ($q \nmid f$). Aus Dimensionsgründen schneiden Q und X sich eigentlich, $C = Q \cap X$ ist ein vollständiger Durchschnitt mit $\deg(C) = 2t$, $p_a(C) = t^2 - 2t + 1$. Es gilt natürlich $Y \subset C$. Aus dem Vergleich der Hilbertpolynome von Y und C folgen dann die Aussagen i) und ii).

Bemerkung 2.11. a) Die Bedingung $h^1(\mathcal{O}_{Y\cap H}(e)) = 0$ ist sicher dann erfüllt, wenn Y keine ebenen Komponenten besitzt. Für glatte, zusammenhängende, nicht entartete Kurven ist die Aussage von Satz 2.9 wohlbekannt [3].

b) Die Fälle $p_a = \frac{1}{4}d^2 - d + 1 - \frac{1}{4}\varepsilon$, $d = 1, 2, 3$, sind bereits in 2.6 und 2.7 behandelt worden.

3. Klassifikation einiger extremer Fälle

In diesem Abschnitt klassifizieren wir torsionsfreie Garben mit Spaltungstyp (1), für die in (10) und (11) Gleichheit gilt.

LEMMA 3.1. Sei F eine torsionsfreie Garbe vom Rang r auf $\mathbb{P}^3$ mit Spaltungstyp (1) und homologischer Dimension $\mathrm{hd}(F) \leq 1$. Dann gilt $h^0(F) \leq r - a$ und es besteht eine exakte Sequenz

$$(31) \qquad 0 \longrightarrow H^0(F) \otimes \mathcal{O} \xrightarrow{ev} F \longrightarrow F' \longrightarrow 0;$$

F' ist torsionsfrei mit $\mathrm{hd}(F') \leq 1$ und $d(F') \leq 1$.

Beweis. Sei $L \subset H \subset \mathbb{P}^3$ eine generische Flagge. Es gilt

$$h^0(F_L(-1)) = h^0(F_H(-1)) = h^0(F(-1)) = 0.$$

Die Restriktionsabbildung $\mathrm{res}_L\colon H^0(F) \longrightarrow H^0(F_L)$ ist deshalb injektiv, es folgt $h^0(F) \leq r - a$. Der Auswertungsmorphismus $ev\colon H^0(F) \otimes \mathcal{O} \longrightarrow F$

ist injektiv und nullstellenfrei auf L. Der Cokern F' ist also singularitätenfrei in Codimension 1, wegen $hd(F') \leq 1$ also torsionsfrei.

Wir untersuchen nun torsionsfreie Garben F mit Spaltungstyp (1) und $\chi(F(-2)) = \binom{m-a_1+1}{2} + a_1$. Zunächst betrachten wir den Fall $a = 0$.

PROPOSITION 3.2. *Sei F eine torsionsfreie Garbe auf $\mathbb{P}^3$ mit trivialem Spaltungstyp und $\chi(F(-2)) = \binom{c_2+1}{2}$. Dann gilt $F \cong \mathcal{O}^{\oplus r}$, oder $c_2 \geq 1$ und F ist durch eine Extension*

$$(32) \qquad 0 \longrightarrow \mathcal{O}^{\oplus(r-1)} \longrightarrow F \longrightarrow I_Y \longrightarrow 0$$

gegeben, wobei $Y \subset \mathbb{P}^3$ eine ebene Kurve vom Grad c_2 ist.

Beweis. Nach Proposition 1.1 ist $\bar{s}_F = 0$, also $hd(F) \leq 1$. Aus (2) folgt $m = c_2$, das Spektrum ist $k_F = (-m, \ldots, -1)$, also $h^2(F) = \binom{m-1}{2}$. Daraus bekommt man $h^0(F) \geq \chi(F) - h^2(F) = r - 1$, nach Lemma 3.1 also $h^0(F) \in \{r-1, r\}$. Ist $h^0(F) = r$, so folgt sofort $F = \mathcal{O}^{\oplus r}$. Ist $h^0(F) = r - 1$, so erhält man

$$0 \longrightarrow \mathcal{O}^{\oplus(r-1)} \longrightarrow F \longrightarrow I_Y \longrightarrow 0$$

mit einer Macauley Kurve Y vom Grad d mit $p_a = \frac{1}{2}(d-1)(d-2)$. Die Behauptung folgt nun aus Proposition 2.6.

Für $a \neq 0$ gilt

PROPOSITION 3.3. *Sei F eine torsionsfreie Garbe auf $\mathbb{P}^3$ mit Spaltungstyp* (1), $a \neq 0$ *und*

$$\chi(F(-2)) = \binom{c_2 - \binom{-c_1}{2} + 2}{2} - 1.$$

Dann gilt $F \cong \mathcal{O}^{\oplus(r-a)} \oplus \mathcal{O}(-1)^{\oplus a}$ oder $c_2 - \binom{-c_1}{2} \geq 1$ und F ist durch eine Extension

$$(33) \qquad 0 \longrightarrow \mathcal{O}^{\oplus(r-a)} + \mathcal{O}(-1)^{\oplus(a-1)} \longrightarrow F \longrightarrow I_Y(-1) \longrightarrow 0$$

gegeben, wobei Y eine ebene Kurve vom Grad $c_2 - \binom{-c_1}{2}$ ist.

Beweis. Es gilt nach (2) $m = c_2 - \binom{-c_1}{2}$. Wieder folgt $\bar{s}_F = 0$, das Spektrum von F liegt fest: $k_F = (-1-m, \ldots, -2)$. Man sieht, daß F genau $r - a$ Schnitte hat und bekommt $0 \longrightarrow \mathcal{O}^{\oplus(r-a)} \longrightarrow F \longrightarrow F'(-1) \longrightarrow 0$. Hier ist F' eine torsionsfreie Garbe, wie wir sie in Proposition 3.2 klassifiziert haben.

Wir betrachten nun die torsionsfreien Garben F, für die in (11) Gleichheit gilt. Sei m durch (2) definiert. Gilt $m \geq 3$, $\chi(F(-2)) = \binom{m-a_1}{2} + 2$, so ist $\bar{s}_F = 0$ oder $\bar{s}_F = m - 2$. Im zweiten Fall hat F

die Darstellung $0 \to F' \to F \to Q \to 0$, wobei F' eine der in 3.2 und 3.3 beschrieben Garben ist, während Q endlichen Träger der Länge $m - 2$ hat. Wir beschränken uns also auf Garben mit homologischer Dimension ≤ 1.

PROPOSITION 3.4. Sei F eine torsionsfreie Garbe auf $\mathbb{P}^3$ mit trivialem Spaltungstyp, $hd(F) \leq 1$, $c_2 \geq 3$ und $\chi(F(-2)) = \binom{c_2}{2} + 2$. Dann existiert eine Erweiterung

$$(34) \qquad 0 \to \mathcal{O}^{\oplus(r-1)} \to F \to I_Y \to 0$$

mit einer Macauley Kurve Y vom Grad $d = c_2$ mit $p_a = \frac{1}{2}(d-2)(d-3)$.

Beweis. Wie oben erhält man aus dem Spektrum $k_F = (-c_2+1, \ldots, -2, -2, -1)$: $h^0(F) = r-1$.

PROPOSITION 3.5. Sei F eine torsionsfreie Garbe auf $\mathbb{P}^3$ mit Spaltungstyp (1), $a \neq 0$, $hd(F) \leq 1$ und

$$\chi(F(-2)) = \binom{c_2 - \binom{-c_1}{2} + 1}{2} + 2.$$

Dann existiert eine Erweiterung

$$(35) \qquad 0 \to \mathcal{O}^{\oplus(r-a)} \oplus \mathcal{O}(-1)^{\oplus(a-1)} \to F \to I_Y(-1) \to 0$$

mit einer Macauley Kurve Y vom Grad $d = c_2 - \binom{-c_1}{2}$ und $p_a = \frac{1}{2}(d-2)(d-3)$.

Beweis. Das Spektrum ist $k_F = (-m, \ldots, -3, -3, -2)$ und es folgt $h^0(F) = r - a$.

Wir haben nun die Garben F mit $hd(F) \leq 1$ klassifiziert, für die in (10) und (11) Gleichheit besteht.

Mit ähnlichen Methoden läßt sich noch der Bereich darunter von $-|\bar{k}^2|$ bis $-|\bar{k}^1|$ behandeln. Es geht also jetzt um torsionsfreie Garben F mit $hd(F) \leq 1$, generischen Spaltungstyp (1) und

$$(36) \qquad \binom{m-a_1+1}{2} + a_1 - 2(m-3) < \chi(F(-2)) < \binom{m-a_1+1}{2} + a_1 - m + 2.$$

LEMMA 3.6. Sei F eine torsionsfreie Garbe vom Rang r auf $\mathbb{P}^3$ mit Spaltungstyp (1) und homologischer Dimension $hd(F) \leq 1$. Es gelte (36). Dann gilt für das Spektrum von F:

$$(37) \qquad k_F = (a_1 - m + 1, \ldots, a_1 - 2, a_1 - 1, a_1 - 1 + b)$$

mit einer ganzen Zahl $b \geq 0$. Es ist

$$(38) \qquad b = \binom{m-a_1}{2} + 1 - \chi(F(-2)).$$

B e w e i s. Es gilt (vgl. 1.8): $\binom{m-a_1+1}{2} + a_1 - 2(m-3) = -|\overline{k}^2|$, für alle möglichen Spektren k_F mit $k_1 = a_1 - m + 2$ gilt also $-|k_F| \leq -|\overline{k}^2|$. Das Spektrum von F muß also mit $k_1 = a_1 - m + 1$ anfangen. Da es zusammenhängend ist, folgt (37). Die Gleichung (38) für b ist klar.

Sei $S(r,a,m,b)$ die Menge der torsionsfreien Garben F vom Rang r auf $\mathbb{P}^3$ mit $hd(F) \leq 1$, Spaltungstyp (1): $a_F = (-1, 0; a, r-a)$, $0 \leq a < r$ (also $a = -c_1$) und Spektrum (37): $k_F = (a_1 - m + 1, \ldots, a_1 - 2, a_1 - 1, a_1 - 1 + b)$. Es gilt also $m = c_2 - \binom{a}{2}$ und $a_1 = 0$ für $c_1 = 0$, $a_1 = -1$ für $c_1 \neq 0$. Durch r, a, m und b sind die Chernklassen von F offensichtlich festgelegt. Wir haben gezeigt, daß alle Garben aus 3.6 zu $S(r, a, m, b)$ gehören. Für diese Garben gilt

(39) $\quad 0 \leq b \leq m - 6.$

LEMMA 3.7. Für eine Garbe $F \in S(r, a, m, b)$ gilt

(40) $$h^0(F) - h^1(F) = \begin{cases} r - a - b & \text{für } a \neq 0, \\ r - 2 - b & \text{für } a = 0. \end{cases}$$

B e w e i s. Ist $k_F = (k_1, \ldots, k_m)$ das Spektrum von F, so folgt

(41) $$h^0(F) - h^1(F) = \chi(F) - h^2(F) = r - a - \Sigma_{k_i \geq -1}(k_i + 2).$$

LEMMA 3.8. Sei $F \in S(r, a, m, b)$. Dann gilt entweder
i) $h^0(F) = r - a$, $a \neq 0$, und es gibt eine Extension

$$0 \to \mathcal{O}^{\oplus(r-a)} \to F \to F'(-1) \to 0$$

mit $F' \in S(a, 0, m, b)$, oder
ii) $h^0(F) < r - a$, und es existiert eine Extension

$$0 \to \mathcal{O}^{\oplus h^0(F)} \to F \to F' \to 0$$

mit $F' \in S(r - h^0(F), a, m, b)$ und $h^0(F') = 0$.

B e w e i s. Dies folgt sofort aus Lemma 3.1.

Wir brauchen also nur die Mengen $S_0(r,a,m,b) = \{F \in S(r,a,m,b) \mid h^0(F) = 0\}$ zu untersuchen. Ist $F \in S_0(r,a,m,b)$, so folgt aus (40): $r \leq a + b$ für $a \neq 0$ und $r \leq b + 2$ für $a = 0$. Im folgenden sei stets $m > 4$. Nach [12] existiert dann ein Reduktionsschritt

(42) $$0 \to F' \to F \to I_{Z,H_0}(1 - m + a_1) \to 0.$$

Hierbei ist F' torsionsfrei, $hd(F') \leq 1$, $h^0(F') = 0$; H_0 ist eine Ebene, $Z \subset H_0$ ein 0-dimensionales abgeschlossenes Unterschema der Länge $l_Z = \chi(\mathcal{O}_Z)$. Für den Spaltungstyp F' gibt es zwei Möglichkeiten:

$$\text{i) } a_{F'} = (-1, 0; a+1, r-a-1), \text{ ii) } a_{F'} = (-2, -1, 0; 1, a-1, r-a) \quad (a \neq 0).$$

Seien $c_i' = c_i(F')$ die Chernklassen von F', $k_{F'} = (k_1', \ldots, k_{m'}')$ das Spektrum von F'. Man erhält

$$c_1' = c_1 - 1 = -a', \; a' = a + 1; \quad c_2' = a_1 + 1 + \binom{a'}{2};$$

$$c_3' = 2 - 2b + 2l_Z - a'c_2' + 2\binom{a'+1}{3}.$$

Für die Länge m' von $k_{F'}$ gilt nach [12] $m' \leq a_1 + 1$.

Bemerkung 3.9. Es sind folgende vier Fälle zu betrachten:

1) $a = 0$, $r = 1$,
2) $a = 0$, $r > 1$,
3) $0 < a < r - 1$,
4) $a = r - 1$; $r \geq 2$.

Wir geben nur die Resultate an:

1) Hier ist $F' = I_L(-1)$, $L \subset \mathbb{P}^3$ eine Gerade, es gilt $l_Z = b + 1$, F ist die Idealgarbe einer Macauley Kurve Y vom Grad $d = c_2$. Man bekommt eine exakte Sequenz

$$0 \longrightarrow \mathcal{O}_L(b) \longrightarrow \mathcal{O}_Y \longrightarrow \mathcal{O}_C \longrightarrow 0$$

mit einer ebenen Kurve $C \subset H_0$ vom Grad $d - 1$. Für $b = 0$ können L und C disjunkt sein. Für $b > 0$ gilt $L \subset C$ und Y entsteht aus C durch Verdopplung längs L (vgl. [13]).

2) Man bekommt $l_Z \leq b$. Sei $B = b - l_Z$, also $k_{F'} = (B - 1)$.

i) Ist $B = 0$, so hat F' Rang $r = 2$ und die Auflösung

$$0 \longrightarrow \mathcal{O}(-2) \longrightarrow \mathcal{O}(-1)^{\oplus 3} \longrightarrow F' \longrightarrow 0.$$

ii) Ist $B = 1$, so gilt $r \leq 3$. Man bekommt $F' = \Omega^1(1)$ oder $F' = N \oplus \mathcal{O}(-1)$ mit einer Nullkorrelationsgarbe N, falls $r = 3$ ist. Ist $r = 2$, so erhält man $0 \longrightarrow I_L \longrightarrow F' \longrightarrow m_{x_0}(-1) \longrightarrow 0$, $x_0 \in \mathbb{P}^3$ ein abgeschlossener Punkt, $L \subset \mathbb{P}^3$ eine Gerade.

iii) Die Fälle $B \geq 2$ sind unübersichtlich.

3) Diese Fälle treten nicht auf. Es wäre nämlich $m' = 0$, also $k_{F'} = \emptyset$, was schnell zu einem Widerspruch führt. Für $0 < a < r - 2$ gilt also $S_0(r, a, m, b) = \emptyset$.

4) Sei jetzt $r \geq 2$, $a = r - 1$. Durch Normieren erhält man aus dem Reduktionsschritt (42):

(43) $\quad 0 \to F''(-1) \to F \to I_{Z,H_0}(-m) \to 0.$

Die Chernklassen c_i'' von F'' sind $c_1''=0$, $c_2''=0$, $c_3''=2(l_Z+1-b)$. Hat F'' trivialen Spaltungstyp, so folgt $F'' \cong \mathcal{O}^{\oplus r}$ Ist der Spaltungstyp nicht trivial, so folgt $k_{F''}=\emptyset$, $c_3''=0$, $l_Z=b-1$ und F' ist gegeben durch eine Extension $0 \to \mathrm{Coker}\,\Phi \to F'' \to \mathrm{Ker}\,\Psi \to 0$ mit einem Epimorphismus $\Psi\colon \mathcal{O}(-1)^{\oplus B} \to \mathcal{O}^{\oplus A}$ und einem Monomorphismus $\Phi\colon \mathcal{O}(-1)^{\oplus B} \to \mathcal{O}^{\oplus(A+r)}$. Hier sind A, B ganze Zahlen ≥ 0. Ist $A>0$, so gilt $A+2 \leq B \leq A+r-1$. Insbesondere ist für $r \leq 3$ nur $(A,B)=(0,0)$ oder $(A,B)=(0,1)$ möglich.

KOROLLAR 3.10. Es gibt keine torsionsfreien Garben F vom Rang r auf $\mathbb{P}^3$ mit $\mathrm{hd}(F) \leq 1$, $h^0(F)=0$, Spaltungstyp $a_F=(-1,0;a,r-a)$, $0<a<r-1$ und

$$\binom{m-a_1+1}{2}+a_1-2(m-3) < \chi(F(-2)) < \binom{m-a_1+1}{2}+a_1-m+2.$$

KOROLLAR 3.11. Es gibt keine torsionsfreien Garben F vom Rang r auf $\mathbb{P}^3$ mit $\mathrm{hd}(F) \leq 1$, $h^0(F)=0$, Spaltungstyp $a_F=(-1,0;r-1,1)$, $r \geq 2$ und $\binom{m-a_1+1}{2}+a_1-m+1=\chi(F(-2))$.

Bemerkung 3.12. Für reflexive Garben F gilt statt (39) die schärfere Ungleichung $b \leq 2-a_1$. Man bekommt so weitere Lücken (vgl. [2]) für stabile reflexive Garben.

Bemerkung 3.13. Folgende extreme Fälle von Garben können vollständig behandelt werden:

1. $a=0$, $b=0$, $r=2$, d.h.: $c_1=0$, $c_3=c_2^2-c_2+2$, $r=2$;
2. $a=0$, $b=1$, $r=3$, d.h.: $c_1=0$, $c_3=c_2^2-c_2$, $r=3$;
3. $0<a=r-1$, $b=1$, $r \geq 2$, d.h.: $c_1=1-r$, $r \geq 2$ und
$$c_3=c_2^2-(r^2-2r)c_2+\binom{r}{3}+\binom{r-1}{2}^2.$$

Dies wird im nächsten Abschnitt geschehen, wo wir die vollständigen Maruyamaschemata der stabilen Garben dieser Typen bestimmen.

4. Kompaktifizierung extremer Modulräume

In diesem Abschnitt sei k stets ein algebraisch abgeschlossener Körper der Charakteristik 0, F eine torsionsfreie Garbe auf $\mathbb{P}^3_k$. F ist (semi-)stabil, wenn für alle echten, kohärenten Untergarben $0 \neq G \subset F$ gilt

(44) $\quad \chi(G(l))/\mathrm{rg}(G) \underset{(\leq)}{<} \chi(F(l))/\mathrm{rg}(F)$ für $l \gg 0$.

Bemerkung 4.1. Ist Q eine kohärente Garbe auf $\mathbb{P}^3$ mit $m = \dim(\operatorname{supp} Q)$, so gilt: $\chi(F(l)) = a\binom{l+m}{m}$ + niedere Terme mit einer ganzen Zahl $a > 0$. F ist also genau dann (semi-)stabil, wenn für alle kohärenten Untergarben $G \subset F$ mit $0 < rg(G) < rg(F)$ und torsionsfreien Cokern F/G (44) gilt: Sei ${}^r\overline{M}_{\mathbb{P}^3}(c_1, c_2, c_3)$ das Projektive Modulschema [7] der S-Äquivalenzklassen semi-stabiler, torsionsfreier Garben F vom Rang r auf $\mathbb{P}^3$ mit den Chernklassen $c_i = c_i(F)$. Es gilt also

$$(45)\quad \chi(F(l)) = r\binom{l+3}{3} + c_1\binom{l+2}{2} - \left(c_2 - \binom{-c_1}{2}\right)(l+1) + \frac{1}{2}c_3 - \frac{1}{2}c_1c_2 - c_2 - \binom{-c_1}{3}.$$

THEOREM 4.2. Für $c_2 \geq 6$ ist das Modulschema ${}^2\overline{M}_{\mathbb{P}^3}(0, c_2, c_2^2 - c_2 + 2)$ irreduzibel, rational, glatt von der Dimension $c_2^2 + 2c_2 + 5$. ${}^2\overline{M}_{\mathbb{P}^3}(0, c_2, c_2^2 - c_2 + 2)$ ist isomorph zu einem projektiven Bündel über $\mathbb{P}^3 \times \mathbb{P}^3$ und besitzt eine universelle Familie.

Bemerkung 4.3. Ist $[F] \subset {}^2\overline{M}_{\mathbb{P}^3}(0, c_2, c_2^2 - c_2 + 2)$, so gilt

$$(46)\quad \chi(F(l))/rg(F) = \binom{l+3}{3} - \frac{1}{2}c_2(l+1) + \frac{1}{4}(c_2^2 - 3c_2 + 2).$$

Nach Maruyama [7] ist ${}^2\overline{M}_{\mathbb{P}^3}(0, c_2, c_2^2 - c_2 + 2)$ ein feiner Modulraum, falls

$$\delta(H) := ggT(2, c_2, \tfrac{1}{2}(c_2 - 1)(c_2 - 2)) = 1$$

ist. Es gilt

$$ggT(2, c_2, \tfrac{1}{2}(c_2 - 1)(c_2 - 2)) = \begin{cases} 2 & \text{für } c_2 \equiv 2\ (4), \\ 1 & \text{sonst.} \end{cases}$$

Wir sehen also, daß die Bedingung $\delta(H) = 1$ zwar hinreichend, aber nicht notwendig dafür ist, daß der entsprechende Modulraum fein ist.

Bemerkung 4.4. Die stabilen, reflexiven Garben vom Rang 2 auf $\mathbb{P}^3$ mit $c_1 = 0$, $c_3 = c_2^2 - c_2 + 2$ sind schon in [2] und [9] klassifiziert worden. ${}^2\overline{M}_{\mathbb{P}^3}(0, c_2, c_2^2 - c_2 + 2)$ ist die Kompaktifizierung dieser Modulschemata.

Beweis von Theorem 4.2. Aus $\chi(\mathcal{O}(l)) = \binom{l+3}{3}$ und (46) folgt $h^0(F) = 0$, falls F semi-stabil ist. Ferner ist $hd(F) \leq 1$, denn wäre $\bar{s}_F \geq 0$, so wäre

$$\chi(F(-2)) + \bar{s}_F \geq \binom{c_2+1}{2} - c_2 + 2,$$

also $k_F = (-c_2, \ldots, -1)$ oder $k_F = (-c_2 + 1, \ldots, -2, -2, -1)$. Aus (41) würde $h^0(F) > 0$ folgen. F hat trivialen Spaltungstyp [10], es besteht nach § 3 die exakte Sequenz

(42) $0 \longrightarrow F' \longrightarrow F \longrightarrow \mathcal{O}_{H_0}(1-c_2) \longrightarrow 0,$

wobei F' die Auflösung

(47) $0 \longrightarrow \mathcal{O}(-2) \longrightarrow \mathcal{O}(-1)^{\oplus 3} \longrightarrow F' \longrightarrow 0$

hat. F' ist reflexiv oder isomorph zu $I_L \oplus \mathcal{O}(-1)$ für eine Gerade L. Wäre $F' \cong I_L \oplus \mathcal{O}(-1)$, so wäre $I_L \subset F$ ein Widerspruch zu Semi-Stabilität von F. Also ist F' reflexiv Sei nun umgekehrt eine nicht-triviale Extension (42) mit einer reflexiven Garbe F' mit Auflösung (47) gegeben. Wir zeigen, daß F dann torsionsfrei und stabil ist. Sei $T = \mathrm{Tors}(F) \neq 0$ die Torsionsuntergarbe von F; $\tilde{F} = F/T$. Wir bekommen ein kommutatives Diagramm

$$\begin{array}{ccccccccc}
 & & & & 0 & & 0 & & \\
 & & & & \downarrow & & \downarrow & & \\
 & & & & T & = & I_{Z,H_0}(1-c_2-d) & & \\
 & & & & \downarrow & & \downarrow & & \\
0 & \longrightarrow & F' & \longrightarrow & F & \longrightarrow & \mathcal{O}_{H_0}(1-c_2) & \longrightarrow & 0 \\
 & & \| & & \downarrow & & \downarrow & & \\
0 & \longrightarrow & F' & \longrightarrow & \tilde{F} & \longrightarrow & \mathcal{O}_W(1-c_2) & \longrightarrow & 0 \\
 & & & & \downarrow & & \downarrow & & \\
 & & & & 0 & & 0 & &
\end{array}$$

$Z \subset H_0$ 0-dimensional, $W \subset H_0$ eine Kurve vom Grad d mit 0-dimensionaler Komponente Z. $\tilde{F}$ hat den Spaltungstyp $(-1, 0)$, es folgt

$$0 \leq c_2(\tilde{F}) = c_2(F') - d = 1 - d,$$

also $d \in \{0,1\}$. Wäre $d = 0$, so $Z = \emptyset$, $T = \mathcal{O}_H(1-c_2)$ und die Extension (42) würde spalten. Wäre $d = 1$, also $c_2(\tilde{F}) = 0$, so bekäme man $k_{\tilde{F}} = \emptyset$, so bekäme man $k_{\tilde{F}} = \emptyset$, also $h^1(\mathcal{O}_W(1-c_2)) = 0$. Es würde folgen $h^1(\mathcal{O}_L(1-c_2)) = 0$, wobei L die 1-dimensionale Komponente von W ist. Das ist für $c_2 > 3$ aber unmöglich. Daher ist F torsionsfrei.

Um die Stabilität von F zu beweisen, braucht man nur Idealgarben $I_Y \subset F$ mit torsionsfreien Quotienten $I_{Y'}$ zu untersuchen (Y 2-codimensional). Man bekommt das folgende Diagramm:

$$\begin{array}{ccccccccc}
 & & 0 & & 0 & & 0 & & \\
 & & \downarrow & & \downarrow & & \downarrow & & \\
0 & \longrightarrow & I_X(-1) & \longrightarrow & I_Y & \longrightarrow & I_{Z,H_0}(1-c_2-\tilde{d}) & \longrightarrow & 0 \\
 & & \downarrow & & \downarrow & & \downarrow & & \\
0 & \longrightarrow & F' & \longrightarrow & F & \longrightarrow & \mathcal{O}_{H_0}(1-c_2) & \longrightarrow & 0 \\
 & & \downarrow & & \downarrow & & \downarrow & & \\
0 & \longrightarrow & I_{Y'} & \longrightarrow & I_{Y'} & \longrightarrow & \mathcal{O}_W(1-c_2) & \longrightarrow & 0 \\
 & & \downarrow & & \downarrow & & \downarrow & & \\
 & & 0 & & 0 & & 0 & &
\end{array} \qquad (48)$$

Da F' reflexiv ist, gilt $I_X(-1) = \mathcal{O}(-1)$, nach Definition von F' also $I_{X'} = I_L$ für eine Gerade $L \subset \mathbb{P}^3$. Sei $d = c_2(I_Y)$, $d' = c_2(I_{Y'})$. Es gilt

$$c_2 = d + d'.$$

Die untere Zeile von (48) liefert $d' = 1 - \tilde{d}$, also $\tilde{d} \leq 1$. Ist $\tilde{d} = 1$, so $d' = 0$ und $k_{I_{Y'}} = \emptyset$, also $h^1(I_{Y'}(-1)) = 0$. Es folgt $h^1(\mathcal{O}_W(-c_2)) = 0$; da W eine Gerade enthält, müßte $c_2 \leq 1$ sein. Ist $\tilde{d} = 0$, so ist $W = Z$ 0-dimensional. Es folgt $d = c_2 - 1$ und für $l \gg 0$

$$\begin{aligned} \chi(I_Y(l)) &= \binom{l+3}{3} - (c_2 - 1)(l+1) + \chi(I_Y(-1)) \\ &< \binom{l+3}{3} - \frac{1}{2} c_2 (l+1) + \frac{1}{2}\chi(F(-1)) = \frac{1}{2}\chi(F(l)). \end{aligned}$$

F ist also stabil. Wir haben bisher gezeigt, daß F genau dann ein Element in ${}^2\overline{M}_{\mathbb{P}^3}(0, c_2, c_2^2 - c_2 + 2)$ definiert, wenn es eine Extension

$$0 \longrightarrow F' \longrightarrow F \longrightarrow \mathcal{O}_{H_0}(1 - c_2) \longrightarrow 0$$

mit einer – eindeutig bestimmten – Ebene $H_0 \subset \mathbb{P}^3$ und einer reflexiven Garbe F' mit Auflösung (47) gibt. Das Modulschema ${}^2\overline{M}_{\mathbb{P}^3}(-1, 1, 1)$ der Garben vom Typ F' ist vollständig und wird durch die Zuordnung $[F'] \longmapsto \operatorname{sing}(F')$ isomorph zu $\mathbb{P}^3$. Wir setzen nun

$$\overline{M} = {}^2\overline{M}_{\mathbb{P}^3}(0, c_2, c_2^2 - c_2 + 2), \quad \overline{M}' = {}^2\overline{M}_{\mathbb{P}^3}(-1, 1, 1).$$

Sei $\underline{F}'$ die universelle Garbe auf $\mathbb{P}^3 \times \overline{M}'$, $\underline{H} \subset \mathbb{P}^3 \times \mathbb{P}^{3\vee}$ die universelle Ebene, $\underline{H} = \{(x, H) \in \mathbb{P}^3 \times \mathbb{P}^{3\vee} \mid x \in H\}$. Wir haben die Projektionen

$$\begin{array}{ccccc} \mathbb{P}^3 & \xleftarrow{\;p_1\;} & \mathbb{P}^3 \times \overline{M}' \times \mathbb{P}^{3\vee} & & \\ & \swarrow^{p_{12}} & \downarrow^{p_{23}} & \searrow^{p_{13}} & \\ \mathbb{P}^3 \times \overline{M}' & & \overline{M}' \times \mathbb{P}^{3\vee} & & \mathbb{P}^3 \times \mathbb{P}^{3\vee} \end{array}$$

Wir setzen

$$\underline{E} := \underline{\operatorname{Ext}}^1_{p_{23}}(p_{13}^* \mathcal{O}_{\underline{H}} \otimes p_1^* \mathcal{O}_{\mathbb{P}^3}(1 - c_2), p_{12}^* \underline{F}').$$

$\underline{E}$ ist lokal frei vom Rang $c_2^2 + 2c_2$ auf $\overline{M}' \times \mathbb{P}^{3\vee}$ und verträglich mit Basiswechsel. Sei $X = \mathbb{P}(\underline{E}^\vee)$, $p\colon X \longrightarrow \overline{M}' \times \mathbb{P}^{3\vee}$ die Bündelprojektion. Der universelle Quotient $\mathcal{O}_{\mathbb{P}(\underline{E}^\vee)}(1)$ von $p^*\underline{E}^\vee$ definiert einen Schnitt $s \in H^0(p^*\underline{E} \otimes \mathcal{O}_{\mathbb{P}(\underline{E}^\vee)}(1))$ ohne Nullstellen. Es gilt aber

$$\begin{aligned} H^0(p^*\underline{E} \otimes \mathcal{O}_{\mathbb{P}(\underline{E}^\vee)}(1)) \cong \operatorname{Ext}^1((\mathrm{id}_{\mathbb{P}^3} \times p)^*(p_{13}^* \mathcal{O}_{\underline{H}} \otimes p_1^* \mathcal{O}_{\mathbb{P}^3}(1 - c_2), \\ (\mathrm{id}_{\mathbb{P}^3} \times p)^* p_{12}^* \underline{F}') \otimes q^* \mathcal{O}_{\mathbb{P}(\underline{E}^\vee)}(1), \end{aligned}$$

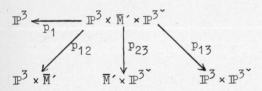

wobei $q: \mathbb{P}^3 \times X \to X$ die Projektion bezeichnet. Der Schnitt s definiert eine universelle Extension über $\mathbb{P}^3 \times X$:

$$0 \to \tilde{\underline{F}}' \otimes q^* \mathcal{O}_{\mathbb{P}(E^\vee)}(1) \to \underline{F} \to \mathcal{O}_{\tilde{\underline{H}}} \otimes \mathcal{O}_{\mathbb{P}^3_X}(1-c_2) \to 0.$$

Hierbei ist $\tilde{\underline{F}}' = (id_{\mathbb{P}^3} \times p)^* p_{12}^* \underline{F}'$ und $\tilde{\underline{H}} = (id_{\mathbb{P}^3} \times p)^{-1} p_{13}^{-1} \underline{H} \subset \mathbb{P}^3_X$. Mit Standardargumenten folgt $X \cong {}^2\overline{M}_{\mathbb{P}^3}(0, c_2, c_2^2 - c_2 + 2)$. Als nächstes bestimmen wir den Modulraum der semi-stabilen torsionsfreien Garben vom Rang 3 auf $\mathbb{P}^3$ mit $c_1 = 0$ und $c_3 = c_2^2 - c_2$. Der offene Teil der Bündel ist schon in [14] untersucht worden.

<u>THEOREM 4.5</u>. <u>Für $c_2 \geq 6$ ist das Modulschema ${}^3\overline{M}_{\mathbb{P}^3}(0, c_2, c_2^2 - c_2)$ irreduzibel, glatt und rational von der Dimension $\frac{1}{2}c_2(3c_2 + 7) + 3$. ${}^3\overline{M}_{\mathbb{P}^3}(0, c_2, c_2^2 - c_2)$ ist isomorph zu einem projektiven Bündel über $\mathbb{P}^3$ und besitzt eine universelle Familie</u>.

<u>Beweis</u>. Es ist $\chi(F(1)) = 3\binom{1+3}{2} - c_2(1+1) - \binom{c_2}{2} - c_2$. Ist F semi-stabil, so folgt $h^0(F) = 0$. Es gibt eine Ebene $H' \subset \mathbb{P}^3$ mit $h^0(F_{H'}) \neq 0$, da nach [12] sonst $c_3 \leq c_2^2 - 3c_2 + 6$ gelten müßte. Anderseits ist – Grauert-Mülich – $h^0(F_H(-1)) = 0$ für generische Ebenen. Ein Schnitt $s_{H'} \in H^0(F_{H'})$ liefert eine exakte Sequenz $0 \to \mathcal{O}_{H'} \to F_{H'} \to Q_{H'} \to 0$ mit einer torsionsfreien, semistabilen Garbe $Q_{H'}$. $Q_{H'}$ und damit auch $F_{H'}$ haben also trivialen Spaltungstyp. Ferner gilt $hd(F) \leq 1$, denn aus $\bar{s}_F > 0$ und $-|k_F| \geq \binom{c_2}{2} + 1$ würde folgen, daß k_F eines der folgenden Spektren wäre:

$$k_F = (-c_2+1, \ldots, -1, -1),\ (-c_2+1, \ldots, -2, -1),$$
$$\text{oder } (-c_2, \ldots, -2, -1).$$

In allen drei Fälle würde $h^0(F) > 0$ gelten. Wir können also die Ergebnisse von Abschnitt 3 anwenden und bekommen eine Extension $0 \to F' \to F \to \mathcal{O}_{H_0}(1 - c_2) \to 0$ mit $F' = \Omega^1_{\mathbb{P}^3}(1)$ oder $F' = N \oplus \mathcal{O}(-1)$. Der Fall $F' = N \oplus \mathcal{O}(-1)$ kommt nicht vor, weil F semi-stabil ist. Wir haben also stets die Darstellung

$$(49) \qquad 0 \to \Omega^1_{\mathbb{P}^3}(1) \to F \to \mathcal{O}_{H_0}(1 - c_2) \to 0.$$

Sei umgekehrt eine nicht-triviale Extension (49) gegeben. Ähnlich wie in 4.2 sieht man, daß F dann torsionsfrei ist. Wir zeigen die Semistabilität: Sei dazu $G \subset F$ eine Untergarbe mit $0 < rg\, G < 3$ und torsionsfreiem Quotienten F/G. Wir setzen $G' = G \cap \Omega^1_{\mathbb{P}^3}(1)$ und erhalten folgendes Diagramm:

$$\begin{array}{ccccccccc}
 & & 0 & & 0 & & 0 & & \\
 & & \downarrow & & \downarrow & & \downarrow & & \\
0 & \longrightarrow & G' & \longrightarrow & G & \longrightarrow & I_{Z,H_0}(1-c_2-\tilde{d}) & \to & 0 \\
 & & \downarrow & & \downarrow & & \downarrow & & \\
0 & \longrightarrow & \Omega^1_{\mathbb{P}^3}(1) & \longrightarrow & F & \longrightarrow & \mathcal{O}_{H_0}(1-c_2) & \longrightarrow & 0 \\
 & & \downarrow & & \downarrow & & \downarrow & & \\
0 & \longrightarrow & Q' & \longrightarrow & Q & \longrightarrow & \mathcal{O}_W(1-c_2) & \longrightarrow & 0 \\
 & & \downarrow & & \downarrow & & \downarrow & & \\
 & & 0 & & 0 & & 0 & &
\end{array}$$

Wir brauchen nur den Fall $c_1(G) = 0$ zu betrachten. Dann ist $I_{Z,H_0}(1 - c_2 - \tilde{d}) \neq 0$ und $c_1(G') = -1$. Weil Q' torsionsfrei ist, ist G' reflexiv. Ist $\operatorname{rg}(G') = 1$, so $G' = \mathcal{O}(-1)$ und Q' eine Nullkorelationsgarbe. Ist $\operatorname{rg}(G')=2$, so ist $Q' = I_{Y'}$ eine Idealgarbe eines abgeschlossenen Unterschemas Y' mit $\dim Y' \leq 1$. In jedem Fall hat Q' trivialen Spaltungstyp, daher auch Q. Es folgt $c_2(Q') \geq 0$, $c_2(Q) \geq 0$. Anderseits ist

$$c_2(Q) = c_2(Q') - \tilde{d}.$$

Aus $1 = c_2(\Omega^1_{\mathbb{P}^3}(1)) = c_2(Q') + c_2(G')$ und $c_2(G') \geq 0$ folgt $c_2(Q') \leq 1$, d.h.: $0 \leq c_2(Q) \leq 1$ und deshalb $-c_2(G)/\operatorname{rg} G < -\frac{1}{3}c_2$. F ist also stabil. Wir betrachten wieder die universelle Ebene $\underline{H} = \{(x,H) \in \mathbb{P}^3 \times \mathbb{P}^{3^\vee} \mid x \in H\}$ und setzen

$$\underline{E} := \underline{\operatorname{Ext}}^1_{p_2}(\mathcal{O}_{\underline{H}} \otimes p_1^* \mathcal{O}_{\mathbb{P}^3}(1-c_2),\ p_1^* \Omega^1_{\mathbb{P}^3}(1)),$$

wobei p_i, $i = 1, 2$, die Projektionen auf den i-ten Faktor von $\mathbb{P}^3 \times \mathbb{P}^{3^\vee}$ sind. Wie in [14] sieht man, daß $X = \mathbb{P}(\underline{E}^\vee)$ isomorph zu ${}^3\overline{M}_{\mathbb{P}^3}(0, c_2, c_2^2 - c_2)$ ist und eine universelle Familie besitzt.

Als letztes untersuchen wir die semi-stabilen, torsionsfreien Garben F vom Rang $r \geq 2$ mit $c_1 = 1 - r$, generischen Spaltungstyp (1) und $c_3 = c_2^2 - (r^2 - 2r)c_2 + \binom{r}{3} + \binom{r-1}{2}^2$. Für eine Garbe F mit diesen Invarianten sind die Begriffe semi-stabil, μ-semi-stabil und μ-stabil äquivalent. Sei $c_2 - \binom{r-1}{2} \geq 6$ und

$$\overline{M}_0 \subset {}^r\overline{M}_{\mathbb{P}^3}(1 - r, c_2, c_2^2 - (r^2 - 2r)c_2 + \binom{r}{3} + \binom{r-1}{2}^2)$$

der offene Teil der Garben F mit generischen Spaltungstyp $a_F = (-1, \ldots, -1, 0)$. Für $r = 2, 3$ ist dies stets erfüllt, $\overline{M}_0$ ist dann das ganze Modulschema.

<u>THEOREM 4.6</u>. <u>Für</u> $c_2 - \binom{r-1}{2} \geq 6$ <u>ist</u> $\overline{M}_0$ <u>eine Zusammenhangskomponente von</u> ${}^r\overline{M}_{\mathbb{P}^3}(1 - r, c_2, c_2^2 - (r^2 - 2r)c_2 + \binom{r}{3} + \binom{r-1}{2}^2)$. $\overline{M}_0$ <u>ist ein Grassmannbündel über</u> $\mathbb{P}^3$.

B e w e i s. Für Garben F mit $[F] \subset \overline{M}_0$ gilt $\mathrm{hd}(F) \leq 1$. Man bekommt – nach Abschnitt 3 – eine Darstellung $0 \to F''(-1) \to F \to \mathcal{O}_{H_0}(-m) \to 0$ mit $m = c_2 - \binom{r-1}{2}$. Hier ist F'' durch eine Extension $0 \to F_2 \to F'' \to F_1 \to 0$ gegeben, wobei F_i als Kern und Cokern trivialer Bündel definiert sind:

$$0 \to F_1 \to \mathcal{O}(-1)^{\oplus B} \to \mathcal{O}^{\oplus A} \to 0, \quad 0 \to \mathcal{O}(-1)^{\oplus B} \to \mathcal{O}^{\oplus (A+r)} \to F_2 \to 0.$$

Da F μ-stabil ist, folgt aus $F_2(-1) \subset F$:

$$-1 + B/(A - B + r) = \mu(F_2(-1)) \leq \mu(F) = -1 + 1/r,$$

also $rB \leq A - B + r$. Dies ist nur für $A = B = 0$ möglich, wir erhalten $F'' = \mathcal{O}^{\oplus r}$, jedes F mit $[F] \in \overline{M}_0$ ist also von der Form

(50) $\quad 0 \to \mathcal{O}(-1)^{\oplus r} \to F \to \mathcal{O}_{H_0}(-m) \to 0.$

Sei umgekehrt F durch eine nicht-triviale Extension (50) gegeben. Offensichtlich hat F dann generischen Spaltungstyp. Wie oben sieht man, daß F auch torsionsfrei ist. Wir zeigen, daß F genau dann semistabil ist, wenn F nicht von der Form

(51) $\quad F \cong \mathcal{O}(-1)^{\oplus s} \oplus G, \ 1 \leq s < r,$

ist. Ist F semi-stabil, so gibt es kein G mit $F \cong \mathcal{O}(-1)^{\oplus s} \oplus G$, da sonst $\mu(G) > \mu(F)$ wäre. Sei F nun nicht μ-stabil.

Es gibt dann eine torsionsfreie Garbe $G \subset F$ vom Rang s mit torsionsfreiem Quotienten $Q = F/G$ und $c_1(G) > s(1/r - 1)$. Wir betrachten wieder das Diagramm

$$\begin{array}{ccccccccc}
 & & 0 & & 0 & & 0 & & \\
 & & \downarrow & & \downarrow & & \downarrow & & \\
0 & \to & G' & \to & G & \to & I_{Z,H_0}(-m-\tilde{d}) & \to & 0 \\
 & & \downarrow & & \downarrow & & \downarrow & & \\
0 & \to & \mathcal{O}(-1)^{\oplus r} & \to & F & \to & \mathcal{O}_{H_0}(-m) & \to & 0 \\
 & & \downarrow & & \downarrow & & \downarrow & & \\
0 & \to & Q' & \to & Q & \to & Q_W(-m) & \to & 0 \\
 & & \downarrow & & \downarrow & & \downarrow & & \\
 & & 0 & & 0 & & 0 & &
\end{array}$$

Es ist $I_{Z,H_0}(-m-\tilde{d}) \neq 0$, also $c_1(G) = c_1(G') + 1$. Da $\mathcal{O}(-1)^{\oplus r}$ μ-semistabil ist, gilt $c_1(G') \leq -s$. Insgesamt folgt $s/r - s < c_1(G) \leq 1 - s$. Das ist nur für $c_1(G) = 1 - s$, also $c_1(Q') = -(r - s)$ möglich. $Q'(1)$ ist dann global erzeugt mit $c_1(Q'(1)) = 0$, deshalb trivial [10]. Die untere Zeile spaltet wegen $\dim W \leq 1$, da Q torsionsfrei ist, muß $\mathcal{O}_W(-m)$

$=0$ sein, es folgt $Q = \mathcal{O}(-1)^{\oplus(r-s)}$ und die Sequenz $0 \to G \to F \to \mathcal{O}(-1)^{\oplus(r-s)} \to 0$ spaltet. Man kann die Bedingung $F \not\cong G + \mathcal{O}(-1)^{\oplus s}$ auch so ausdrücken: Eine nicht triviale Extension ε: $0 \to \mathcal{O}(-1)^{\oplus r} \to F \to \mathcal{O}_{H_0}(-m) \to 0$ definiert genau dann eine semi-stabile Garbe F, wenn die duale Abbildung $\mathcal{O}^{\oplus r} \xrightarrow{\varepsilon^*(1)} \mathrm{Ext}^1(\mathcal{O}_{H_0}(1-m), \mathcal{O}) = \mathcal{O}_{H_0}(m)$ einen Monomorphismus $H^0(\varepsilon^*(1))$: $H^0(\mathcal{O}^{\oplus r}) \to H^0(\mathcal{O}_{H_0}(m))$ induziert. Wie in [11] folgt daraus, daß $\overline{M}_0$ isomorph zum Grassmannbündel p: $X \to \mathbb{P}^3$ der "r-dimensionalen Untervektorräume in $H^0(\mathcal{O}_{H_0}(m))$" ist.

Insbesondere ist $\overline{M}_0 = \emptyset$ für $r > \binom{c_2 - \binom{r-1}{2} + 2}{2}$. Es ist naheliegend, zu fragen, ob die Inklusion $\overline{M}_0 \subset {}^r\overline{M}_{\mathbb{P}^3}(1-r, c_2, c_2^2 - (r^2-2r)c_2 + \binom{r}{3} + \binom{r-1}{2}^2)$ echt ist.

Literatur

[1] BARTH, W. und G. ELENCWAJG: Concernant la cohomologie des fibrés algébriques stables sur $\mathbb{P}_n(\mathbb{C})$. LNM 683, Springer (1978), 1-24.

[2] CHANG, M.-Ch.: Stable rank 2 reflexive sheaves on $\mathbb{P}^3$ with large c_3. Crelles J. 343 (1983), 99-107.

[3] GRUSON, L. und C. PESKINE: Genre des courbes de l'espace projectif. Algebraic Geometry. LNM 687, Springer (1978).

[4] HARTSHORNE, R.: Algebraic Geometry. Graduate Texts in Math. 52, Springer (1977).

[5] ——: Stable reflexive sheaves. Math. Ann. 254 (1980), 121-176.

[6] ——: Stable reflexive sheaves II. Invent. Math. 66 (1982), 165-190.

[7] MARUYAMA, M.: Moduli of stable sheaves II. J. Math. Kyoto Univ. 18 (1978), 557-614.

[8] OKONEK, C.: Reflexive Garben auf $\mathbb{P}^4$. Math. Ann. 260 (1982), 211-237.

[9] ——: Moduli extremer reflexiver Garben auf $\mathbb{P}^n$. Crelles J. 338 (1983), 138-194.

[10] ——, M. SCHNEIDER und H. SPINDLER: Vector bundles on complex projective spaces. Birkhäuser (1980).

[11] —— und H. SPINDLER: Reflexive Garben vom Rang $r > 2$ auf $\mathbb{P}^n$. Crelles J. 344 (1983), 38-64.

[12] —— und ——: Das Spektrum torsionsfreier Garben I. Manuscripta Math. 47, 187-228 (1984).

[13] SAUER, T.: Nonstable reflexive sheaves on $\mathbb{P}^3$. Trans. Amer. Math. Soc. 281 (1984), 633-655.

[14] SPINDLER, H.: Die Modulräume stabiler 3-Bündel auf $\mathbb{P}^3$ mit den Chernklassen $c_1 = 0$, $c_3 = c_2^2 - c_2$. Math. Ann. 256 (1981), 133-143.

Mathematisches Institut der
Georg-August Universität
Bunsenstraße 3-5, D-3400 Göttingen, BRD

QUASI-REGULAR BOUNDARY AND STOKES' FORMULA FOR A SUB-ANALYTIC LEAF

Wiesław Pawłucki (Kraków)

Contents

page

Summary. The aim of this paper is to prove some differential properties of sub-analytic leaves related to the conditions of Whitney, and to derive Stokes' formula for sub-analytic leaves.

Introduction

The aim of this paper is to prove some differential properties of sub-analytic leaves and, as an application, to get Stokes' formula for sub-analytic leaves.

Let X be a real n-dimensional vector space. A subset $A \subset X$ is called semi-analytic if for any $a \in X$ there exist a neighbourhood U of the point a in X and analytic functions g_{ij}, f_i, $i = 1,\ldots,r$; $j = 1,\ldots,s$, defined on U, such that

$$A \cap U = \bigcup_{i=1}^{r} \Big(\bigcap_{j=1}^{s} \{g_{ij} > 0\} \cap \{f_i = 0\} \Big).$$

A bounded subset $E \subset X$ is called sub-analytic if there exist a real finite dimensional vector space Y and a bounded semi-analytic set $A \subset X \times Y$ such that $E = p(A)$, where $p : X \times Y \to X$ is the natural projection. A subset $E \subset X$ is called sub-analytic if for any $a \in X$ the germ E_a is the germ of a bounded sub-analytic set in X.

The class of sub-analytic sets in X is closed with respect to the union and intersection of locally finite families and to the difference of sets. This class has the following topological properties. The closure and the interior of a sub-analytic set are sub-analytic. The family of connected components of a sub-analytic set is locally finite and these components are also sub-analytic. If E is a sub-analytic set in X, then the set of all regular points of E (i.e. such points $a \in E$ that the germ E_a is the germ of an analytic submanifold) is sub-analytic.

For a sub-analytic set $E \subset X$ we define the <u>dimension</u> of E as $\max\{\dim \mathfrak{S} |$ where $\mathfrak{S}$ is an analytic submanifold, $\mathfrak{S} \subset E\}$. The dimension has the following properties:

1) If $X = U \oplus V$ and $p : X \to U$ is the natural projection, then $\dim p(E) \leq \dim E$.

2) $\dim \overline{E} = \dim E$.

3) $\dim E < \dim X$ if and only if E is nowhere dense in X.

4) If $E \neq \emptyset$ and $E' = \{x \in E | x$ is a regular point of dimension equal to $\dim E\}$, then $\dim(E \setminus E') < \dim E$.

5) If $E \neq \emptyset$, then $\dim(\overline{E} \setminus E) < \dim E$.

Let X and Y be two vector spaces (from now on all the vector spaces considered will be real and finite dimensional). A mapping $f : D \to Y$, $D \subset X$, is called <u>sub-analytic</u> if f, considered as a subset of $X \times Y$, is sub-analytic. If K, L, and M are bounded sub-analytic sets in X, Y, and Z, respectively, and if $f : K \to L$ and $g : L \to M$ are sub-analytic mappings, then $g \circ f$ is a sub-analytic mapping. The sum, product, and quotient of any two real bounded sub-analytic functions are sub-analytic.

A set $\Gamma \subset X$ is called a <u>semi-(sub)-analytic leaf</u> if it is a semi-(sub)-analytic set and at the same time an analytic submanifold of X.

The notions of semi-(sub)-analytic sets and their properties can be naturally generalized to the case when X is an analytic manifold with a countable basis of the topology. The proofs of the above mentioned properties can be found in [2], [3], [7] (cf. also [6] and [12]).

Let X be a C^1-manifold of dimension n, M - a C^1-submanifold of X of dimension m, $m \geq 1$, and let $a \in \overline{M} \setminus M$. We say that a is a <u>regular boundary point for</u> M if the germ M_a is C^1-diffeomorphic with the germ at 0 of the set $\{x \in \mathbb{R}^n | x_m > 0, \ x_j = 0; \ j = m+1, \ldots, n\}$; in other words, if there exists a map $f : U \to \mathbb{R}^n$, such that $a \in U$,

$f(U \cap \bar{M})$ is an open subset of $\{x_m \geq 0, \; x_j = 0; \; j = m+1,\dots,n\}$ and $f(U \cap (\bar{M} \setminus M)) \subset \{x_j = 0; \; j = m,\dots,n\}$. One can easily check that if this condition is fulfilled, then the germ $(\bar{M} \setminus M)_a$ is a germ of a manifold of class C^1 and dim. $m-1$. We call a point a <u>quasi-regular boundary point</u> for M if there exists a neighbourhood V of the point a in X such that $V \cap (\bar{M} \setminus M)$ is a C^1-submanifold of dimension $m-1$, $V \cap M$ has only a finite number of topological components $(\Lambda_i)_i$ and for any i: $a \in \bar{\Lambda}_i \setminus \Lambda_i$, and a is a regular boundary point for Λ_i.

For convenience sake, we recall Whitney's regularity conditions (cf. [8] and [13]). Let X be a vector space and $G_p(X)$ (or for short G_p) the Grassman manifold of p-dimensional vector subspaces of X. In particular, by G_1 (or P) we will denote the projective space over X. Let N and M be two C^1-submanifolds of X such that $N \subset \bar{M} \setminus M$, and let $k = \dim N < l = \dim M$. Let a be a point of the manifold N.

<u>Condition (A)</u>. If (x_ν) is a sequence of points belonging to M convergent to a, $W \in G_l$, and a sequence of the tangent spaces $T_{x_\nu} M$ is convergent to W (in G_l), then $T_a N \subset W$.

<u>Condition (B)</u>. If (x_ν) and (y_ν) are two sequences convergent to a, and such that $x_\nu \in M$ and $y_\nu \in M$ (for any ν), and if $\lambda \in P$ and $W \in G_1$ are such that $\mathbb{R}(x_\nu - y_\nu)$ tends to λ and $T_{x_\nu} M$ tends to W, then $\lambda \subset W$.

The conditions (A) and (B) are invariant with respect to C^1-diffeomorphisms, so they can be generalized to the case of X being any C^1-manifold. The condition (B) implies the condition (A).

1. Quasi-regular boundary points and (B)-Whitney's condition

Let us assume that X is an n-dimensional analytic manifold. Let M be a sub-analytic leaf in X of dimension m, $m \geq 1$, and N the set of all $(m-1)$-regular points of the set $\bar{M} \setminus M$, i.e. $N = \{x \in \bar{M} \setminus M \mid$ the germ $(\bar{M} \setminus M)_x$ is the germ of an analytic submanifold of dim. $m-1\}$. The set N is also a sub-analytic leaf (cf. [7], [12]).

<u>THEOREM 1.1</u>. <u>Let</u> a <u>be a point of</u> N. <u>The point</u> a <u>is a quasi-regular boundary point for</u> M <u>if and only if there exists a neighbourhood</u> U <u>of</u> a <u>in</u> N <u>such that the pair</u> (N,M) <u>fulfils the condition</u> (B) <u>at any point</u> $x \in U$.

In the proof of Theorem 1.1 we shall need the following two theorems.

THEOREM 1.2. Let E be a vector space and let $\phi,\psi : A \to \mathbb{R}$, $A \subset E$, be two sub-analytic functions with compact graphs. Additionally, let us assume that $\{\psi = 0\} \subset \{\phi = 0\}$. Then, there exist constants $K, \alpha > 0$ such that $|\phi(x)| \leq K \cdot |\psi(x)|^{\alpha}$ for any $x \in A$.

THEOREM 1.3. Let Λ be a sub-analytic leaf in a vector space E, $\dim \Lambda = p$. Let τ be a mapping $\Lambda \ni x \to T_x \Lambda \in G_p(E)$. Then τ is a sub-analytic mapping.

The proof of Theorem 1.2 is contained in [8] and that of Theorem 1.3 in [4]. The following is an immediate corollary to Theorem 1.3.

THEOREM 1.4. If Ω is an open subset of $\mathbb{R}^p$ and $f : \Omega \to \mathbb{R}$ is an analytic and sub-analytic bounded function, then the derivatives $\partial f/\partial x_k$ are sub-analytic functions.

Proof of Theorem 1.1. Without the loss of generality we can assume that X is a neighbourhood of 0 in $\mathbb{R}^n$, $a = 0$, N is an open neighbourhood of 0 in $Y = \mathbb{R}^m \times 0$, and $M \cap Y = \emptyset$. It is obvious that if 0 is a quasi-regular boundary point for M, then there exists a neighbourhood U of 0 in Y such that the pair (N,M) satisfies the condition (B) at any point $x \in U$. Let us assume that the pair (N,M) satisfies the condition (B) at any point $y \in N$. We take a mapping

$$h : \mathbb{R}^n \ni (u_1,\dots,u_n) \to (u_1,\dots,u_{m-1},(\sum_{i=m}^{n} u_i^2)^{\frac{1}{2}}) \in R^m .$$

There exists an open neighbourhood U of 0 in $\mathbb{R}^n$ such that $h|M \cap U : M \cap U \to \mathbb{R}^m$ is a local analytic isomorphism. In fact, if such U dit not exist, it would be possible to choose two sequences (x_ν), (v_ν) such that $x_\nu \in M$, $v_\nu \in T_{x_\nu}M$, $|v_\nu| = 1$, $(d_{x_\nu}h)(v_\nu) = 0$ (for any ν) and $x_\nu \to 0$. Let x'_ν be the orthogonal projection of x_ν on Y. If necessary, by choosing subsequences, we can assume that $T_{x_\nu}M \to W$, $v_\nu \to u$ and $(x_\nu - x'_\nu)/|x_\nu - x'_\nu| \to w$, for some $W \in G_m$, $u,w \in \mathbb{R}^n$. Since

$$(d_{x_\nu}h)(v_\nu) = (e_1\ v_\nu,\dots,e_{m-1}\ v_\nu,\ \frac{x_\nu - x'_\nu}{|x_\nu - x'_\nu|} \cdot v_\nu) = 0,$$

then u is orthogonal to $Y + \mathbb{R}w$. Because of the condition (A) we have $Y \subset W$ and because of the condition (B) $w \in W$. Hence $W = Y + \mathbb{R}w$. Therefore $u \in W$ and u is orthogonal to W, so we get a contradiction.

Since any neighbourhood of 0 in $\mathbb{R}^n$ contains a neighbourhood of the form $h^{-1}(G\times[0,\delta))$, where G is a neighbourhood of 0 in Y and $\delta>0$, then there exists an open ball B in Y with the centre 0 and $\delta>0$ such that for $U=h^{-1}(B\times(0,\delta))$, $h|M\cap U: M\cap U\to B\times(0,\delta)$ is a local analytic isomorphism. We can assume that $M\cap U$ is a closed subset of U, so the mapping $h|M\cap U: M\cap U\to B\times(0,\delta)$ is proper. Therefore, it is a finite covering mapping (cf. [8], p. 21). Since the set $B\times(0,\delta)$ is homeomorphic with a ball, for any connected component Γ of the set $M\cap U$, $h|\Gamma$ is an analytic isomorphism of Γ on $B\times(0,\delta)$ (cf. [8], p. 23). The set $M\cap U$ has a finite number of connected components (Γ_j), $j=1,\dots,s$, and Γ_j are sub-analytic. Let us fix j and consider the analytic isomorphism $f=(h|\Gamma_j)^{-1}$ of $B\times(0,\delta)$ on Γ_j. For any $(y,t)\in B\times(0,\delta)$, $f(y,t)=(y,\phi_m(y,t),\dots,\phi_n(y,t))$, where ϕ_i are some analytic and sub-analytic bounded functions defined on $B\times(0,\delta)$. The mapping can be extended to a homeomorphism $\tilde{f}$ of $B\times\times[0,\delta)$ on $\overline{\Gamma}_j\cap U=\Gamma_j\cup B$, such that $\tilde{f}(y,0)=y$ for $y\in B$.

For any $(y,t)\in B\times(0,\delta)$,

$$\text{(I)}\qquad t=\Big[\sum_{i=m}^{n}(\phi_i(y,t))^2\Big]^{\frac12}.$$

Differentiating the above relation with the respect to t, we get

$$\text{(II)}\qquad t=\sum_{i=m}^{n}\phi_i(y,t)\cdot\frac{\partial\phi_i}{\partial t}(y,t).$$

For any $(y,t)\in B\times(0,\delta)$, the vector

$$\frac{\partial f}{\partial t}(y,t)=\Big(0,\dots,0,\frac{\partial\phi_m}{\partial t}(y,t),\dots,\frac{\partial\phi_n}{\partial t}(y,t)\Big)$$

is a non-zero vector of the tangent space $T_{f(y,t)}\Gamma_j$ and is orthogonal to Y.

Because of the condition (B), for any $b\in B$ the angle between the directions of the vectors $(0,\dots,0,\phi_m(y,t),\dots,\phi_n(y,t))$ and

$$\Big(0,\dots,0,\frac{\partial\phi_m}{\partial t}(y,t),\dots,\frac{\partial\phi_n}{\partial t}(y,t)\Big)$$

tends to 0, when (y,t) tends to $(b,0)$.

Due to the equalities (I) and (II), the scalar product of the unit vectors of these vectors is equal to

$$\Big[\sum_{i=m}^{n}\Big(\frac{\partial\phi_i}{\partial t}(y,t)\Big)^2\Big]^{\frac12}.$$

Thus, the above means that

$$\text{(III)} \qquad \sum_{i=m}^{n} \left(\frac{\partial \phi_i}{\partial t}(y,t)\right)^2 \to 1$$

when $(y,t) \to (b,0)$.

In particular, the derivatives $\partial\phi_i/\partial t$ are bounded on any set of the form $B' \times (0,\delta')$, where $\overline{B'} \subset B$ and $0 < \delta' < \delta$. Owing to the condition (B), for any $b \in B$ the difference of the unit vectors of the vectors

$$(0,\ldots,0,\phi_m(y,t),\ldots,\phi_n(y,t))$$

and

$$(0,\ldots,0,\frac{\partial \phi_m}{\partial t}(y,t),\ldots,\frac{\partial \phi_n}{\partial t}(y,t))$$

tends to 0, when (y,t) tends to $(b,0)$.

Because of (I), (II), and the above we get

$$\text{(IV)} \qquad \frac{\partial \phi_i}{\partial t}(y,t) - \frac{\phi_i(y,t)}{t} \to 0$$

when $(y,t) \to (b,0)$ and $i = m,\ldots,n$. From this fact it follows that the functions $\phi_i(y,t)t^{-1}$ are bounded on any set of the form $B' \times (0,\delta')$, where $\overline{B'} \subset B$ and $0 < \delta' < \delta$. Let B' be an open neighbourhood of 0 in Y such that $\overline{B'} \subset B$ and let $0 < \delta' < \delta$. Let us define the functions γ_i as follows:

$$\gamma_i(y,t) = \frac{\partial \phi_i}{\partial t}(y,t) - \frac{\phi_i(y,t)}{t}$$

for $(y,t) \in B' \times (0,\delta')$ and $i = m,\ldots,n$. The functions γ_i are sub-analytic because of Theorem 1.4 and $\gamma_i(y,t) \to 0$ when $t \to 0$, uniformly with respect to $y \in B'$.

It follows from Theorem 1.2 that there exist constants $K,\alpha > 0$ such that

$$|\gamma_i(y,t)| \leq Kt^{\alpha};$$

hence

$$|\gamma_i(y,t)|t^{-1} \leq Kt^{\alpha-1}$$

for $(y,t) \in B' \times (0,\delta')$. For any $y \in B'$, the function $s \to \gamma_i(y,s)s^{-1}$ is integrable on the interval $(0,\delta')$ and the function

$$B' \times [0,\delta') \ni (y,t) \to \int_0^t \gamma_i(y,s)s^{-1}\,ds$$

is continuous. Let

$$\psi_i(y,t) = \phi_i(y,t)t^{-1} - \int_0^t \psi_i(y,s)s^{-1}\,ds$$

for $i = m,\dots,n$. It is easily checked that $\partial\psi_i/\partial t \equiv 0$ on $B' \times (0,\delta')$, so the functions $\psi_i(y,t) = \psi_i(y)$ do not depend on t. These functions are continuous on B'. Therefore, for any $b \in B'$ there exist the limits

$$\text{(V)} \qquad \lim_{(y,t)\to(b,0)} \frac{\phi_i(y,t)}{t} = \psi_i(b)$$

and

$$\text{(VI)} \qquad \lim_{(y,t)\to(b,0)} \frac{\partial f}{\partial t}(y,t) = (0,\dots,0,\ \psi_m(b),\dots,\ \psi_n(b)).$$

Since the vectors $t^{-1}\cdot(0,\dots,0,\phi_m(y,t),\dots,\phi_n(y,t))$ are unit vectors, the vectors $(0,\dots,0,\psi_m(b),\dots,\psi_n(b))$ are unit as well.

Now, let $b \in B'$, (x_ν) be a sequence of points of Γ_j convergent to b, $W \in G_m$, and let the sequence $T_{x_\nu}M$ be convergent to W. Then, W must contain the unit vector $(0,\dots,0,\psi_m(b),\dots,\psi_n(b))$ and also $W \supset Y$ (because of the condition (A)). Therefore, for any $b \in B'$ there exists

$$\text{(VII)} \qquad \lim_{x\to b} T_x\Gamma_j = Y + \mathbb{R}(0,\dots,0,\ \psi_m(b),\dots,\ \psi_n(b)).$$

It follows from the equality (I) that

$$\text{(VIII)} \qquad 0 = \sum_{i=m}^{n} \phi_i(y,t)\frac{\partial\phi_i}{\partial y_k}(y,t)$$

for $(y,t) \in B \times (0,\delta)$ and $k = 1,\dots,m-1$. Let us fix a point $b \in B'$. Let a sequence $((y_\nu,t_\nu))$ of points of the set $B \times (0,\delta)$ be convergent to $(b,0)$ and let the sequence of unit vectors

$$\frac{\partial f}{\partial y_k}(y_\nu,t_\nu)/\left|\frac{\partial f}{\partial y_k}(y_\nu,t_\nu)\right|$$

be convergent to a unit vector $v = (v_1,\dots,v_n)$. Owing to (V), (VII), and (VIII), we have $v \in Y$. Hence,

$$\sum_{i=m}^{n} \left(\frac{\partial\phi_i}{\partial y_k}(y,t)\right)^2 / \left[1 + \sum_{i=m}^{n} \left(\frac{\partial\phi_i}{\partial y_k}(y,t)\right)^2\right] \to 0$$

when $(y,t) \to (b,0)$ and finally,

$$\frac{\partial \phi_i}{\partial y_k}(y,t) \to 0$$

when $(y,t) \to (b,0)$, $i = m,\dots,n$, $k = 1,\dots,m-1$. Therefore,

$$\text{(IX)} \qquad \lim_{(y,t)\to(b,0)} \frac{\partial f}{\partial y_k}(y,t) = (0,\dots,0,\underset{(k)}{1},0,\dots,0),$$

where $k = 1,\dots,m-1$.

From the equalities (IX) and (VI), and from the fact that $\tilde{f}$ is a homeomorphism, it follows that the restriction $\tilde{f}|B' \times [0,\delta')$ is a C^1-embedding. Thus, 0 is a regular boundary point for $f(B' \times (0,\delta'))$. Since $f(B' \times (0,\delta')) = \Gamma_j \cap h^{-1}(B' \times (0,\delta'))$, then 0 is a regular boundary point for Γ_j. This completes the proof of the theorem.

Now, we give two examples showing that the assumption of sub-analyticity in Theorem 1.1 cannot be dropped.

Examples:

1) Let $M = \{(x,y) \in \mathbb{R}^2 \mid 0 < x < 1, \ y = x \cdot \sin\sqrt{|\ln x|}\}$ and $N = \{(0,0)\}$. At (0,0) the pair (N,M) fulfils the condition (B), but this point is not boundary quasi-regular for M.

2) Let $M = \{(x,y,z) \in \mathbb{R}^3 \mid y > 0, \ z = \exp((x^2+1)\ln y)\}$ and $N = \{(x,y,z) \in \mathbb{R}^3 \mid y = z = 0\}$. At (0,0,0) the pair (N,M) fulfils the condition (B), but this point is not boundary quasi-regular for M.

2. Quasi-regular boundary of a sub-analytic leaf

In this section we introduce the notion of the quasi-regular boundary and prove some of its properties.

Let X be a real n-dimensional analytic manifold having a countable basis of topology. Let M be a sub-analytic leaf in X of dimension m, $m \geq 1$, and let N be a sub-analytic leaf consisted of all (m-1)-regular points of the set $\overline{M} \setminus M$. We shall call the set $\Sigma = \{x \in N \mid x$ is a quasi-regular boundary point for $M\}$ the _quasi-regular boundary of the leaf_ M.

THEOREM 2.1. _If_ Σ _is non-empty, then it is a sub-analytic leaf of constant dimension_ $m-1$.

Proof. It is sufficient to show that the set Σ is sub-analytic. We may assume that X is a vector space.

Let $Z = \{(\lambda, W) \in P \times G_m |, \lambda \subset W\}$. Let us take the mapping

$$\Phi : M \times N \ni (x,y) \mapsto (\mathbb{R} \cdot (x-y), T_xM) \in P \times G_m .$$

It follows from Theorem 1.3 that this mapping is sub-analytic. Let $\bar{\Phi}$ be the closure of Φ considered as a subset of $X^2 \times P \times G_m$. Let $E = \{x \in N | \ ((x,x) \times P \times G_m) \cap \bar{\Phi} \subset (x,x) \times Z\}$. It is precisely the set of all points $x \in N$, at which the pair (N,M) satisfies the condition (B). One can easily see that the set E is sub-analytic. Since $\Sigma = \mathrm{int}_N E$ (cf. Theorem 1.1), the set Σ is sub-analytic. This completes the proof.

R e m a r k. If the leaf M is semi-analytic, then Σ is also a semi-analytic leaf (cf. [8] and Theorem 1.1).

THEOREM 2.2. *The set* Σ *is dense in* N.

To prove Theorem 2.2 we need some lemmas.

LEMMA 2.3. *Let* X *and* Y *be two vector spaces,* E *- a sub-analytic subset of* $X \times Y$, *and* $\pi : X \times Y \to X$ *- the natural projection. Let us assume that* $\dim \pi_E^{-1}(x) \geq r$ *for any* $x \in \pi(E)$. *Then,* $\dim E \geq \geq r + \dim \pi(E)$.

P r o o f of Lemma 2.3. According to Lemmas A and B of [2], E can be represented as a union $E = \bigcup_j \Gamma_j$ of a countable family of semi-analytic leaves such that for any j the mapping $\pi|\Gamma_j$ has constant rank and for any number d the set $\bigcup\{\Gamma_j | \ \mathrm{rank}\, \pi|\Gamma_j = d\}$ is sub-analytic. The set $H = \bigcup\{\pi(\Gamma_j) | \ \mathrm{rank}\, \pi|\Gamma_j < \dim \pi(E)\}$ is sub-analytic in X, of dimension smaller than $\dim \pi(E)$ (cf. [2], Prop.1). Therefore, there exists a point $a \in \pi(E) \setminus H$. Then, $(\pi|E)^{-1}(a) = \bigcup\{(\pi|\Gamma_j)^{-1}(a) | \ \mathrm{rank}\, \pi|\Gamma_j = \dim \pi(E)\}$. There exists i such that $\dim(\pi|\Gamma_i)^{-1}(a) \geq r$ and rank $\pi|\Gamma_i = \dim \pi(E)$ (cf. [2], Prop. 1). Thus $\dim E \geq \dim \Gamma_i = \dim(\pi|\Gamma_i)^{-1}(a) + \mathrm{rank}(\pi|\Gamma_i) \geq r + \dim \pi(E)$. This completes the proof of the lemma.

LEMMA 2.4. *Let us assume that* Ω *is an open, nonempty subset of* $\mathbb{R}^{m-1} = \mathbb{R}^{m-1} \times 0 \subset \mathbb{R}^n$ *and* M *is a sub-analytic leaf in* $\mathbb{R}^n$ *of dimension* m, *which is bounded and such that* $\Omega \subset \bar{M} \setminus M$. *Let* $\pi : (x_1, \dots, x_n) \mapsto (x_1, \dots, x_m)$ *be the natural projection. Let us suppose that the set* $F = \{x \in M | \ \mathrm{rank}\, \pi|M < m\}$ *is nowhere dense in* M. *Then there exist a point* $c \in \Omega$ *and its neighbourhood* Q *in* $\mathbb{R}^m = \mathbb{R}^m \times 0 \subset \mathbb{R}^n$ *such that: 1)* $Q = B \times (-\delta, \delta)$, *where* B *is an open ball in* $\mathbb{R}^{m-1}$ *with centre* c, $\delta > 0$; *2) each topological component* Γ *of the set* $M \cap (Q \times \mathbb{R}^{n-m})$, containing in its closure the point c, is of the form $\Gamma = \{(v,\phi(v)) | v \in \Delta\}$, where $\Delta = B \times (0,\delta)$ or $\Delta = B \times (-\delta,0)$ and $\phi : \Delta \to \mathbb{R}^{n-m}$ is an analytic mapping such that $\phi(x_1,\dots,x_m)$ tends to 0, uniformly with respect to $(x_1,\dots,x_{m-1}) \in B$, when $x_m \to 0$.

Proof of Lemma 2.4. Theorem 1.3 assures that the set F is sub-analytic and since the set F is nowhere dense in M, $\dim F \leq m-1$. Thus, $\dim(\bar{F} \setminus F) \leq m-2$ and $\Omega \cap \bar{F}$ is nowhere dense in Ω. Additionally, for $E = M \cap \pi^{-1}(R^{m-1})$ we have $\dim \bar{E} \leq m-1$, so $\Omega \cap \bar{E}$ is nowhere dense in Ω.

Let us take $c \in \Omega \setminus (\bar{E} \cup \bar{F})$. There exists a neighbourhood $U = B \times (-\delta_m,\delta_m) \times \dots \times (-\delta_n,\delta_n)$ of c in $\mathbb{R}^n$ such that:

I. B is an open ball in $\mathbb{R}^{m-1}$ with the centre c, $B \subset \Omega$, $\delta_j > 0$,

II. $U \cap (\bar{M} \setminus M) = B$;

III. $U \cap \bar{E} = \emptyset$,

IV. $U \cap \bar{F} = \emptyset$.

Let $Q = \pi(U)$. Owing to II and III, $M \cap U$ is a closed subset of $U \setminus \pi^{-1}(B)$; hence the mapping $\pi | M \cap U : M \cap U \to Q \setminus B$ is proper. Since, because of IV, it is also a local analytic isomorphism, this mapping is a finite covering. As the components of the set $Q \setminus B$ are homeomorphic with a ball in $\mathbb{R}^m$, when Γ is a component of the set $M \cap U$, then $\Gamma = \{(v,\phi(v)) |, v \in \Delta\}$, where Δ is one of the two components of $Q \setminus B$ and $\phi : \Delta \to R^{n-m}$ is an analytic mapping. As II implies, $\phi(x_1,\dots,x_m) \to 0$ uniformly with respect to $(x_1,\dots,x_{m-1})$ when $x_m \to 0$.

Proof of Theorem 2.2. We can assume that X is an open subset of a vector space V and N is a nonempty, open subset of a vector subspace W of dim. $m-1$. It suffices to show that $\Sigma \neq \emptyset$. Let us take the mapping $\tau : M \ni x \to T_x M \in G_m(V)$. Because of Theorem 1.3 and Lemma 2.3, there exists an open, nonempty subset $\Omega \subset N$ such that the set $\bar{\tau} \cap \cap (\{x\} \times G_m(V))$ is finite for any $x \in \Omega$ ($\bar{\tau}$ denotes the closure of τ in $V \times G_m(V)$). Let $a \in \Omega$ and Z be the vector subspace of dimension $n-m$ such that $Z \cap W = 0$, $Z \cap T = 0$, for any T with the property $(a,T) \in \bar{\tau}$. Let U be an m-dimensional subspace of V such that $Z \cap U = 0$ and $W \subset U$. Introducing an affine coordinate system we can suppose that $V = \mathbb{R}^n$, $W = \mathbb{R}^{m-1} = \mathbb{R}^{m-1} \times 0 \subset \mathbb{R}^n$, $U = \mathbb{R}^m = \mathbb{R}^m \times 0 \subset \mathbb{R}^n$, $Z = \mathbb{R}^{n-m} = 0 \times \mathbb{R}^{n-m}$, and $a = 0$.

By taking, if necessary, a smaller Ω, we can assume that $Z \cap T = 0$, when $(x,T) \in \bar{\tau}$, $x \in \Omega$. Then there exist a point c and a neighbourhood as in Lemma 2.4. If Γ is a topological component of the set $M \cap (Q \times \mathbb{R}^{n-m})$ and $c \in \bar{\Gamma}$, then $\Gamma = \{(v,\phi(v)) \mid v \in \Delta\}$. The above properties imply that the derivative $d\phi$ can be extended to a continuous mapping on the set $\Delta \cup B$, so ϕ can be extended to a C^1-mapping on $\Delta \cup B$. Hence, $c \in \Sigma$. The proof of Theorem 2.2 is completed.

Now, we are going to present another proof of Theorem 2.2, based on Lemma 2.4 and the following version of the Puiseux theorem.

THEOREM 2.5. *Let* X, Y *be two vector spaces and* Λ *- a sub-analytic bounded* m-*dimensional leaf in* X. *Let us assume that* $f : \Lambda \times (0,\delta) \to Y$, *where* $\delta > 0$, *is an analytic sub-analytic and bounded mapping. Then, there exist a set* E *and an integer* $k > 0$ *such that:*

1) E *is a closed subset of* Λ, *sub-analytic in* X;

2) $\dim E \leq m - 1$;

3) *for any* $a \in \Lambda \setminus E$, *the mapping* $(x,t) \mapsto f(x,t^k)$ *has an analytic extension on a neighbourhood of the point* $(a,0)$ *in* $\Lambda \times \mathbb{R}$.

The proof can be found in [9].

Proof of Theorem 2.2. We can assume that X is an open subset in $\mathbb{R}^n$ and that $\Omega = N$ and M satisfy the assumptions of Lemma 2.4. Thus, there exist a point c and its neighbourhood Q fulfilling the conditions 1) and 2) of Lemma 2.4. Moreover, due to Theorem 2.5, we can assume that, for any mapping ϕ of the condition 2), the mapping

$$\Delta \ni (x_1,\ldots,x_{m-1},x_m) \mapsto \phi(x_1,\ldots,x_{m-1},\varepsilon \cdot x_m^k),$$

where k is a positive even integer and $\varepsilon = 1$ for $\Delta = B \times (0,\delta)$, and $\varepsilon = -1$ for $\Delta = B \times (-\delta,0)$, can be extended to an analytic mapping on Q. This implies that the partial derivatives $\partial\phi/\partial x_\nu$, $\nu = 1,\ldots,m-1$, can be extended to continuous functions on $\Delta \cup B$.

Let $\phi = (\phi_1,\ldots,\phi_{n-m})$. Let us take the expansions

$$\phi_j(x_1,\ldots,x_{m-1},\varepsilon \cdot x_m^k) = \sum_{i=p_j}^{\infty} \alpha_{ij}(x_1,\ldots,x_{m-1}) \cdot x_m^i,$$

where $\alpha_{p_j j} \not\equiv 0$ on B. Of course, $p_j > 0$ since $\phi_j \equiv 0$ on B. If $\alpha_{p_j j}(b) \neq 0$, then there exists the limit

$$\lim_{x \to b} \frac{\partial \phi_j}{\partial x_m}(x_1,\ldots,x_{m-1},\varepsilon x_m^k) = \lim_{x \to b} \frac{\varepsilon}{k} \cdot x_m^{1-k} \cdot \frac{\partial \psi_j}{\partial x_m}(x_1,\ldots,x_m),$$

where $\psi_j(x_1,\ldots,x_m) = \phi_j(x_1,\ldots,x_{m-1},\varepsilon x_m^k)$. This limit is finite if

$p_j \geq k-1$ and infinite if $p_j < k-1$. Therefore there exists an open and dense subset G of B such that the partial derivatives $\partial\phi_j/\partial x_m$ can be extended to continuous functions with values in $\mathbb{R} \cup \{-\infty,+\infty\}$ on the set $\Delta \cup B$. It follows that any point $b \in G$ is a regular boundary point for Γ. Thus there is a point $z \in B$, which is a quasi-regular boundary point for M.

3. Stokes' formula for a sub-analytic leaf

The aim of this section is to prove a version of Stokes' formula for a sub-analytic leaf. This problem for semi-analytic leaves was considered in [1] and [5]. The treatment presented here is based on Prof. S. Łojasiewicz's lectures on analysis given at the Jagiellonian University. The definitions concerning differential forms on manifolds and their integration have been taken from Prof. de Rham's work [10].

Let X be an Euclidean, n-dimensional vector space and M - a C^1-submanifold of X, of dimension p, $p \geq 1$. For any $k = 0,\dots,p$, an <u>even differential k-form</u> on M is a mapping α in which to each point $x \in M$ corresponds a k-linear antisymmetric form $\alpha(x)$ on the tangent space T_xM. An <u>odd differential k-form</u> on M is a mapping β in which to each pair (x,o), where $x \in M$ and o is an orientation of the tangent space T_xM, corresponds a k-linear antisymmetric form $\beta(x,o)$ on T_xM, and additionally, $\beta(x,-o) = -\beta(x,o)$ for any pair (x,o).

Since even differential forms are the usual differential forms, in the remaining part of the paper we shall drop "even". In particular, odd differential 0-forms on M are real functions defined on the set of pairs (x,o), where $x \in M$ and o is an orientation of T_xM, which are odd with respect to o. The orientability of the manifold M is equivalent to the existence of a nowhere vanishing continuous odd differential 0-form on M. An <u>Euclidean density</u> on M is an odd p-form σ_M on M, defined in the following way: $\sigma_M(x,o)$ is a p-linear antisymmetric form on T_xM such that $\sigma_M(x,o)(e_1,\dots,e_p) = 1$, where $(e_1,\dots,e_p)$ is an orthonormal basis of T_xM and $(e_1,\dots,e_p) \in o$. By putting $\mu_M(E) = \int_E \sigma_M$, for any Lebesgue measurable subset E of M, we get the <u>p-dimensional Lebesgue measure</u> on M. We say that the manifold M <u>has the property</u> (C_p) if for any open, bounded subset G of X, there exists a constant $L \geq 0$ such that for any open ball B_r of radius r, contained in G, we have $\mu_M(M \cap B_r) \leq Lr^p$.

Let us assume that M is oriented and $N \subset \bar{M} \setminus M$ is a C^1-submanifold of dimension $p-1$, and any point $a \in N$ is a regular boundary point for M. For any $a \in N$, there exists a diffeomorphism $f : U \to \mathbb{R}^n$ defined on a neighbourhood of a in X and such that $f(\bar{M} \cap U)$ is an open subset of the set $\{x_1 \leq 0,\ x_j = 0_j,\ j = p+1,\dots,n\}$ and $f(U \cap N) \subset \subset \{x_1 = x_j = 0;\ j = p+1,\dots,n\}$. Let us consider the spaces

$$\{x \in \mathbb{R}^n \mid\ x_1 = 0;\quad j = p+1,\dots,n\}$$

and

$$\{x \in \mathbb{R}^n \mid\ x_1 = x_j = 0;\quad j = p+1,\dots,n\}$$

with the orientations defined by the identifications with $\mathbb{R}^p$ and $\mathbb{R}^{p-1}$, respectively. There exists precisely one orientation on N such that if $a \in N$, f is a diffeomorphism as above, and f_M is orientation-preserving, then f_N is orientation-preserving as well.

We shall call this orientation the <u>induced</u> <u>orientation</u> on N (by M). To this orientation corresponds certain odd differential 0-form ε on N (of the absolute value 1) which will be called the <u>induced</u> 0-<u>form</u> on N (by M).

Now, let us assume that M is oriented and N is a C^1-submanifold of X, of dimension $p-1$, $N \subset \bar{M} \setminus M$, and each point $a \in N$ is a quasi-regular boundary point for M. We define ε, the induced 0-form on N by M, as follows. Let $a \in N$ and W be a neighbourhood of a in X such that $(\bar{M} \setminus M) \cap W = N \cap W$, $M \cap W$ has only a finite number of topological components $(\Lambda_i)_i$, and each point $x \in N \cap W$ is a regular boundary point for all Λ_i. If we denote by ε_i the 0-form induced on $N \cap W$ by Λ_i, then we can put $\varepsilon = \Sigma_i\, \varepsilon_i$ on $N \cap W$. Using a suitable partition of unity, we easily get the following version of Stokes' theorem.

<u>THEOREM 3.1</u>. <u>Let</u> M <u>be</u> <u>an</u> <u>oriented</u> <u>submanifold</u> <u>and</u> <u>let</u> Σ <u>be</u> <u>a</u> C^1-<u>submanifold</u> <u>of</u> X, <u>of</u> <u>dimension</u> $p-1$, <u>such</u> <u>that</u> <u>each</u> <u>point</u> a <u>is</u> <u>a</u> <u>quasi</u>-<u>regular</u> <u>boundary</u> <u>point</u> <u>for</u> M. <u>Let</u> ε <u>be</u> <u>the</u> <u>induced</u> 0-<u>form</u> <u>on</u> Σ <u>by</u> M <u>and</u> α - <u>a</u> <u>differential</u> (p-1)-<u>form</u> <u>of</u> <u>the</u> <u>class</u> C^1 <u>on</u> X, <u>such</u> <u>that</u> $\bar{M} \cap \operatorname{supp} \alpha$ <u>is</u> <u>compact</u> <u>and</u> $(\bar{M} \setminus M) \cap \operatorname{supp} \alpha \subset \Sigma$. <u>Then</u>

$$\int (d\alpha)_M = \int \varepsilon\, \alpha_\Sigma .$$

Let us assume that the submanifold M is oriented. Then, an (even) differential p-form $\tilde{\sigma}_M$ corresponds in a natural way to the Euclidean density σ_M. If α is a differential p-form on M, then $\alpha = \phi \cdot \tilde{\sigma}_M$, where ϕ is a real function uniquely determined by the form α. The form α is integrable on M if and only if ϕ is integrable with respect to the measure μ_M. If the above conditions are satisfied, then

$$\int\alpha = \int\phi \, d\, \mu_M.$$

Let us denote by $A_k(X)$ the space of k-linear antisymmetric forms on X $(k = 0,\dots,n)$ with the norm

$$|u| = \sup\{|u(x_1,\dots,x_k)| \mid |x_i| \leq 1, \ i = 1,\dots,k\}.$$

Let α be a differential continuous p-form on X and ϕ - a function on M such that $\alpha_M = \phi \cdot \tilde{\sigma}_M$. Then $|\phi(x)| \leq |\alpha(x)|$ for $x \in M$, i.e. $|\phi| \leq |\alpha|_M$. Therefore, if α_M is integrable, then

$$|\int\alpha_M| \leq \int|\alpha|_M \, d\, \mu_M.$$

LEMMA 3.2. *Let us assume that the submanifold M has the property (C_p) and α is a differential $(p-1)$-form of class C^1 on X such that $\bar{M} \cap \operatorname{supp}\alpha$ is compact. Let F be a compact subset of X such that its $(p-1)$-dimensional Hausdorff measure is 0. Let G be a neighbourhood of F in X. Then, for any $\varepsilon > 0$ there exists a function $\psi : X \to [0,1]$ of class C^1 such that:*

1) *$\operatorname{supp}\psi$ is compact and contained in G,*
2) *$\psi = 1$ in a neighbourhood of F,*
3) *$|\int(d(\psi\alpha))_M| \leq \varepsilon$.*

Proof. It is possible to assume that G is bounded. Let ρ be a C^1-function on X such that $0 \leq \rho \leq 1$, $\rho(x) = 0$ when $|x| \geq 1$ and $\rho(x) = 1$ when $|x| \leq 1/2$. Moreover, let us take a function

$$\gamma : [0,+\infty) \to [0,1]$$

of class C^1 such that $\gamma(0) = 0$ and $\gamma(t) = 1$ for $t \geq 1$. Let ϕ be a function on M such that $(d\alpha)_M = \phi \cdot \tilde{\sigma}_M$. Since $\int|\phi| d\mu_M < +\infty$, it follows from the absolute continuity of the integral that there exists a constant $\eta_1 > 0$ such that if E is a measurable subset of M and $\mu_M(E) < \eta_1$, then

$$\int_E |\phi| d\mu_M \leq \varepsilon/2.$$

Let $0 < \eta < \min(4^{1-p} \cdot \operatorname{dist}(F, X \setminus G), 1)$. From the assumption on F it follows that

$$F \subset \bigcup_{i=1}^{\infty} F_i$$

and

$$\sum_{i=1}^{\infty} r_i^{p-1} < \eta,$$

where $r_i = \operatorname{diam} F_i$. Moreover, we can assume that each F_i contains a certain point a_i of F. Let us denote $Q_i = \{|x - a_i| < 2r_i\}$, $B_i = \{|x - a_i| < 4r_i\}$ for $i = 1,2,\ldots$. For each i we have $F_i \subset Q_i \subset B_i \subset \subset G$. Since

$$F \subset \bigcup_{i=1}^{\infty} Q_i,$$

there exists k such that $F \subset \bigcup_{i=1}^{k} Q_i$. Let us define a function ψ as follows:

$$\psi(x) = \gamma\Big(\sum_{i=1}^{k} \rho\Big(\frac{x - a_i}{4r_i}\Big)\Big)$$

for $x \in X$. Then, $\operatorname{supp}\psi \subset \bigcup_{i=1}^{k} B_i \subset G$ and $\psi = 1$ on $\bigcup_{i=1}^{k} Q_i$.

Let $N = M \cap \big(\bigcup_{i=1}^{k} B_i\big)$. Thus,

$$\Big|\int (d(\psi\alpha))_M\Big| \le \Big|\int ((d\psi) \wedge \alpha)_M\Big| + \Big|\int (\psi d\alpha)_M\Big|$$

$$\le \int |(d\psi) \wedge \alpha|_N \, d\mu_M + \int |\phi| \, d\mu_M \le C_1 \cdot \int |d\psi|_M \, d\mu_M + \int |\phi|_N \, d\mu_M,$$

where the constant C_1 depends only on α and G. We have

$$|d\psi(x)| \le C_2 \cdot \sum_{i=1}^{k} \frac{1}{r_i} \Big|d\rho\Big(\frac{x - a_i}{4r_i}\Big)\Big|$$

for $x \in X$, where the constant C_2 depends only on γ. Therefore

$$\int |d\psi|_M \, d\mu_M \le C_2 \sum_{i=1}^{k} \frac{1}{r_i} \cdot \int \Big|d\rho\Big(\frac{x - a_i}{4r_i}\Big)\Big|_M \, d\mu_M$$

$$\le C_2 \cdot C_3 \cdot \sum_{i=1}^{k} \frac{1}{r_i} \cdot \mu_M(M \cap B_i),$$

where the constant C_3 depends only on ρ. Owing to the property (C_p), we have $\mu_M(M \cap B_i) \le C_4 \cdot r_i^p$, where the constant C_4 depends only on M and G.

Then,

$$C_1 \cdot \int |d\psi|_M \, d\mu_M \leq C_1 C_2 C_3 C_4 \sum_{i=1}^{k} r_i^{p-1} \leq C'\eta,$$

where $C' = C_1 C_2 C_3 C_4$. Additionally, since $\eta < 1$,

$$\mu_M(N) \leq \sum_{i=1}^{k} \mu_M(M \cap B_i) \leq C_4 \cdot \sum_{i=1}^{k} r_i^p \leq C_4 \cdot \sum_{i=1}^{k} r_i^{p-1},$$

so $\mu_M(N) \leq C_4 \cdot \eta$. If the constant η satisfies the inequalities $C' \cdot \eta \leq \varepsilon/2$ and $C_4 \cdot \eta \leq \eta_1$, then $|\int (d(\psi\alpha)_M| \leq \varepsilon$. This completes the proof of the lemma.

<u>THEOREM 3.3</u>. *Let M be a C^1-submanifold of X, of dimension p, oriented and having the property (C_p). Let Σ be a C^1-submanifold of X, of dimension $p-1$, which consists of quasi-regular boundary points for M. Let ε be the induced 0-form on Σ by M. Let α be a differential $(p-1)$-form on X of class C^1 such that $\bar{M} \cap \operatorname{supp} \alpha$ is compact, the set $(\bar{M} \setminus (M \cup \Sigma)) \cap \operatorname{supp} \alpha$ has the $(p-1)$-dimensional Hausdorff measure equal to 0, and the form $\varepsilon \cdot \alpha_\Sigma$ is integrable on Σ. Then*

$$\int (d\alpha)_M = \int \varepsilon \cdot \alpha_\Sigma .$$

Proof. Let $F = (\bar{M} \setminus (M \cup \Sigma)) \cap \operatorname{supp} \alpha$. Let us take a decreasing sequence (G_k) $(k = 1,2,\dots)$ of open sets in X such that

$$F = \bigcap_{k=1}^{\infty} G_k .$$

For any k there exists a C^1-function $\psi_k : X \to [0,1]$ such that $\operatorname{supp} \psi_k \subset G_k$, $\psi_k = 1$ in a neighbourhood of F and $|\int (d(\psi_k \alpha))_M| < 1/k$. Since

$$\int (d((1-\psi_k)\alpha))_M = \int \varepsilon ((1-\psi_k)\alpha)_\Sigma ,$$

we get the required equality by passing to the limit.

To apply Theorem 3.3 to a sub-analytic leaf we need the following theorem proved in [11].

<u>THEOREM 3.4</u>. *Let E be a bounded sub-analytic set in X. Then*

$$E = \bigcup_{i=1}^{s} W_i ,$$

where each of the sets W_i is of the form $W_i = \{u + h_i(u) \mid , u \in \Omega_i\}$, and $h_i : \Omega_i \to V_i$ is an analytic sub-analytic mapping defined on an open subset Ω_i of U_i, where U_i is a vector subspace of X, V_i its orthogonal complement and $d\phi_i$ is bounded on Ω_i.

LEMMA 3.5. If M is a p-dimensional bounded sub-analytic leaf in X, then M satisfies the condition (C_p) and $\mu_M(M) < +\infty$.

It is an easy corollary to Theorem 3.4.

LEMMA 3.6. If E is a sub-analytic subset of X and $\dim E < m$, then its m-dimensional Hausdorff measure is equal to 0.

One can easily obtain it from Theorem 3.4.

THEOREM 3.7 (Stokes' formula for a sub-analytic leaf). Let M be a p-dimensional, oriented and sub-analytic leaf in X and let Σ be the quasi-regular boundary of M, and ε - the induced 0-form on Σ by M. Let α be a differential $(p-1)$-form of class C^1 on X, such that the set $\bar{M} \cap \operatorname{supp} \alpha$ is compact. Then

$$\int (d\alpha)_M = \int \varepsilon \alpha_\Sigma .$$

Proof. It follows from Theorems 2.1 and 2.2 that the set $\bar{M} \setminus (M \cup \Sigma)$ is sub-analytic of dimension smaller than $p-1$. Lemma 3.6 assures that its $(p-1)$-dimensional Hausdorff measure is 0. Since Σ is a sub-analytic leaf, the family of its topological components is locally finite. Thus $|\varepsilon|$ is bounded on any bounded subset of X. From that and Lemma 3.5 it follows that $\varepsilon \alpha_\Sigma$ is integrable on Σ. An application of Theorem 3.3 completes the proof.

References

[1] BUNGART, L., Stokes' Theorem on real analytic varieties, Proc. Nat. Acad. U.S.A. 54 (1965), 343-344.

[2] DENKOWSKA, Z., S. ŁOJASIEWICZ, and J. STASICA, Certaines propriétés élémentaires des ensembles sous-analytiques, Bull. Acad. Polon. Sci., Sér. Sci. Math. Astr. Phys. 27 (1979), 529-535.

[3] ——, ——, and ——, Sur le théorème du complémentaire pour les ensembles sous-analytiques, Bull. Acad. Polon. Sci., Sér. Sci. Math. Astr. Phys. 27 (1979), 537-539.

[4] —— and K. WACHTA, La sous-analycité de l'application tangente, Bull. Acad. Polon. Sci. Ser. Sci. Math. Astr. Phys. 30 (1982), 329-331.

[5] HERRERA, M., Integration on a semi-analytic set, Bull. Soc. Math. France 94 (1966), 141-180.

[6] HIRONAKA, H., Subanalytic sets, in: Number theory, algebraic geomet-

ry and commutative algebra, in honour of Y. Akizuki; Kinokuniya, Tokyo 1973, pp. 453-493.

[7] KURDYKA, K., Regular points of sub-analytic sets, Thesis, Jagiellonian University, Cracow.

[8] ŁOJASIEWICZ, S., Ensembles semi-analytiques, I.H.E.S., Bures-sur-Yvette 1965.

[9] PAWŁUCKI, W., Le théorème de Puiseux pour une application sous-analytique, Bull. Acad. Polon. Sci., to appear.

[10] de RHAM, G., Variétés différentiables, Hermann, Paris 1960.

[11] STASICA, J., Whitney's condition for sub-analytic sets, Zeszyty Naukowe Uniwersytetu Jagiellońskiego. Prace Matematyczne N° 23, Cracow 1982.

[12] TAMM, M., Subanalytic sets in the calculus of variation, Acta Math. 146 (1981), 1-2.

[13] WHITNEY, H., Tangents to an analytic variety, Ann. of Math. (2) 81 (1965).

Institute of Mathematics
Jagiellonian University
Reymonta 4
PL-30-059 Kraków, Poland

ON THE CONTINUITY OF THE MINIMA OF VARIATIONAL INTEGRALS IN ORLICZ-SOBOLEV SPACES

Giovanni Porru* (Cagliari)

Contents page

Summary. In this paper I investigate the minima of the variational integral $\int_\Omega f(x,u,\nabla u)dx$, where $f(x,u,p) : \Omega \times R^1 \times R^n \to R^1$ satisfies the following condition: $M(|p|) - K \leq f(x,u,p) \leq aM(|p|) + K$ with $K \geq 0$, $a \geq 1$, and $M(t)$ - an N-function. I prove the continuity of such minima when $M(t)$ satisfies the Δ_2 condition and the following further condition: there exists an m_ε with $\lim_{\varepsilon\to 0^+} \varepsilon m_\varepsilon = 0$ for which $M(\beta t) \leq$ $\leq m_\varepsilon \beta^{n-\varepsilon} M(t)$ if $0 < \beta < 1$ and $t > 0$.

Introduction. Let us consider the variational integral

$$(0.1) \qquad F(u,\Omega) = \int_\Omega f(x,u,Du)dx,$$

where Ω is a bounded domain of R^n $(n \geq 2)$ and $f(x,u,\xi) : \Omega \times R^1 \times R^n \to$ $\to R$ is a Carathéodory function such that there exist $K \geq 0$, $a \geq 1$, and $p > 1$, for which

$$(0.2) \qquad |\xi|^p - K \leq f(x,u,\xi) \leq a|\xi|^p + K, \quad |\xi| = (\sum_{i=1}^n \xi_i^2)^{\frac{1}{2}}.$$

A classical problem is to study the minima for $F(u,\Omega)$ among all functions $u_0 + v$, where u_0 is a fixed function of $W^{1,p}(\Omega)$ and v

---

* The author is a member of the GNAFA of the CNR.

belongs to $W_0^{1,p}(\Omega)$. For the existence of such minima (for which the main hypothesis is the convexity of $f(x,u,\xi)$ with respect to ξ) we refer to [5]. When $p > n$, a minimum function of $F(u,\Omega)$ is trivially Hölder continuous by the Sobolev Theorem. Also for $1 < p \leq n$, a function minimizing $F(u,\Omega)$ is Hölder continuous - for the proof (and for a complete bibliography) we refer to [5] and [2].

Now, instead of (0.2) let us consider

(0.3) $\quad M(|\xi|) - K \leq f(x,u,\xi) \leq a\,M(|\xi|) + K,$

where $M(t) : [0,+\infty) \to R^+$ is an <u>N-function</u>, that is a continuous convex function such that

$$\lim_{t \to 0^+} [M(t)/t] = 0, \qquad \lim_{t \to +\infty} [M(t)/t] = +\infty.$$

Correspondently to the N-function $M(t)$, the <u>Orlicz class</u> $K_M(\Omega)$ is the set of all (equivalence classes of) measurable functions u defined on Ω and satisfying

$$\int_\Omega M(|u|)\,dx < +\infty.$$

$K_M(\Omega)$ is always a convex set of functions but it may not be linear. The <u>Orlicz space</u> $L_M(\Omega)$ is defined to be the linear hull of the Orlicz class $K_M(\Omega)$. $L_M(\Omega)$ is ([3]) a Banach space equipped with the norm

$$||u||_M = \inf\{k > 0 : \int_\Omega M(|u(x)|/k)\,dx \leq 1\}.$$

The <u>Orlicz-Sobolev</u> space $W_M^1(\Omega)$ consists of those (equivalence classes of) functions $u \in L_M(\Omega)$ whose distributional derivatives $\partial u/\partial x_i$, $i = 1,\dots,n$, also belong to $L_M(\Omega)$. Even $W_M^1(\Omega)$ is a Banach space with respect to the norm

$$||u||_{W_M^1} = ||u||_M + \sum_{i=1}^{n} ||\partial u/\partial x_i||_M .$$

Obviously, $W_M^1(\Omega)$ is a natural space for the investigation of the minima for $F(u,\Omega)$ under the hypothesis (0.3). The existence problem is implicitly discussed in [5]. The aim of this paper is to investigate the continuity of the minima when $M(t)$ is such that:

(0.4) $\quad$ if $\alpha > 1$, $M(\alpha t) \leq \ell(\alpha) M(t)$;

(0.5) $\quad$ if $0 < \beta < 1$, $\forall \varepsilon > 0$ $M(\beta t) \leq m_\varepsilon \beta^{n-\varepsilon} M(t)$, with $\lim_{\varepsilon \to 0} \varepsilon m_\varepsilon = 0$.

Hypothesis (0.4) is usually called Δ_2 condition. When it is satisfied, the Orlicz class $K_M(\Omega)$ coincides with $L_M(\Omega)$ and is separable ([3]). Hypothesis (0.5) is rather restrictive. It is (trivially) satisfied for $M(t) = t^p$ with $p \geq n$, but this case does not interest us. It is not verified for $M(t) = t^p$ with $p < n$. Nevertheless, there exist functions $M(t)$ such that $\lim_{t\to\infty}[M(t)/t^n] = 0$, $\lim_{t\to\infty}[M(t)/t^{n-\varepsilon}] = +\infty$ for every $\varepsilon > 0$, and which satisfy (0.5). One such function is

$$M(t) = \begin{cases} t^n & \text{if } 0 < t < 1, \\ t^n(\log et)^s & \text{if } 1 \leq t, \quad -1 < s < 0. \end{cases}$$

Obviously, the latter N-function satisfies also (0.4).

From now on, we denote by C constants independent of ε.

1. Preliminary lemmas

LEMMA 1.1. *Let $M(t)$ be an N-function satisfying (0.4) and (0.5). Let Σ be an unitary sphere of R^n and let $u(x) \in W^1_M(\Sigma)$. If $\nabla_\Sigma u$ denotes the tangential component on Σ of the gradient ∇u, then we have for $0 \leq \varepsilon \leq \frac{1}{2}$:*

$$\int_\Sigma M(|u - u_\Sigma|)d\Sigma \leq C\, m_\varepsilon \int_\Sigma M(|\nabla_\Sigma u|)d\Sigma, \tag{1.1}$$

where u_Σ is the integral mean of u on Σ.

Proof. Let $u \in C^1(\Sigma)$ be a function such that $u_\Sigma = 0$. If $(r, \theta_1, \ldots, \theta_{n-1})$ are the polar coordinates with origin at the center of Σ, we may suppose that $u(1, 0, \theta_2, \ldots, \theta_{n-1}) = 0$. Consequently, we have

$$|u| = |u(1,0,\theta_2,\ldots,\theta_{n-1})| \leq \int_0^\pi |(\partial/\partial t)u(1,t,\theta_2,\ldots,\theta_{n-1})|dt$$

$$\leq \int_0^\pi |\nabla_\Sigma u|(\sin t)^{s_\varepsilon}(\sin t)^{-s_\varepsilon}dt,$$

with $s_\varepsilon = (n-2)/(n-1-\varepsilon)$.

In virtue of (0.4) and Jensen's inequality we find

$$M(|u|) \leq \ell(\mu_\varepsilon)M\Big(\mu_\varepsilon^{-1}\int_0^\pi |\nabla_\Sigma u|(\sin t)^{s_\varepsilon}(\sin t)^{-s_\varepsilon}dt\Big)$$

$$\leq \ell(\mu_\varepsilon)\,\mu_\varepsilon^{-1}\int_0^\pi M(|\nabla_\Sigma u|(\sin t)^{s_\varepsilon})(\sin t)^{-s_\varepsilon}dt,$$

where $\mu_\varepsilon = \int_0^\pi (\sin t)^{-s_\varepsilon}\, dt$.

The quantity $\ell(\mu_\varepsilon)/\mu_\varepsilon$ is increasing with respect to ε. As we are interested in little values of ε, say in $0 \le \varepsilon \le \frac{1}{2}$, we may suppose that $\ell(\mu_\varepsilon)/\mu_\varepsilon \le C = \ell(\mu_{\frac{1}{2}})/\mu_{\frac{1}{2}}$. By using (0.5) we obtain

$$M(|u|) \le C\, m_\varepsilon \int_0^\pi M(|\nabla_\Sigma u|)(\sin t)^{n-2}\, dt.$$

After integration on Σ we find

$$\int_\Sigma M(|u|)d\Sigma \le C\, m_\varepsilon \int_\Sigma M(|\nabla_\Sigma u|)d\Sigma.$$

Now, let $u \in W^1_M(\Sigma)$ and let $\overset{\nu}{u} \in C^1(\Sigma)$ be a sequence of functions with $\overset{\nu}{u}_\Sigma = 0$, $\upsilon = 1,2,\ldots$, and such that

$$\lim_{\nu\to\infty} \| \overset{\nu}{u} - u + u_\Sigma \|_{W^1_M(\Sigma)} = 0.$$

For each ν we have

$$(1.2) \qquad \int_\Sigma M(|\overset{\nu}{u}|)d\Sigma \le C\, m_\varepsilon \int_\Sigma M(|\nabla_\Sigma \overset{\nu}{u}|)d\Sigma.$$

Since convergence in the norm of L_M implies mean convergence (see [3], p. 75), we get

$$\lim_{\nu\to\infty} \int_\Sigma M(|\overset{\nu}{u} - u + u_\Sigma|)d\Sigma = 0.$$

For a suitable subsequence of $\overset{\nu}{u}$ (that we denote again by $\overset{\nu}{u}$), we have

$$\lim_{\nu\to\infty} M(|\overset{\nu}{u}|) = M(|u - u_\Sigma|) \quad \text{a.e. on } \Sigma.$$

Because of Fatou's Lemma we obtain

$$(1.3) \qquad \int_\Sigma M(|u - u_\Sigma|)d\Sigma \le \liminf_{\nu\to\infty} \int_\Sigma M(|\overset{\nu}{u}|)d\Sigma.$$

If $\xi, \eta \in R^n$, then (0.4) implies

$$M(|\xi + \eta|) \le \ell(2)/2\, (M(|\xi|) + M(|\eta|)).$$

Thus, if $\xi = \nabla_\Sigma \overset{\nu}{u} - \nabla_\Sigma u$ and $\eta = \nabla_\Sigma u$, we get

$$M(|\nabla_\Sigma \overset{\nu}{u}|) \le \ell(2)/2\, M(|\nabla_\Sigma \overset{\nu}{u} - \nabla_\Sigma u|) + \ell(2)/2\, M(|\nabla_\Sigma u|).$$

From the latter, as $\overset{\nu}{u} \to u - u_\Sigma$ in $W^1_M(\Sigma)$, we derive

(1.4) $$\liminf_{\nu\to\infty} \int_\Sigma (M(|\nabla_\Sigma \overset{\nu}{u}|)d\Sigma \le \ell(2)/2 \int_\Sigma M(|\nabla_\Sigma u|)d\Sigma.$$

Now, (1.2), (1.3), and (1.4) give (1.1).

LEMMA 1.2. _Let_ $M(t)$ _be an_ N-_function satisfying_ (0.4) _and_ (0.5), _let_ F _be a compact set contained in_ Ω, _and let_ $u(x)$ _be a function belonging to_ $W^1_M(\Omega)$. _Suppose that for given_ $C \ge 2$, _for some_ $\varepsilon < 1/Cm_\varepsilon = \delta_\varepsilon$, _and for an arbitrary ball_ B_q (_of radius_ q) _contained in_ F _there exists a constant_ A _such that_

(1.5) $$\int_{B_q} M(|\nabla u|)dx \le A\, q^{\delta_\varepsilon}.$$

Then, if $B_{\frac{3}{2}\rho} \subset F$ _and if_ x', x'' _belong to the ball_ $B_{\frac{1}{2}\rho}$ _concentric to_ $B_{\frac{3}{2}\rho}$, _we have_

$$M(|u(x') - u(x'')|) \le L(\varepsilon)\rho^{\delta_\varepsilon - \varepsilon}.$$

Proof. It is sufficient to prove the assertion for $u(x) \in C^1(\Omega)$. Arguing as in [4] (p. 57) we find

$$C|u(x') - u(x'')| \le 1/\rho^n \int_0^\rho \psi_1(r)\, r^{n-1}\, dr + 1/\rho^n \int_0^\rho \psi_2(r)\, r^{n-1}\, dr,$$

where

$$\psi_i(r) = \int_0^r dq \int_{\Sigma_i} |\nabla u| d\Sigma, \quad i = 1,2,$$

Σ_1 (resp. Σ_2) being the unit sphere with center x' (resp. x''). In virtue of (0.4) and Jensen's inequality we find

(1.6) $$M(|u(x') - u(x'')|) \le C/\rho^n \int_0^\rho M(\psi_1(r))r^{n-1}dr + C/\rho^n \int_0^\rho M(\psi_2(r))r^{n-1}dr.$$

In order to evaluate the first integral at the right-hand of (1.6), we define

$$\phi(q) = \int_{B_q} |\nabla u| dx,$$

where B_q is the ball of center x' and radius q. If $\gamma = (n - \delta_\varepsilon)/(n - \varepsilon)$, in virtue of (0.5) and Jensen's inequality we get

$$M(q^\gamma q^{-n} \phi(q)) \le C\, m_\varepsilon\, q^{\gamma(n-\varepsilon)-n} \int_{B_q} M(|\nabla u|)dx.$$

By using (1.5), the latter gives

(1.7) $$M(q^\gamma\, q^{-n} \phi(q)) \le C\, A\, m_\varepsilon\, q^{\gamma(n-\varepsilon)-n+\delta_\varepsilon}.$$

Since

$$\phi'(q)\, q^{1-n} = \int_\Sigma |\nabla u| d\Sigma,$$

we obtain

$$\psi_1(r) = \int_0^r \phi'(q)\, q^{1-n}\, dq.$$

Integration by parts yields

$$\psi_1(r) = (n-1) \int_0^r \phi(q)\, q^{-n}\, dq + \phi(r)\, r^{1-n}.$$

Thus, by using (0.4) and (0.5) we find

$$M(\phi_1(r)) \leq C[M(\int_0^r \phi(q)\, q^{\gamma-n}\, q^{-\gamma}\, dq)$$
$$+ M(r^{1-\gamma}\, \phi(r)\, r^{\gamma-n})]$$
$$\leq C\, m_\varepsilon [r^{(1-\gamma)(n-\varepsilon)-1+\gamma}\, (1-\gamma)\ell(1/(1-\gamma)) \int_0^r M(\phi(q) q^{\gamma-n}) q^{-\gamma}\, dq$$
$$+ r^{(1-\gamma)(n-\varepsilon)}\, M(\phi(r)\, r^{\gamma-n})].$$

If we use (1.7), after some simplification we obtain

$$M(\psi_1(r)) \leq A\, C\, m_\varepsilon^2 [\ell(n/(\delta_\varepsilon - \varepsilon)) + 1]\, r^{\delta_\varepsilon - \varepsilon}.$$

The same estimate holds for $M(\phi_2(r))$. Applying it to (1.6), the Lemma follows easily.

2. Main result

We say that the function $u \in W^1_M(\Omega)$ is a <u>minimum</u> for $F(u,\Omega)$ if

$$F(u,\Omega) \leq F(u+v,\Omega)$$

for every $v \in \overset{\circ}{W}{}^1_M(\Omega)$ (= completion of $C^\infty_o(\Omega)$ with respect to the W^1_M norm).

<u>THEOREM 2.1. Let</u> $M(t)$ <u>be an</u> N-<u>function satisfying</u> (0.4) <u>and</u> (0.5), <u>and let</u> $u \in W^1_M(\Omega)$ <u>be a minimum for</u> $F(u,\Omega)$. <u>Then, if</u> (0.3) <u>holds</u>, $u(x)$ <u>is locally continuous on</u> Ω.

<u>Proof</u>. We follow [5] (p. 105). Let B_r be a ball of center x_o and radius r, contained in Ω. If

$$D(r) = \int_{B_r} M(|\nabla u|)dx,$$

we have, almost everywhere,

$$D'(r) = \int_{\partial B_r} M(|\nabla u|)\, dS,$$

where ∂B_r is the boundary of B_r. Since $|\nabla u| \geq \frac{1}{r}|\nabla_\Sigma u|$ (Σ is the unit sphere concentric to B_r), we also have

$$(2.1) \qquad D'(r) \geq \int_{\partial B_r} M(|\nabla_\Sigma u|/r)\, dS = r^{n-1} \int_\Sigma M(|\nabla_\Sigma u|/r)\, d\Sigma.$$

After integration over B_r of the left-hand of (0.3), with $\xi = \nabla u$, we obtain

$$(2.2) \qquad D(r) \leq \int_{B_r} f(x,u,\nabla u) dx + \omega_n r^n K,$$

where ω_n is the measure of the unit ball. Let $z(\rho,\Sigma)$ ((ρ,Σ) are polar coordinates with center x_o) be a function such that $z(\rho,\Sigma) = u(r,\Sigma)$ if $\rho \geq r$. By integration over B_r of the right-hand of (0.3) (with $\xi = \nabla z$), we find

$$\int_{B_r} f(x,z,\nabla z) dx \leq a \int_{B_r} M(|\nabla z|) dx + \omega_n r^n K.$$

From the latter and (2.2) we derive

$$(2.3) \qquad D(r) \leq a \int_{B_r} M(|\nabla z|) dx + 2\omega_n r^n K.$$

We take

$$z(\rho,\Sigma) = \begin{cases} z_o + \frac{\rho}{r}[u(r,\Sigma) - z_o] & \text{if } \rho < r \\ u(\rho,\Sigma) & \text{if } \rho \geq r \end{cases}$$

with $z_o = \int_\Sigma u(r,\Sigma) d\Sigma$.

Since

$$|\nabla z|^2 = |u(r,\Sigma) - z_o|^2/\tau^2 + |\nabla_\Sigma u|^2/\tau^2,$$

then

$$|\nabla z| \leq |u(r,\Sigma) - z_o|/r + |\nabla_\Sigma u|/r.$$

As $M(t)$ is convex and (0.4) holds, we find

$$M(|\nabla z|) \leq \ell(2)/2[M(|u(r,\Sigma) - z_o|/r) + M(|\nabla_\Sigma u|/r)],$$

$$\int_\Sigma M(|\nabla z|) d\Sigma \leq C \int_\Sigma M(|u(r,\Sigma) - z_o|/r) d\Sigma + C \int_\Sigma M(|\nabla_\Sigma u|/r) d\Sigma.$$

From the latter and the Lemma 1.1 we obtain

$$\int_\Sigma M(|\nabla z|)d\Sigma \leq C\, m_\varepsilon \int_\Sigma M(|\nabla_\Sigma u|/r)d\Sigma,$$

from which, by using (2.3) and (2.1), we have

$$D(r) \leq a \int_0^r \rho^{n-1}\, d\rho \int_\Sigma M(|\nabla z|)d\Sigma + 2\omega_n r^n K \leq C\, m_\varepsilon\, r^n \int_\Sigma M(|\nabla_\Sigma u|/r)d\Sigma$$

$$+ 2\omega_n r^n K \leq C\, m_\varepsilon\, (rD'(r) + r^n).$$

Finally, by integration, for $0 < q < R$ we get

$$D(q) \leq [R^n/(n-\delta_\varepsilon) + D(R)](\tfrac{q}{R})^{1/C\, m_\varepsilon}. \tag{2.4}$$

If F is a compact set contained in Ω and if we consider a ball B_q contained in F, then (2.4) gives

$$\int_{B_q} M(|\nabla u|)dx \leq A\, q^{\delta_\varepsilon},$$

where $\delta_\varepsilon = 1/C\, m_\varepsilon$ and A depends only on a, K, n, Ω, dist. $(F,\partial\Omega)$, and $\int_\Omega M(|\nabla u|)dx$. Now, by Lemma 1.2, the theorem follows.

References

[1] ADAMS, R.A., Sobolev spaces, Academic Press, New York 1975.

[2] GIAQUINTA, M. and E. GIUSTI, On the regularity of the minima of variational integrals, in print.

[3] KRASNOSEL'SKII, M.A. and Y.B. RUTICKIĬ, Convex functions and Orlicz spaces, P. Noordhoff LTD, Groningen 1961.

[4] LADYZHENSKAJA, O.A. and N.N. URAL'TSEVA, Linear and quasilinear elliptic equations, Academic Press, New York 1968.

[5] MORREY, C.B., Multiple integrals in the calculus of variations, Springer-Verlag, Berlin 1968.

Dipartimento di Matematica
Università di Cagliari
via Ospedale 72
I-09100 Cagliari, Italy

GEOMETRIES OF THE PROJECTIVE MATRIX SPACE

Binyamin Schwarz and Abraham Zaks (Haifa)

Contents | page

Summary. In our paper on matrix Möbius transformations (Comm. Algebra 9 (19) (1981), 1913-1968) we introduced the one-dimensional left-projective space over the complex $n \times n$ matrices $P = P_1(M_n(\mathbb{C}))$. For $n = 1$ this space is the projective complex line $P_1(\mathbb{C})$ and so is homeomorphic to the Riemann sphere. We studied the topology of P and the projective mappings of P onto itself. Here we present some generalizations of certain parts of the Euclidean, the spherical and the non-Euclidean geometry from the scalar to the multidimensional case.

1. Introduction

We use, as far as possible, the terminology and notation of Caratheodory's book [2].

We state the algebraic results needed to settle the geometric assertions. The detailed study will appear elsewhere.

We outline first the basic notions and results of [10], on which the present work relies; when doing this we also describe our notation.

Complex $n \times n$ matrices are denoted by capital Roman letters $P = (p_{ik})_1^n$; $P^* = (\overline{p}_{ki})_1^n$ is the conjugate transpose of $P = (p_{ik})_1^n$, and

$|P|$ denotes the determinant of P. Diagonal matrices $L = (\ell_i \delta_{ik})_1^n$ are denoted by $L = \{\ell_1, \dots, \ell_n\}$. For Hermitian matrices $H = H^*$, $H > 0$ (< 0) means H is positive (negative) definite. Complex $n \times 2n$ matrices are always written in block form $(P_1\,P_2)$, where P_1 and P_2 are $n \times n$ matrices. The set of all $n \times 2n$ matrices of rank n is denoted by $C_o(2n^2)$. Two matrices $(P_1\,P_2)$ and $(\tilde{P}_1\,\tilde{P}_2)$ of $C_o(2n^2)$ are (left- or row-) equivalent - if there exists a nonsingular $n \times n$ matrix R such that

$$(1.1)\qquad (\tilde{P}_1\,\tilde{P}_2) = R(P_1\,P_2), \quad |R| \neq 0.$$

The corresponding equivalence classes are the points of the projective space P. The point corresponding to the matrix $(P_1\,P_2) \in C_o(2n^2)$ is denoted by $\underline{P}$, and we write $\underline{P} = f(P_1\,P_2)$ and $(P_1\,P_2) \in f^{-1}[\underline{P}]$. f is the standard map from $C_o(2n^2)$ to the space P and the topology of P is the quotient topology relative to f and the usual topology of $C_o(2n^2)$. P is a connected and compact Hausdorff space whose topology has a countable base. Thus P is metrizable but, for $n > 1$, no concrete metric was given in [10].

A point $\underline{P}$ is called <u>finite</u> if, for any $(P_1\,P_2) \in f^{-1}[\underline{P}]$, $|P_2| \neq 0$. For such a point there exists a unique canonical matrix $(P\,I) \in f^{-1}[\underline{P}]$, $(I = (\delta_{ik})_1^n)$. $\underline{0} = f(0\,I)$ is called the <u>origin</u> or the <u>south pole</u>. The particular infinite point $\underline{N} = f(I\,0)$ is called the <u>north pole</u>.

Complex $2n \times 2n$ matrices are denoted as follows: $\underset{\sim}{P}$, $\underset{\sim}{S}$ etc. Whenever such a matrix is written in block form each of the four blocks will be an $n \times n$ matrix. Projectivities of the space P are given by nonsingular $2n \times 2n$ matrices $\underset{\sim}{S}$ operating on the $n \times 2n$ matrices corresponding to the points of P. For given

$$\underset{\sim}{S} = \begin{bmatrix} A & C \\ B & D \end{bmatrix}, \quad |\underset{\sim}{S}| \neq 0,$$

and arbitrary $\underline{P} = f(P_1\,P_2)$ we set $(P_1^S\,P_2^S) = (P_1\,P_2)\underset{\sim}{S} = (P_1A + P_2B \;\; P_1C + P_2D)$ and call $\underline{P}^S = f(P_1^S\,P_2^S)$ <u>the map of $\underline{P}$ under the projectivity</u> S. Note that if $\underline{P}$ is a finite point, $\underline{P} = f(P\,I)$, and if its map $\underline{P}^S$ is also finite, then the canonical matrix corresponding to $\underline{P}^S$ is $(P^S\,I) = ((PC + D)^{-1}(PA + B)\,I)$. The matrix Möbius transformation

$$(1.2)\qquad P^S = (PC + D)^{-1}(PA + B), \quad \begin{vmatrix} A & C \\ B & D \end{vmatrix} \neq 0$$

is thus the restriction of the projectivity S to the finite points of the space. The projectivities of P onto itself form a group under composition. Every projectivity is a homeomorphism of the space onto itself. Two nonsingular matrices $\underset{\sim}{S}_1$ and $\underset{\sim}{S}_2$ correspond to the same projectivity if and only if there exists a scalar $s \neq 0$ such that $\underset{\sim}{S}_1 = s\underset{\sim}{S}_2$. For $n = 1$, $z = f(\zeta_1 \zeta_2) = \zeta_1/\zeta_2$ and (1.2) reduces to the classical Möbius transformation

$$z^S = \frac{az+b}{cz+d}, \qquad \begin{vmatrix} a & c \\ b & d \end{vmatrix} \neq 0.$$

For a different approach to matrix Möbius transformations see e.g. [4, 9, 11].

Throughout the present paper $\| \ \|$ will denote the spectral norm. The singular values $\ell_i \geq 0$, $i = 1,\dots,n$ of an $n \times n$ matrix A are the square roots of the eigenvalues of A^*A, $A^*A \geq 0$, and $\|A\| = \max_i \ell_i$. If $|A| \neq 0$, then $A^*A > 0$ and hence all $\ell_i > 0$. For unitary matrices U and V and for arbitrary A, the matrices A, UA, AV, and UAV have the same singular values, hence $\|A\| = \|UA\| = \|AV\| = \|UAV\|$ [6, 8]. We mention the following known factorization result [3, 8].

PROPOSITION 1.1. *For any* $n \times n$ *matrix* A *there exist unitary matrices* U *and* V *and a diagonal matrix* $L = \{\ell_1,\dots,\ell_n\}$, $\ell_i \geq 0$, $i = 1,\dots,n$, *such that*

(1.3) $\quad A = ULV.$

The elements ℓ_i *of* L *are the singular values of* A.

This factorization is not unique. The singular values $\ell_1,\dots,\ell_n$ of A can be ordered arbitrarily, and if L is given there is still some freedom for the choice of the unitary factors.

2. Euclidean geometry

Let $\underline{P}$ and $\underline{Q}$ be finite points and let $(P\,I)$ and $(Q\,I)$ be the corresponding canonical matrices. We define the *Euclidean distance* $d(\underline{P},\underline{Q})$ by

(2.1) $\quad d(\underline{P},\underline{Q}) = \|P - Q\|.$

This clearly defines a metric for the set of finite points of P and we call the resulting metric space the *Gaussian plane*. The projectivities of P keeping this distance invariant are characterized as follows.

THEOREM 2.1. *The projectivity* $\underset{\sim}{S}$ *keeps the Euclidean distance invariant if and only if the corresponding matrices* S *are of the form*

$$(2.2) \qquad \underset{\sim}{S} = s\begin{bmatrix} U_1 & 0 \\ U_2 P_0 & U_2 \end{bmatrix}, \quad U_1 U_1^* = U_2 U_2^* = I,$$

where $s \neq 0$ *is an arbitrary complex number.*

For the group of these *Euclidean motions* the following holds.

THEOREM 2.2. *Let the finite points* $\underline{P}_\nu = f(P_\nu I)$, $\underline{Q}_\nu = f(Q_\nu I)$, $\nu = 1,2$, *be given. There exists an Euclidean motion mapping simultaneously* $\underline{P}_1$ *into* $\underline{P}_2$ *and* $\underline{Q}_1$ *into* $\underline{Q}_2$ *if and only if the two matrices* $Q_1 - P_1$ *and* $Q_2 - P_2$ *have the same set* $(\ell_1,\ldots,\ell_n)$ *of singular values.*

In the scalar case a circle in the Gaussian plane partitions this plane into three sets: the circle γ, its inside the disc Δ^-, and its outside Δ^+. In the multidimensional case the Gaussian plane, and later on the Riemann sphere and the unit disk, will be partitioned into $\binom{n+2}{2}$ sets. These sets are defined by the rank and the signature of certain Hermitian forms. We rely on the following obvious assertion. Let $\underset{\sim}{H} = \underset{\sim}{H}^*$ be a given $2n \times 2n$ matrix and let $\underline{P}$ be a given point of the projective space. For every $(P_1\, P_2) \in f^{-1}[\underline{P}]$, the rank and the signature of the Hermitian matrix

$$(2.3) \qquad H(P_1,P_2) = (P_1\, P_2)\underset{\sim}{H}(P_1\, P_2)^*$$

depends only on $\underset{\sim}{H}$ and on the point $\underline{P}$ and not on the choice of the matrix $(P_1\, P_2) \in f^{-1}[\underline{P}]$.

For the Gaussian plane we consider only forms in P; i.e. we set $H(P) = (P\, I)\underset{\sim}{H}(P\, I)^*$. The *Euclidean circle* $\gamma_e(\underline{0},r)$ with center at the origin and radius $r > 0$ and the corresponding partition are defined by

$$(2.4) \qquad \underset{\sim}{H}_e(r) = \begin{bmatrix} I & 0 \\ 0 & -r^2 I \end{bmatrix}, \quad r > 0.$$

We set $H_e(P,r) = (P\, I)\underset{\sim}{H}_e(r)(P\, I)^* = PP^* - r^2 I$ and define

$$\gamma_e(\underline{0},r) = \{\underline{P} : \underline{P} = f(P\, I),\quad H_e(P,r) = 0\},$$

$$(2.4') \qquad \Delta_e^-(\underline{0},r) = \{\underline{P} : \underline{P} = f(P\, I),\quad H_e(P,r) < 0\},$$

$$\Delta_e^+(0,r) = \{\underline{P} : \underline{P} = f(P\,I), \quad H_e(P,r) > 0\}.$$

For $n \geq 2$ the $\binom{n+2}{2} - 3$ remaining sets of the partition of the plane are defined by the rank and signature of $H_e(P,r)$. Let $\underline{P} = f(P\,I)$ be a finite point for which $|P| \neq 0$. Then we define

$$(2.5) \qquad \check{\underline{P}} = f((P^*)^{-1} I)$$

and call $\check{\underline{P}}$ the point <u>inverse</u> to $\underline{P}$ (with respect to the <u>unit circle</u> $\gamma_e(\underline{0},1)$). $\gamma_e(\underline{0},r)$, its inside and its outside may be characterized as follows.

<u>PROPOSITION 2.3. For any given</u> r, $r > 0$, <u>the following holds for the sets defined by</u> (2.4'):

$$(2.6) \qquad \begin{aligned} &\gamma_e(\underline{0},r) = \{\underline{P} : \underline{P} = f(P\,I), \ P = rU, \ UU^* = I\}, \\ &\gamma_e(\underline{0},r) = \{\underline{P} : d(\underline{0},\underline{P}) = 1/d(\underline{0},\check{\underline{P}}) = r\}, \\ &\Delta_e^-(\underline{0},r) = \{\underline{P} : d(\underline{0},\underline{P}) < r\}, \\ &\Delta_e^+(\underline{0},r) = \{\underline{P} : d(\underline{0},\check{\underline{P}}) < 1/r\}. \end{aligned}$$

In the first equation the unitary $n \times n$ matrices serve as parameter. It follows that the points of each circle depend on n^2 real parameters. This holds true for all circles and straight lines of the three geometries. The Euclidean circle $\gamma_e(\underline{P}_o,r)$ with center $\underline{P}_o = f(P_o I)$ and radius r, $r > 0$, and the sets of the corresponding partition are defined as the map of $\gamma_e(\underline{0},r)$ and the corresponding sets under a motion (2.2) with arbitrary unitary U_1 and U_2 and given P_o. These sets are determined by the rank and signature of $\tilde{H}_e(P,r) = (P\,I)\tilde{\underset{\sim}{H}}_e(r) \times (P\,I)^*$ where $\tilde{\underset{\sim}{H}}_e(r) = \underset{\sim}{S}^{-1} \underset{\sim}{H}_e(r)(\underset{\sim}{S}^*)^{-1}$. It follows that $\tilde{H}_e(P,r) = (P - P_o) \times (P - P_o)^* - r^2 I$, and $\gamma_e(\underline{P}_o,r)$, $\Delta_e^-(\underline{P}_o,r)$ and $\Delta_e^+(\underline{P}_o,r)$ may again be characterized by distances.

The <u>finite Hermitian</u> line ℓ_e is given by $\ell_e = \{\underline{P} : \underline{P} = f(P\,I), P = P^*\}$. The corresponding Hermitian matrix is

$$(2.7) \qquad \underset{\sim}{K} = i\begin{bmatrix} 0 & -I \\ I & 0 \end{bmatrix};$$

it induces the form $K(P) = (P\,I)\underset{\sim}{K}(P\,I)^*$. The <u>straight lines</u> of the Gaussian plane are obtained from ℓ_e by Euclidean motions; $\underset{\sim}{K}$ is replaced by $\underset{\sim}{K}^s = \underset{\sim}{S}^{-1}\underset{\sim}{K}(\underset{\sim}{S}^*)^{-1}$.

3. Spherical geometry

For the Gaussian plane we used the canonical matrices $(P\,I)$. In the present section we use $n \times 2n$ matrices belonging to a compact subset K of $C_0(2n^2)$. We define $(P_1\,P_2) \in K$ if

$$(3.1)\qquad P_1P_1^* + P_2P_2^* = I.$$

PROPOSITION 3.1 [10, p. 1928]. *Let* P *be a given point of* $P_1(M_n(\mathbb{C}))$. *There exists a matrix* $(P_1\,P_2)$ *such that* $(P_1\,P_2) \in f^{-1}[\underline{P}]$ *and* $(P_1\,P_2) \in K$. *Moreover,* $(\tilde{P}_1\,\tilde{P}_2) \in f^{-1}[\underline{P}]$ *and* $(\tilde{P}_1\,\tilde{P}_2) \in K$ *if and only if*

$$(3.2)\qquad (\tilde{P}_1\,\tilde{P}_2) = U(P_1\,P_2), \quad UU^* = I.$$

Now, we complete each $n \times 2n$ matrix of K to a unitary $2n \times 2n$ matrix $\underset{\sim}{P}$.

PROPOSITION 3.2. *Let* P_1 *and* P_2 *be* $n \times n$ *matrices satisfying* (3.1). *Then there exist* $n \times n$ *matrices* P_3 *and* P_4 *such that the* $2n \times 2n$ *matrix*

$$(3.3)\qquad \underset{\sim}{P} = \begin{bmatrix} P_1 & P_2 \\ P_3 & P_4 \end{bmatrix}$$

is unitary. The completing matrix $(P_3\,P_4)$ *is unique up to a premultiplication by a unitary* $n \times n$ *matrix.*

For a given point $\underline{P}$ the corresponding unitary matrix $\underset{\sim}{P}$, such that $\underline{P} = f(P_1\,P_2)$, is thus not uniquely given. If also

$$(3.3')\qquad \underset{\sim}{\tilde{P}} = \begin{bmatrix} \tilde{P}_1 & \tilde{P}_2 \\ \tilde{P}_3 & \tilde{P}_4 \end{bmatrix}$$

is unitary and satisfies $f(\tilde{P}_1\,\tilde{P}_2) = \underline{P}$, then $\underset{\sim}{P}$ and $\underset{\sim}{\tilde{P}}$ are *block-equivalent*; i.e. they satisfy

$$(3.4)\qquad \begin{bmatrix} \tilde{P}_1 & \tilde{P}_2 \\ \tilde{P}_3 & \tilde{P}_4 \end{bmatrix} = \begin{bmatrix} U_1 & 0 \\ 0 & U_2 \end{bmatrix} \begin{bmatrix} P_1 & P_2 \\ P_3 & P_4 \end{bmatrix}, \qquad U_1U_1^* = U_2U_2^* = I.$$

For a given $\underline{P}$ let $\underset{\sim}{P}$ be a corresponding unitary matrix. While $\underset{\sim}{P}$ is given only up to block-equivalence, clearly the point $\hat{\underline{P}} = f(P_3 P_4)$ depends only on $\underline{P}$. We call $\hat{\underline{P}}$ the <u>antipode</u> of the given point $\underline{P}$. For $n = 1$ and $z \neq \infty$ we may choose

$$(3.5) \qquad \underset{\sim}{P} = \begin{bmatrix} z/(1+|z|^2)^{1/2} & 1/(1+|z|^2)^{1/2} \\ -1/(1+|z|^2)^{1/2} & \bar{z}/(1+|z|^2)^{1/2} \end{bmatrix}$$

and obtain $\hat{z} = -1/\bar{z}$. This motivates our terminology. For arbitrary n, $n \geq 1$, the unit matrix $I = \begin{bmatrix} I & 0 \\ 0 & I \end{bmatrix}$ corresponds to the antipodal pair $(\underline{N}, \underline{0})$.

The following structure theorem for unitary $2n \times 2n$ matrices may be of independent interest.

THEOREM 3.3. *Let $\underset{\sim}{P}$ be a given unitary $2n \times 2n$ matrix. There exist four unitary $n \times n$ matrices U_1, U_2, V_1, V_2 and two diagonal matrices L_1 and L_2 satisfying*

$$(3.6) \qquad L_1 = \{\ell_1, \ldots, \ell_n\}, \quad L_2 = \{\ell_1', \ldots, \ell_n'\}, \quad \ell_i \geq 0, \quad \ell_i' \geq 0, \quad \ell_i^2 + \ell_i'^2 = 1, \quad i = 1, \ldots, n,$$

such that

$$(3.7) \qquad \underset{\sim}{P} = \begin{bmatrix} P_1 & P_2 \\ P_3 & P_4 \end{bmatrix} = \begin{bmatrix} U_1 & 0 \\ 0 & U_2 \end{bmatrix} \begin{bmatrix} L_1 & L_2 \\ -L_2 & L_1 \end{bmatrix} \begin{bmatrix} V_1 & 0 \\ 0 & V_2 \end{bmatrix} = \begin{bmatrix} U_1L_1V_1 & U_1L_2V_2 \\ -U_2L_2V_1 & U_2L_1V_2 \end{bmatrix}.$$

The singular values of the blocks P_1 and P_4 are the elements ℓ_i of L_1, and the singular values of P_2 and P_3 are the elements ℓ_i' of L_2. The factorization (3.7) is not unique. The order of the singular values of P_1 as elements of L_1 can be prescribed, and if L_1 is given there is still some freedom for the choice of the unitary factors U_1, U_2, V_1 and V_2. It follows from Theorem 3.3 that, for unitary $2n \times 2n$ matrices (3.3), $\|P_\nu\| \leq 1$, $\nu = 1, \ldots, 4$, $\|P_1\| = \|P_4\|$, $\|P_2\| = \|P_3\|$ and $\|P_1\|^2 + \|P_2\|^2 \geq 1$.

Let $\underline{P}$ and $\underline{Q}$ be two given points of the projective space, and let $\underset{\sim}{P}$, given by (3.3), and $\underset{\sim}{Q} = \begin{bmatrix} Q_1 & Q_2 \\ Q_3 & Q_4 \end{bmatrix}$ be corresponding unitary matrices. The matrix

$$(3.8)\quad \underset{\sim}{R} = \begin{bmatrix} R_1 & R_2 \\ R_3 & R_4 \end{bmatrix} = \underset{\sim}{P}\underset{\sim}{Q}^*$$

is also unitary. The norms $\|R_\nu\|$, $\nu = 1,\ldots,4$ do not change if $\underset{\sim}{P}$ and $\underset{\sim}{Q}$ are replaced by block equivalent matrices (cf. (3.4)). Hence $\|R_2\| = \|R_3\|$ depends only on the points $\underline{P}$ and $\underline{Q}$. We define the <u>chordal distance</u> $\chi(\underline{P},\underline{Q})$ by

$$(3.9)\quad \chi(\underline{P},\underline{Q}) = \|R_2\| = \|P_1 Q_3^* + P_2 Q_4^*\|$$

and obtain easily the following.

<u>THEOREM 3.4</u>. <u>The function $\chi(\underline{P},\underline{Q})$ defines a metric for the space</u> P.

For $n = 1$ this reduces to the usual chordal distance $\chi(z,w)$ between two points on the Riemann sphere (with diameter 1) (cf. (3.5)). We also mention that the metric topology of P, corresponding to χ, is the topology built in [10]; i.e. the quotient topology relative to f and the topology of $C_o(2n^2)$ (cf. [5]). The projectivities of P keeping χ invariant ("<u>rotations of the sphere</u>") are characterized as follows.

<u>THEOREM 3.5</u>. <u>The projectivity $\underset{\sim}{S}$ keeps the chordal distance invariant if and only if the corresponding matrices $\underset{\sim}{S}$ are of the form $\underset{\sim}{S} = s\underset{\sim}{U}$, $\underset{\sim}{U}\underset{\sim}{U}^* = \underset{\sim}{I}$, where $s \neq 0$ is an arbitrary complex number</u>.

For $n = 1$ this reduces to $z^s = (az + b)/(\bar{b}z - \bar{a})$, $|a|^2 + |b|^2 > 0$, $\underset{\sim}{S} = \begin{bmatrix} a & \bar{b} \\ b & -\bar{a} \end{bmatrix} = (|a|^2 + |b|^2)^{1/2}\underset{\sim}{U}$, $\underset{\sim}{U}\underset{\sim}{U}^* = \underset{\sim}{I}$.

<u>PROPOSITION 3.6</u>. a) <u>For any pair of points</u> (P,Q), $\chi(\underline{P},\underline{Q}) \leq 1$. b) <u>For any antipodal pair</u> $(\underline{P},\hat{\underline{P}})$, $\chi(\underline{P},\hat{\underline{P}}) = 1$. c) <u>Rotations of the sphere map antipodal pairs into antipodal pairs</u>.

We note that, for $n = 1$, part c) of this proposition follows from parts a) and b), as in this scalar case the antipode $\hat{\underline{P}}$ is the only point at maximal distance from the given point $\underline{P}$. For $n > 1$ this is not true and there are many points $\underline{Q}$, in addition to the antipode $\hat{\underline{P}}$, such that $\chi(\underline{P},\underline{Q}) = 1$. As distances are invariant under rotations, we consider only the points at maximal distance from the origin $\underline{0} = f(0\ I)$. The set of infinite points $P_\infty = \{\underline{P} = \underline{P} : \underline{P} = f(P_1\ P_2),\ |P_2| = 0\}$ was considered in [10] and it was shown that this set is closed and nowhere dense in P. Now, we obtain the following.

PROPOSITION 3.7. a) $P_\infty = \{\underline{P} : \chi(\underline{0},\underline{P}) = 1\}$. b) The only point at chordal distance 1 from every infinite point is the origin $\underline{0}$.

The following holds for the group of the spherical rotations.

THEOREM 3.8. Let the points $\underline{P}$, $\underline{Q}$, $\underline{P}_0$ and $\underline{Q}_0$ be given and let $\underset{\sim}{P}$, $\underset{\sim}{Q}$, $\underset{\sim}{P}_0$ and $\underset{\sim}{Q}_0$ be corresponding unitary matrices. There exists a rotation of the sphere mapping simultaneously $\underline{P}$ into $\underline{P}_0$ and $\underline{Q}$ into $\underline{Q}_0$ if and only if the two blocks R_1 and R_1^o of the matrices

$$(3.8') \quad \underset{\sim}{R} = \begin{bmatrix} R_1 & R_2 \\ R_3 & R_4 \end{bmatrix} = \underset{\sim}{P}\underset{\sim}{Q}^*, \quad \underset{\sim}{R}_o = \begin{bmatrix} R_1^o & R_2^o \\ R_3^o & R_4^o \end{bmatrix} = \underset{\sim}{P}_o \underset{\sim}{Q}_o^*$$

have the same set $(\ell_1,\ldots,\ell_n)$ of singular values.

Let

$$(3.10) \quad \underset{\sim}{H}_s(r) = \begin{bmatrix} (1-r^2)I & 0 \\ 0 & -r^2 I \end{bmatrix}, \quad 0 < r < 1.$$

For arbitrary $(P_1\, P_2) \in C_o(2n^2)$, not necessarily in K, we set

$$(3.10') \quad H_s(P_1,P_2,r) = (P_1\, P_2)\underset{\sim}{H}_s(r)(P_1\, P_2)^* = (1-r^2)P_1P_1^* - r^2P_2P_2^*.$$

The spherical circle $\gamma_s(\underline{0},r)$ with center at $\underline{0}$ and radius $r, 0 < r < 1$, and the corresponding spherical disks $\Delta_s^-(\underline{0},r)$ and $\Delta_s^+(0,r)$ are defined as follows.

$$\gamma_s(\underline{0},r) = \{\underline{P} : \underline{P} = f(P_1\, P_2),\quad H_s(P_1,P_2,r) = 0\},$$

$$(3.11) \quad \Delta_s^-(\underline{0},r) = \{\underline{P} : \underline{P} = f(P_1\, P_2),\quad H_s(P_1,P_2,r) < 0\},$$

$$\Delta_s^+(\underline{0},r) = \{\underline{P} : \underline{P} = f(P_1\, P_2),\quad H_s(P_1,P_2,r) > 0\},$$

and the remaining sets of the partition of the space are defined by the rank and the signature of $H_s(P_1,P_2,r)$. The spherical circle and the corresponding disks may be characterized as follows.

PROPOSITION 3.9. For any given $r, 0 < r < 1$, the following holds for the sets defined by (3.11):

$$\gamma_s(\underline{0},r) = \{\underline{P} : \underline{P} = f(P_1\,P_2), P_1 = rU_1, P_2 = (1-r^2)^{1/2}U_2, U_1U_1^* = U_2U_2^* = I\},$$

$$\gamma_s(\underline{0},r) = \{\underline{P} : \chi(\underline{0},\underline{P}) = r,\ \chi(\underline{N},\underline{P}) = (1-r^2)^{1/2}\},$$

$$(3.12)\quad \Delta_s^-(\underline{0},r) = \{\underline{P} : \chi(\underline{0},\underline{P}) < r\},$$

$$\Delta_s^+(\underline{0},r) = \{\underline{P} : \chi(\underline{N},\underline{P}) < (1-r^2)^{1/2}\}.$$

For $n \geq 2$ we have, in general, strict inequality in $\chi^2(\underline{0},\underline{P}) + \chi^2(\underline{N},\underline{P}) \geq 1$. Equality holds only if $\underline{P} = \underline{0}$, or $\underline{P} = \underline{N}$, or if $P \in \gamma_s(\underline{0},r)$, $r = \chi(\underline{0},\underline{P})$. Let $(\underline{M},\hat{\underline{M}})$ be a pair of antipodal points and let

$$\underset{\sim}{M} = \begin{bmatrix} M_1 & M_2 \\ M_3 & M_4 \end{bmatrix}$$

be a corresponding unitary matrix $\underline{M} = f(M_1\,M_2)$, $\hat{\underline{M}} = f(M_3\,M_4)$. Let U_1 and U_2 be arbitrary unitary matrices and set

$$(3.13)\quad \underset{\sim}{S} = \begin{bmatrix} U_1 & 0 \\ 0 & U_2 \end{bmatrix}\begin{bmatrix} M_1 & M_2 \\ M_3 & M_4 \end{bmatrix}.$$

For the corresponding rotations S we have $\underline{0}^S = \hat{\underline{M}}$, $\underline{N}^S = \underline{M}$. The spherical circle $\gamma_s(\hat{\underline{M}},r)$ with center $\hat{\underline{M}}$ and radius $r, 0 < r < 1$, and the sets of the corresponding partition are defined as the map of $\gamma_s(\underline{0},r)$ and the corresponding sets under a rotation (3.13). These sets are determined by the rank and signature of $\tilde{H}_s(P_1,P_2,r) = (P_1\,P_2)\,\underset{\sim}{\tilde{H}}_s(r)(P_1\,P_2)^*$ where $\underset{\sim}{\tilde{H}}_s(r) = \underset{\sim}{S}^*\underset{\sim}{H}_s(r)\underset{\sim}{S}$. It follows that

$$\tilde{H}_s(P_1,P_2,r) = (P_1M_1^* + P_2M_2^*)(P_1M_1^* + P_2M_2^*)^* - r^2(P_1P_1^* + P_2P_2^*).$$

The circle $\gamma_s(\hat{\underline{M}},r)$ and the disks $\Delta_s^-(\hat{\underline{M}},r)$ and $\Delta_s^+(\hat{\underline{M}},r)$ may also be characterized by distances.

We add the following remark. The spherical circles with center at $\underline{0}$ (parallels of latitude) are by their definition Euclidean circles with center at $\underline{0}$. Indeed, $\gamma_s(\underline{0},r_s)$ coincides with $\gamma_e(\underline{0},r_e)$ if $r_e^2 = r_s^2/(1-r_s^2)$, $0 < r_s < 1$. For $n=1$ all the other spherical circles are Euclidean ones. For $n \geq 2$ this does not hold true in general.

The equator $\gamma_s(\underline{0},1/\sqrt{2})$ is defined by $\{\underline{P} : \underline{P} = f(P_1\,P_2), P_1P_1^* - P_2P_2^* = 0\}$. Its map under a rotation (3.13) yields the great circle $\gamma_s(\hat{\underline{M}},1/\sqrt{2})$; e.g. for $M = (1/\sqrt{2})\begin{bmatrix} iI & I \\ I & iI \end{bmatrix}$ we obtain the <u>Hermitian great</u>

circle $\ell_s = \{\underline{P} : \underline{P} = f(P_1 P_2), P_2 P_1^* - P_1 P_2^* = 0\}$.

In analogy to the scalar case we define, for any n, the spherical distance $E_s(\underline{P},\underline{Q})$. We set

$$E_s(\underline{P},\underline{Q}) = \arcsin \chi(\underline{P},\underline{Q}). \tag{3.14}$$

Thus, $E_s(\underline{P},\underline{Q})/\chi(\underline{P},\underline{Q}) \to 1$ as $\chi(\underline{P},\underline{Q}) \to 0$. $0 \le \chi(\underline{P},\underline{Q}) \le 1$ implies

$$0 \le E_s(\underline{P},\underline{Q}) \le \pi/2. \tag{3.15}$$

For $n = 1$ $E_s(\underline{P},\underline{Q})$ is the arc length of the geodesic arc on the sphere of diameter 1 connecting $\underline{P}$ and $\underline{Q}$. Theorem 3.4 implies that a) $E_s(\underline{P},\underline{Q}) = 0$ if and only if $\underline{P} = \underline{Q}$, and b) $E_s(\underline{P},\underline{Q}) = E_s(\underline{Q},\underline{P})$. To prove the triangle inequality it sufficies to establish the following special case.

PROPOSITION 3.10. *If* $E_s(\underline{0},\underline{P}) + E_s(\underline{0},\underline{Q}) \le \pi/2$, *then* $E_s(\underline{P},\underline{Q}) \le E_s(\underline{0},\underline{P}) + E_s(\underline{0},\underline{Q})$.

This proposition is equivalent to the following inequality for the spectral norm.

THEOREM 3.11. *Let*

$$L_1 = \{\ell_1,\dots,\ell_n\},\quad L_2 = \{\ell_1',\dots,\ell_n'\},\quad \ell_i \ge 0,\ \ell_i' \ge 0,\quad \ell_i^2 + \ell_i'^2 = 1,\quad i = 1,\dots,n,$$

$$M_1 = \{m_1,\dots,m_n\},\quad M_2 = \{m_1',\dots,m_n'\},\quad m_i \ge 0,\ m_i' \ge 0,\quad m_i^2 + m_i'^2 = 1,\quad i = 1,\dots,n, \tag{3.16}$$

$$\ell = \max_i \ell_i, \qquad m = \max_i m_i,$$

and let U *and* V *be unitary* $n \times n$ *matrices. If* $\ell^2 + m^2 \le 1$, *then*

$$\|L_1 U M_2 + L_2 V M_1\| \le \ell(1 - m^2)^{1/2} + m(1 - \ell^2)^{1/2}.$$

This inequality, which is of independent interest, has been recently proved by London [7]. Indeed, London proves a more general result, by assuming only $\|U\| \le 1$, $\|V\| \le 1$, instead of the above assumption $UU^* = VV^* = I$. Thus, Proposition 3.10 is established and E_s is a distance.

THEOREM 3.12. *The function* $E_s(\underline{P},\underline{Q}) = \arcsin \chi(\underline{P},\underline{Q})$ *defines a metric for the space* P.

For the chordal distance it follows that, if $\underline{M} \neq \underline{P}$ and $\underline{M} \neq \underline{Q}$, then we have strict inequality in the triangle inequality $\chi(\underline{P},\underline{Q}) < \chi(\underline{P},\underline{M}) + \chi(\underline{M},\underline{Q})$. For the spherical distance the situation is different. Theorem 3.12 implies, and indeed is equivalent to, the following statement.

THEOREM 3.13. *Let $\lambda(\Gamma)$ be the arc length of the curve Γ with respect to the chordal distance. Then for any pair of points $(\underline{P},\underline{Q})$ there exists a smooth curve Γ_0 joining $\underline{P}$ to $\underline{Q}$ such that $\lambda(\Gamma_0) = E_s(\underline{P},\underline{Q}) = \min \lambda(\Gamma)$. Here the minimum is taken over all rectifiable curves Γ joining $\underline{P}$ to $\underline{Q}$.*

Thus, E_s is intrinsic in the sense of Alexandrov [1]. The curve Γ_0 may be chosen to lie in a great circle.

4. Non-Euclidean geometry

The unitary $2n \times 2n$ matrices used in the spherical geometry will now be replaced by $\underset{\sim}{J}$-unitary matrices. To define these matrices we use the following $2n \times 2n$ matrix

$$(4.1) \qquad \underset{\sim}{J} = \begin{bmatrix} I & 0 \\ 0 & -I \end{bmatrix},$$

where, as always, each block is a $n \times n$ matrix. The $2n \times 2n$ matrix $\underset{\sim}{P}$ is called *$\underset{\sim}{J}$-unitary* if

$$(4.2) \qquad \underset{\sim}{P}\,\underset{\sim}{J}\,\underset{\sim}{P}^* = \underset{\sim}{J};$$

see [9]. Using the block notation (3.3) for $\underset{\sim}{P}$, we see that $\underset{\sim}{P}$ is $\underset{\sim}{J}$-unitary if and only if the following three relations hold:

$$(4.3) \qquad P_1P_1^* - P_2P_2^* = I,$$

$$(4.4) \qquad P_3P_3^* - P_4P_4^* = -I,$$

$$(4.5) \qquad P_1P_3^* - P_2P_4^* = 0.$$

The $\underset{\sim}{J}$-unitary $2n \times 2n$ matrices form a group. If $\underset{\sim}{P}$ is $\underset{\sim}{J}$-unitary, then so is $\underset{\sim}{P}^*$. The following analogue of Proposition 3.2 holds.

PROPOSITION 4.1. *Let P_1 and P_2 be $n \times n$ matrices satisfying* (4.3). *Then there exist $n \times n$ matrices P_3 and P_4 such that the*

$2n \times 2n$ matrix $\underset{\sim}{P}$, given by (3.3), is $\underset{\sim}{J}$-unitary. The completing matrix $(P_3\, P_4)$ is unique up to a premultiplication by a unitary $n \times n$ matrix.

The structure theorem for these matrices is as follows.

THEOREM 4.2. Let $\underset{\sim}{P}$ be a given $\underset{\sim}{J}$-unitary $2n \times 2n$ matrix. There exist four unitary $n \times n$ matrices U_1, U_2, V_1 and V_2, and two diagonal matrices Λ_1 and Λ_2 satisfying

$$(4.6) \qquad \Lambda_1 = \{\lambda_1', \ldots, \lambda_n'\}, \quad \Lambda_2 = \{\lambda_1, \ldots, \lambda_n\}, \quad \lambda_i' \geq 0, \quad \lambda_i \geq 0, \quad \lambda_i'^2 - \lambda_i^2 = 1, \quad i = 1, \ldots, n$$

such that

$$(4.7) \qquad \underset{\sim}{P} = \begin{bmatrix} P_1 & P_2 \\ P_3 & P_4 \end{bmatrix} = \begin{bmatrix} U_1 & 0 \\ 0 & U_2 \end{bmatrix} \begin{bmatrix} \Lambda_1 & \Lambda_2 \\ \Lambda_2 & \Lambda_1 \end{bmatrix} \begin{bmatrix} V_1 & 0 \\ 0 & V_2 \end{bmatrix} = \begin{bmatrix} U_1\Lambda_1 V_1 & U_1\Lambda_2 V_2 \\ U_2\Lambda_2 V_1 & U_2\Lambda_1 V_2 \end{bmatrix}.$$

As in the unitary case this factorization is not unique. We denote the set of all matrices $(P_3\, P_4)$ satisfying (4.4) by J^-, and the set of all matrices $(P_1\, P_2)$ satisfying (4.3) by J^+. We also set

$$(4.8) \qquad \begin{aligned} \Delta^- &= \{\underline{P} : \underline{P} = f(P_3\, P_4), \quad P_3P_3^* - P_4P_4^* < 0\}, \\ \Delta^+ &= \{\underline{P} : \underline{P} = f(P_1\, P_2), \quad P_1P_1^* - P_2P_2^* > 0\}. \end{aligned}$$

Note that $\Delta^- = \Delta_s^-(\underline{0}, 1/\sqrt{2}) = \Delta_e^-(\underline{0}, 1)$ and $\Delta^+ = \Delta_s^+(\underline{0}, 1/\sqrt{2}) \supset \Delta_e^+(\underline{0}, 1)$. We call Δ^- the unit disk. The following is the analogue of Proposition 3.1.

PROPOSITION 4.3. Let $\underline{P}$ be a given point in the unit disk Δ^-. There exists a matrix $(P_3\, P_4)$ such that $(P_3\, P_4) \in f^{-1}[\underline{P}]$ and $(P_3\, P_4) \in J^-$. Moreover, $(\hat{P}_3\, \hat{P}_4) \in f^{-1}[\underline{P}]$ and $(\hat{P}_3\, \hat{P}_4) \in J^-$ if and only if

$$(3.2') \qquad (\hat{P}_3\, \hat{P}_4) = U(P_3\, P_4), \quad UU^* = I.$$

Let $\underline{P} \in \Delta^-$ be given and let $(P_3\, P_4) \in f^{-1}[\underline{P}]$, $(P_3\, P_4) \in J^-$. We complete $(P_3\, P_4)$ to a $\underset{\sim}{J}$-unitary matrix $\underset{\sim}{P}$. It follows by Propositions 4.1 and 4.3 that this $\underset{\sim}{J}$-unitary matrix $\underset{\sim}{P}$, corresponding to the given point $\underline{P} \in \Delta^-$, is given up to block-equivalence (3.4). We set $\check{\underline{P}} = f(P_1\, P_2)$; the point $\check{\underline{P}} \in \Delta^+$ depends only on $\underline{P} \in \Delta^-$ and we call $\check{\underline{P}}$

the point inverse to $\underline{P}$. For $n=1$, let $z=\zeta_3/\zeta_4$ be in $\Delta^-: |z|<1$. We may set $\zeta_3 = z/(1-|z|^2)^{1/2}$, $\zeta_4 = 1/(1-|z|^2)^{1/2}$. The matrix

$$(4.9)\quad \underset{\sim}{P} = \begin{bmatrix} \zeta_1 & \zeta_2 \\ \zeta_3 & \zeta_4 \end{bmatrix} = \begin{bmatrix} 1/(1-|z|^2)^{1/2} & \bar{z}/(1-|z|^2)^{1/2} \\ z/(1-|z|^2)^{1/2} & 1/(1-|z|^2)^{1/2} \end{bmatrix}$$

is $\underset{\sim}{J}$-unitary and the point $\check{z} = \zeta_1/\zeta_2 = 1/\bar{z}$ is inverse to the given point z.

Equality (2.5) defined the inverse point $\check{\underline{P}}$ for every finite point $\underline{P} = f(P\,I)$, for which $|P| \neq 0$. Every point $\underline{P}$ of Δ^- is finite and if for such a point $\underline{P} = f(P\,I)$ $|P| \neq 0$, then the two definitions yield the same inverse point $\check{\underline{P}}$. However, our present definition defines $\check{\underline{P}}$ for every point $\underline{P} = f(P_3\,P_4)$ of Δ^-; i.e. also in the case when $|P_3| = |P_2| = 0$ and so $\check{\underline{P}}$ is an infinite point of Δ^+.

Let $\underline{P}$ and $\underline{Q}$ be points in Δ^- and let $\underset{\sim}{P}$ and $\underset{\sim}{Q}$ be corresponding $\underset{\sim}{J}$-unitary matrices. We set

$$(4.10)\quad \underset{\sim}{R} = \begin{bmatrix} R_1 & R_2 \\ R_3 & R_4 \end{bmatrix} = \underset{\sim}{P}\,\underset{\sim}{J}\,\underset{\sim}{Q}^*,$$

and define

$$(4.11)\quad \rho(\underline{P},\underline{Q}) = \|R_2\|\,.$$

$\rho(\underline{P},\underline{Q})$ is well defined, i.e. depends only on the points $\underline{P}$ and $\underline{Q}$ in Δ^-. As $\underset{\sim}{R}$ is $\underset{\sim}{J}$-unitary it follows that $\|R_2\| = \|R_3\|$ and $\|R_1\|^2 = 1 + \|R_2\|^2$. $\rho(\underline{P},\underline{Q}) = 0$ if and only if $\underline{P} = \underline{Q}$ and $\rho(\underline{P},\underline{Q}) = \rho(\underline{Q},\underline{P})$. In general, the triangle inequality does not hold for $\rho(\underline{P},\underline{Q})$. For each pair of points $(\underline{P},\underline{Q})$ in Δ^- we define the <u>pseudo-chordal distance</u> $\psi(\underline{P},\underline{Q})$ by

$$(4.12)\quad \psi(\underline{P},\underline{Q}) = \frac{\rho(\underline{P},\underline{Q})}{(1+\rho^2(\underline{P},\underline{Q}))^{1/2}}\,.$$

The triangle inequality holds for ψ and so we obtain

<u>THEOREM 4.4. The function</u> $\psi(\underline{P},\underline{Q})$ <u>defines a metric for</u> Δ^-.

By (4.12) $0 \leq \psi(\underline{P},\underline{Q}) < 1$. It also follows that

$$(4.12') \quad \psi(\underline{P},\underline{Q}) = \frac{\|R_2\|}{\|R_1\|}, \qquad \underset{\sim}{R} = \underset{\sim}{P}\,\underset{\sim}{J}\,\underset{\sim}{Q}^*.$$

For $n=1$ let $\underset{\sim}{P}$ be the matrix (4.9) corresponding to the point z, $|z|<1$ and let $\underset{\sim}{Q}$ be the analogous matrix corresponding to w, $|w|<1$. Then $\psi(z,w) = |z-w|/|1-z\bar{w}|$. This is the classical pseudo-chordal distance [2]. The following is the analogue of Theorems 2.1 and 3.5.

THEOREM 4.5. *The projectivity S maps Δ^- onto itself and keeps the pseudo-chordal distance invariant if and only if the corresponding matrices $\underset{\sim}{S}$ are of the form $\underset{\sim}{S} = s\underset{\sim}{T}$, $\underset{\sim}{T}\underset{\sim}{J}\underset{\sim}{T}^* = \underset{\sim}{J}$, where $s \neq 0$ is an arbitrary complex number.*

We call these projectivities *non-Euclidean motions*; for $n=1$ this is the classical case. The following analogue of Theorems 2.2 and 3.8 holds for the group of non-Euclidean motions.

THEOREM 4.6. *Let the points $\underline{P}$, $\underline{Q}$, $\underline{P}_o$ and $\underline{Q}_o$ in Δ^- be given and let $\underset{\sim}{P}$, $\underset{\sim}{Q}$, $\underset{\sim}{P}_o$ and $\underset{\sim}{Q}_o$ be corresponding $\underset{\sim}{J}$-unitary matrices. There exists a non-Euclidean motion mapping simultaneously $\underline{P}$ into $\underline{P}_o$ and $\underline{Q}$ into $\underline{Q}_o$ if and only if the two blocks R_3 and R_3^o of the matrices*

$$(4.10') \quad \underset{\sim}{R} = \begin{bmatrix} R_1 & R_2 \\ R_3 & R_4 \end{bmatrix} = \underset{\sim}{P}\,\underset{\sim}{J}\,\underset{\sim}{Q}^*, \qquad \underset{\sim}{R}_o = \begin{bmatrix} R_1^o & R_2^o \\ R_3^o & R_4^o \end{bmatrix} = \underset{\sim}{P}_o\,\underset{\sim}{J}\,\underset{\sim}{Q}_o^*$$

have the same set $(\lambda_1,\dots,\lambda_n)$ of singular values.

For $\underline{M}$, $\underline{P}$ and $\underline{Q}$ in Δ^-, $\underline{M} \neq \underline{P}$ and $\underline{M} \neq \underline{Q}$ the strict inequality $\psi(\underline{P},\underline{Q}) < \psi(\underline{P},\underline{M}) + \psi(\underline{M},\underline{Q})$ holds. Thus, ψ is not intrinsic and again it is desirable to introduce a larger intrinsic distance in Δ^-. In analogy to the scalar case, for any pair of points in Δ^-, we define the *non-Euclidean distance* $E_n(\underline{P},\underline{Q})$ by

$$(4.13) \quad E_n(P,Q) = \frac{1}{2}\log\frac{1+\psi(\underline{P},\underline{Q})}{1-\psi(\underline{P},\underline{Q})},$$

$E_n(\underline{P},\underline{Q})/\psi(\underline{P},\underline{Q}) \to 1$ as $\psi(\underline{P},\underline{Q}) \to 0$. The triangle inequality holds for E_n.

THEOREM 4.7. *The function* $E_n(\underline{P},\underline{Q})$ *defines a metric for* Δ^-.

This is equivalent to the following statement.

THEOREM 4.8. *Let* $\lambda(\Gamma)$ *be the arc length of the curve* Γ, $\Gamma \subsetneq \Delta^-$, *with respect to the pseudo-chordal distance. Then for any pair of points* $(\underline{P},\underline{Q})$ *in* Δ^- *there exists a smooth curve* Γ_o, $\Gamma_o \subset \Delta^-$, *joining* $\underline{P}$ *to* $\underline{Q}$ *such that* $\lambda(\Gamma_o) = E_n(\underline{P},\underline{Q}) = \min \lambda(\Gamma)$. *Here the minimum is taken over all rectifiable curves* Γ *in* Δ^- *joining* $\underline{P}$ *to* $\underline{Q}$.

Let

$$(4.14) \quad \underset{\sim}{H}_n(r) = \begin{bmatrix} \cosh^2 rI & 0 \\ 0 & -\sinh^2 rI \end{bmatrix}, \quad 0 < r < \infty.$$

For arbitrary $(P_3\, P_4) \in C_o(2n^2)$ we set

$$(4.14') \quad H_n(P_3,P_4,r) = (P_3\, P_4)\underset{\sim}{H}_n(r)(P_3\, P_4)^* = \cosh^2 r\, P_3P_3^* - \sinh^2 r\, P_4P_4^*.$$

The *non-Euclidean circle* $\gamma_n(\underline{0},r)$ with center at $\underline{0}$ and radius r, $0 < r < \infty$, and the corresponding *non-Euclidean disk* $\Delta_n^-(\underline{0},r)$ are defined as follows.

$$(4.15) \quad \begin{aligned} \gamma_n(\underline{0},r) &= \{\underline{P} : \underline{P} = f(P_3\, P_4),\ H_n(P_3,P_4,r) = 0\}, \\ \Delta_n^-(\underline{0},r) &= \{\underline{P} : \underline{P} = f(P_3\, P_4),\ H_n(P_3,P_4,r) < 0\}. \end{aligned}$$

Both lie in Δ^-; the Hermitian form (4.14') partitions Δ^- into $\binom{n+2}{2}$ sets. For $r_e = \tanh r_n$, $0 < r_n < \infty$, $0 < r_e < 1$, we obtain $\gamma_e(\underline{0},r_e) = \gamma_n(\underline{0},r_n)$ and $\Delta_e^-(\underline{0},r_e) = \Delta_n^-(\underline{0},r_n)$. $\gamma_n(\underline{0},r)$ and $\Delta_n^-(\underline{0},r)$ may be characterized as follows.

PROPOSITION 4.9. *For any* $r, 0 < r < \infty$, *the following holds for the sets defined by* (4.15):

$$(4.16) \quad \begin{aligned} \gamma_n(\underline{0},r) &= \{\underline{P} : \underline{P} = f(P_3\, P_4),\ P_3 = \sinh r\, U_1, P_4 = \cosh r\, U_2,\ U_1U_1^* = \\ &= U_2U_2^* = I\}, \\ \Delta_n^-(\underline{0},r) &= \{\underline{P} : E_n(\underline{0},\underline{P}) < r\}. \end{aligned}$$

Let $(M_3 M_4) \in J^-$ and let $\underline{M} = f(M_3 M_4)$ be the corresponding point in Δ^-. Let

$$\underset{\sim}{M} = \begin{bmatrix} M_1 & M_2 \\ M_3 & M_4 \end{bmatrix}$$

be a $\underset{\sim}{J}$-unitary completion of $(M_3 M_4)$. Let U_1 and U_2 be arbitrary unitary matrices and define $\underset{\sim}{S}$ by (3.13). This non-Euclidean motion maps $\underline{0}$ into $\underline{M}$. We define $\underset{\sim}{\tilde{H}}_n(r) = S^{-1}\underset{\sim}{H}_n(r)(S^{-1})^*$ and set $\tilde{H}_n(P_3,P_4,r) = (P_3\ P_4)\underset{\sim}{\tilde{H}}_n(r)(P_3\ P_4)^* = \cosh^2 r(P_3P_3^* - P_4P_4^*) + (P_3M_3^* - P_4M_4^*)(P_3M_3^* - P_4M_4^*)^*$. The non-Euclidean circle $\gamma_n(\underline{M},r)$ with center $\underline{M} \in \Delta^-$ and radius r, $0 < r < \infty$, and the corresponding disk $\Delta_n^-(\underline{M},r)$ are defined as maps of $\gamma_n(\underline{0},r)$ and $\Delta_n^-(\underline{0},r)$ under this non-Euclidean motion S; these sets are also characterized by the form $\tilde{H}_n(P_3,P_4,r)$.

The <u>Hermitian</u> <u>diameter</u> ℓ_n of Δ^- is defined by

$$\ell_n = \{\underline{P} : \underline{P} \in \Delta^-,\ \underline{P} = f(P_3\ P_4),\ P_4P_3^* - P_3P_4^* = 0\}.$$

Thus, ℓ_n is the set of all points $\underline{P} = f(P_3\ P_4)$ in Δ^- for which $K(P_3\ P_4) = (P_3\ P_4)\underset{\sim}{K}(P_3\ P_4)^* = 0$, where $\underset{\sim}{K}$ is defined by Eq. (2.7). The <u>non-Euclidean</u> <u>straight</u> <u>lines</u> are defined as the maps of ℓ_n under non-Euclidean motions. The corresponding Hermitian forms are determined by $\underset{\sim}{\tilde{K}} = \underset{\sim}{S}^{-1}\underset{\sim}{K}(\underset{\sim}{S}^{-1})^*$. The curve Γ_0 of Theorem 4.8 may be chosen to lie in a non-Euclidean straight line.

References

[1] BUSEMANN, H, Convex surfaces, Interscience, New York 1958.

[2] CARATHEODORY, C., Theory of functions of a complex variable, Vol.1, Chelsea, New York 1954.

[3] GANTMACHER, F.R., The theory of matrices, Vol.1, Chelsea, New York 1959.

[4] HUA, L.K., Geometries of matrices III. Fundamental theorems in the geometries of symmetric matrices, Trans. Amer. Math. Soc. <u>61</u> (1947), 229-255.

[5] KELLEY, J.L., General topology, Van Nostrand, Princeton, N.J. 1955.

[6] LANCASTER, P., Theory of matrices, Academic Press, New York 1969.

[7] LONDON, D., An inequality for the spectral norm of certain matrices, Linear and Multilinear Algebra, <u>14</u>(1983), 37-44.

[8] MARSHALL, A.W. and I. OLKIN, Inequalities: theory of Majorization and the applications, Academic Press, New York 1979.

[9] POTAPOV, V.P., The multiplicative structure of J-contractive

matrix functions, Amer. Math. Soc. Translations, Ser. II 15 (1960), 131-243.

[10] SCHWARZ, B. and A. ZAKS, Matrix Möbius transformations, Comm. Algebra (9) (19) (1981), 1913-1968.

[11] SIEGEL, C.L., Symplectic geometry, Amer. J. Math. 65 (1943), 1-86.

Acknowledgement. This research was supported by the Fund for the Promotion of Research at the Technion.

Technion-Israel Institute of Technology
Department of Mathematics
Haifa, Israel

TEN OPEN PROBLEMS CONNECTED WITH HERMITIAN AND KÄHLER MANIFOLDS

Yum-Tong Siu (Cambridge, MA)

1. If M is a compact complex manifold which admits a Hermitian holomorphic line bundle whose curvature form is everywhere semipositive and is strictly positive on a nonempty open subset G of M, is M Moišeson? The special case when G is dense in M is known as the Grauert-Riemenschneider conjecture. The case where $M-G$ is of measure zero in M was recently proved by Siu. The case when M is Kähler follows from the usual proof of the vanishing theorem of Kodaira and was first observed by Riemenschneider.

2. Let M be a compact complex manifold and L be a Hermitian holomorphic line bundle over M whose curvature form is positive semidefinite everywhere and positive definite at some point. Prove that the infimum of $\lambda(M, L^k)$ for all positive integers k is positive, where $\lambda(M, L^k)$ is the smallest positive eigenvalue of the Laplacian $\bar{\partial}^*\bar{\partial}$ on the Hilbert space of all global L^2 sections of L^k over M. (This statement, if proved, would answer in the affirmative the preceding question.)

3. Is every complex surface admitting a negatively curved Kähler-Einstein metric biholomorphic to a quotient of the ball? The case with additional pinching assumptions was proved by Siu-Yang.

4. Is every noncompact complete Kähler manifold of positive bisectional curvature Stein? (It is known that there are holomorphic functions separating points and giving local coordinates.)

5. Is every n-dimensional complete Kähler manifold of positive sectional curvature biholomorphic to $\mathbb{C}^n$? The case $n=2$ was recently proved by Mok.

6. Do there exist always nonconstant bounded holomorphic functions on simply connected complete Kähler manifold with sectional curvature bounded from above by a negative number?

7. Is every simply connected complete Kähler manifold with negative bisectional curvature Stein? It is not known whether such a manifold is necessarily noncompact.

8. Is every compact Kähler manifold with nonnegative bisectional curvature and positive Ricci curvature biholomorphic to a Hermitian symmetric manifold? The case of complex dimension three was recently proved by Bando.

9. Is every compact Kähler-Einstein manifold with nonnegative bisectional curvature locally symmetric. The case of nonnegative sectional curvature was proved by A. Gray.

10. On an irreducible compact Hermitian symmetric manifold of rank >1, is every Kähler metric with nonnegative bisectional curvature necessarily an invariant metric?

Department of Mathematics
Harvard University
Science Center, One Oxford Street
Cambridge, MA 02138, U.S.A.

SOME OPEN PROBLEMS ON HOLOMORPHIC FUNCTIONS OF ONE VARIABLE

Jan Stankiewicz (Rzeszów)

1. Let $U = \{z: |z| < 1\}$, $St = \{f(z) = z + a_2 z^2 + \dots : \mathrm{Re}[z f'(z)/f(z)] > 0 \text{ in } U\}$, $f_\varepsilon(z) = [f(z) + \varepsilon z]/(1+\varepsilon)$, and

$$N_\delta(f) = \{g(z) = z + b_2 z^2 + \dots : \sum_{n=2}^{\infty} n|a_n - b_n| < \delta\}.$$

In 1892 S. Ruscheweyh proved that if for every ε, $|\varepsilon| < \delta$, a function $f_\varepsilon(z) \in St$, then the whole neighbourhood $N_\delta(f)$ is contained in St. Q.I. Rahman and J. Stankiewicz proved that if, for some positive integer $n \geq 2$ and every ε, $|\varepsilon| < \delta$, a function $f_{n,\varepsilon}(z) = f(z) + z^n \in St$, then for every ε, $|\varepsilon| < \delta$, the function $f_\varepsilon(z) \in St$ and hence $N_\delta(f) \subset St$. For the converse theorem we only have that if for every ε, $|\varepsilon| < \delta$, $f_\varepsilon(z) \in St$, then for every ε, $|\varepsilon| < \delta/n$, we have $f_{n,\varepsilon} \in St$.

CONJECTURE 1 (Rahman, Ruscheweyh, and Stankiewicz). If for some positive integer n and for every ε, $|\varepsilon| < \delta$, a function $f_{n,\varepsilon} \in St$, then $N_{n\delta}(f) \subset St$.

Remark. These problems are also investigated for many other subclasses of univalent functions.

2. We say that f is subordinate to F in $U = \{z: |z| < 1\}$ and write $f \prec F$ if there exists a function $w(z)$, $|w(z)| \leq |z|$ in U, such that $f(z) = F(w(z))$. Analogously, we say that f is majorized by F and write $f \ll F$ if there exists a function $b(z)$, $|b(z)| \leq 1$ in U, such that $f(z) = b(z)F(z)$.

In 1970 M.S. Robertson introduced a concept of quasisubordination $f <\!\prec F$ as follows: $f <\!\prec F \iff \exists_g\ f \ll g$ and $g \prec F$. In 1977 J. Stankiewicz introduced a concept of quasimajorization $f \prec\!< F$ as follows: $f \prec\!< F \iff \exists_h\ f \prec h$ and $h \ll F$. He also proved that $f \prec\!< F \Rightarrow f <\!\prec F$. The converse implication $f <\!\prec F \Rightarrow f \prec\!< F$ is, in general, false (J. Krzyż and J. Stankiewicz). It is known that if $F(z) \equiv z$ then the last implication holds for every holomorphic function $f(z)$.

PROBLEM. It will be very interesting to find some subclasses A and B of holomorphic functions such that for all $f \in A$ and $F \in B$ the identity $f < \prec F \Longleftrightarrow f \prec < F$ holds true.

3. For the functions $f(z) = a_1 z + a_2 z^2 + \dots$, $g(z) = b_1 z + b_2 z^2 + \dots$ we define the convolution (or: the Hadamard product) as follows:

$$(f * g)(z) := a_1 b_1 z + a_2 b_2 z^2 + \dots .$$

Let S^c denote the class of functions of the form $f(z) = a_1 z + a_2 z^2 + \dots$, such that $f(U)$ is a convex domain.

CONJECTURE 2 (H.S. Wilf 1961). $f \prec F$ and $F, G \in S^c \Longrightarrow f * G \prec F * G$.

CONJECTURE 3 (Rahman and Stankiewicz 1980). $f \prec F$, $g \prec G$, and $F, G \in S^c \Longrightarrow f * g \prec F * G$.

Conjecture 2 was proved in 1973 by S. Ruscheweyh and T. Sheil-Small. Conjecture 3 is still open. Rahman and Stankiewicz proved it only under the additional condition that either f or g belongs to the class S^c.

Institute of Mathematics and Physics
Ignacy Łukasiewicz Technical University of Rzeszów
Poznańska 2, PL-35-084 Rzeszów, Poland

A STRENGTHENED CONTOUR-AND-SOLID PROPERTY FOR LIPSCHITZ FUNCTIONS AND EXTENSION OF THE DERIVATIVE TO THE BOUNDARY

Promarz Melikovič Tamrazov (Kiev)

Summary. A strengthened contour-and-solid property for Lipschitz functions is proved as a sharpening of a result of [4]. The property is then applied to generalize a theorem of [3] on continuous extension of the derivative of a function, holomorphic in an open set, to the boundary. (© Институт математики АН УССР, 1983, preprint 83.35).

1. Let $\overline{\mathbb{C}}$ denote the one-point-compactification of the complex plane $\mathbb{C}$. For an open set $G \subset \mathbb{C}$, denote by ∂G and $\overline{\partial G}$ the boundary of G in $\mathbb{C}$ and $\overline{\mathbb{C}}$, respectively. For a function f on $\overline{G}$,[1] we consider the contour and solid moduli of continuity $\omega_{\partial G}(f,\delta)$ and $\omega_{\overline{G}}(f,\delta)$ (cf. [3, 4]). We shall prove the following result which strengthens that part of Theorem 9 in [4] which is connected with Lipschitz functions:

THEOREM 1. *Suppose that G is an open set in $\mathbb{C}$ whose boundary ∂G contains at least two points, and f is a continuous function on $\overline{G}$, holomorphic in G and satisfying the contour Lipschitz condition*

$$\omega_{\partial G}(f,\delta) \le \beta\delta \quad \forall\, \delta > 0. \tag{1}$$

Then the following statements hold true: a) *If G is bounded, then*

$$\omega_{\overline{G}}(f,\delta) \le \delta \sup_{x\in(0,\delta]} \omega_{\partial G}(f,x)/x \quad \forall\, \delta > 0 \tag{2}$$

and

$$|f'(z)| \le \overline{\lim_{x\to 0}}\, \omega_{\partial G}(f,x)/x \quad \forall\, z \in G. \tag{3}$$

b) *If $z=\infty$ is an accumulation point of $\overline{\partial G}$ and the function $(\log|f(z)|)/\log|z|$ is bounded from above as $z \to \infty$ within each unbounded connected component of G, then the estimates* (2) *and* (3) *remain valid*. c) *If $z=\infty$ is an isolated point of $\overline{\partial G}$ and $f(z) = o(|z|^2)$ $(z\to\infty)$, then* (2) *and* (3) *remain valid as well.*

[1] $\overline{G} = G \cup \partial G$

COROLLARY. Suppose that G is an open set in $\mathbb{C}$ whose complement contains at least two points, and f is a continuous function on $\overline{G}$, holomorphic in G and satisfying the condition

$$(4) \qquad \omega_{\partial G}(f,\delta) = o(\delta) \quad (\delta \to 0).$$

Then the following statements hold true: a) If G is bounded, then $f'(z) \equiv 0$ in G and

$$(5) \qquad \omega_{\overline{G}}(f,\delta) = o(\delta) \quad (\delta \to 0).$$

b) If $z = \infty$ is an accumulation point of $\overline{\partial G}$ and the function $(\log|f(z)|)/\log|z|$ is bounded from above as $z \to \infty$ within each unbounded connected component of G, then $f'(z) \equiv 0$ in G and the relation (5) remains valid. c) If $z = \infty$ is an isolated point of $\overline{\partial G}$ and $f(z) = o(|z|^2)$ $(z \to \infty)$, then $f'(z) \equiv 0$ and (5) remains valid as well.

2. We shall also prove the following result which gives a generalization of Theorem 5.1 in [3]: Let $f'_{\partial G}$ and $f'_{\overline{G}}$ denote the contour and solid derivatives of a function f (cf. [3]). The contour derivative may be defined only on the set $(\partial G)_1$ of accumulation points of ∂G. Let further $(\partial G)_*$ denote the set of all irregular boundary points of G with the property that for each of them no portion of ∂G centred at that point is removable for bounded holomorphic functions.

THEOREM 2. Suppose that G is an open set in $\mathbb{C}$ for which $(\partial G)_1$ is nonempty, and f is a continuous function on $\overline{G}$, holomorphic in G and satisfying the condition

$$\omega_{\partial G}(f,\delta) \le \beta\delta \quad \forall\ \delta > 0.$$

If the point $z = \infty$ belongs to a boundary component of a connected component of G and this boundary component does not reduce to a point, we assume additionally that the function $(\log|f(z)|)/\log|z|$ is bounded from above as $z \to \infty$. Then, if on the set $(\partial G)_1$ there exists the contour derivative $f'_{\partial G}$ and it is continuous, then

$$\lim_{\zeta \to z,\ \zeta \in G} f'(\zeta) = f'_{\partial G}(z) \quad \forall\ z \in (\partial G)_1 \setminus (\partial G)_*.$$

3. The proofs will be given with help of the method developed in [3]. We start with a lemma: Let D be a domain in $\overline{\mathbb{C}}$ whose boundary $\overline{\partial D}$ contains at least three points and let $z_0 \in D \cap \mathbb{C}$. Then it is well known that the universal cover $\hat{D}$ of D is conformally equivalent to the unit

disc. Denote by p the projection $\hat{D} \longrightarrow D$, and by σ an arbitrarily chosen conformal homeomorphism of the open disc $|w| < 1$ onto $\hat{D}$, for which $p(\sigma(0)) = z_0$ and $(p \circ \sigma)'(0) > 0$. Let

(6) $\quad R(w) = R_{D,z_0}(w) \underset{\text{def}}{=} p(\sigma(w))$ for $|w| < 1$.

By definition, it is clear that $R(0) = z_0$, $R'(0) > 0$, and that the mapping $R = R_{D,z_0}$ is uniquely determined by D and z_0 as a local homeomorphism; therefore

(7) $\quad R'(w) \neq 0 \;\; \forall\, w\colon |w| < 1.$

LEMMA. <u>Suppose that</u> ∂D <u>is not a set of logarithmic capacity zero. Then the function</u> $R(w)$ <u>has its angular boundary values</u> $R(e^{i\theta})$ <u>almost everywhere on the circle</u> $|z| = 1$ <u>and, for almost all</u> $\theta \in [0, 2\pi]$, <u>these angular boundary values belong to</u> ∂D <u>and are regular boundary points of</u> D.

<u>4</u>. <u>Proof of the lemma</u>. In D there is the generalized Green function $g(\zeta, z)$. It is well known (cf. [2], pp. 336-338) that

(8) $\quad g(R(0), R(w)) = -\log|w\,B(w)|$ for $|w| < 1$,

where
$$B(w) = \prod_{\nu}(|t_\nu|/t_\nu)(t_\nu - w)/(1 - \bar{t}_\nu w)$$

is the Blaschke product with simple zeros t_ν, for which the series $\sum_\nu (1 - |t_\nu|)$ converges. The function B(w) has its angular boundary values a.e. (= almost everywhere) on the circle $|w| = 1$ and they are equal 1 in absolute value. Consequently, the function (8) has its angular boundary values on the boundary set $E \subset \{w\colon |w| = 1\}$ of the total measure and they are equal 0. Therefore, if R(w) has at $w_1 \in E$ its angular boundary value $R(w_1)$, then $R(w_1) \in \overline{\partial D}$ (otherwise we had $g(R(0), R(w_1)) > 0$). Thus we have proved that at almost all boundary points, where the function R(w) has its angular boundary values, they are in $\overline{\partial D}$ and, by the uniqueness theorem for analytic functions, we may suppose that these values belong to ∂D.

If D is a bounded domain, then the function R(w) is bounded and hence it has its angular boundary values a.e. on the circle $|w| = 1$. If $\overline{\mathbb{C}} \setminus D$ contains a nondegenerate connected component K, then there is a conformal homeomorphism $\varphi(z)$ of $\overline{\mathbb{C}} \setminus K$ onto a bounded domain Q such that the inverse mapping $\psi(\zeta) = \varphi^{-1}(\zeta)$ is continuously extendable onto $\overline{Q}$ in the generalized sense and $\psi(\zeta)$ is infinite at most at two points of $\overline{Q}$. Then $R_{D,z_0}(w) = \psi(R_{Q,\varphi(z_0)}(w))$. Moreover, the function $R_{Q,\varphi(z_0)}(w)$ is not identically constant; it is also bounded and hence it has its angular boundary values a.e. on the unit circle; their ψ-images are finite and the function $\psi(\zeta)$ is continuous on $\overline{Q}$ in the generalized sense. Therefore the function $R_{D,z_0}(w)$ has its angular boundary values at any point where $R_{Q,\varphi(z_0)}(w)$ has the same property. This means that in the case in question $R_{D,z_0}(w)$ has its angular boundary values a.e.

It remains to consider the case where $\overline{\mathbb{C}} \setminus D$ is a totally disconnected set. At first we suppose that $z = \infty \in D$.

Denote by $B^*(w)$ the Blaschke product with simple zeros t^*_ν ranging over the whole set of roots of the equation $R(t) = \infty$. Then the function

$$b(w) \underset{\text{def}}{=} \log\{[R(w) - z_0]B^*(w)/w\,B(w)\},$$

by $t^*_\nu \neq 0$, satisfies the conditions

$$b(0) = \log[R'(0)\,B^*(0)/B(0)] \neq \infty,$$

$$b(t^*_\nu) = \log\{[t^*_\nu B(t^*_\nu)]^{-1} \lim_{w \to t^*_\nu} [R(w) - z_0]B^*(w)\} \neq \infty \;\; \forall\, \nu,$$

as the function $R(w)$ has a simple pole at t^*_ν (as it is locally univalent in $|w| < 1$). Thus the function $b(w)$ is holomorphic in $|w| < 1$. On the other hand, the function

$$s(z) \underset{\text{def}}{=} \log|z - z_0| + g(z_0, z) - g(\infty, z)$$

is harmonic and bounded in D; moreover,

$$s(R(w)) = \operatorname{Re}\log\{[R(w) - z_0]B^*(w)/w\,B(w)\} = \operatorname{Re} b(w).$$

Thus $\operatorname{Re} b(w)$ is bounded in $|w| < 1$. Hence the function $b(w)$ has its angular boundary values a.e. on $|w| = 1$. Consequently, also the function

$$[R(w) - z_0]B^*(w)/w\,B(w) = \exp b(w)$$

has its boundary values a.e. on $|w| = 1$. Since both functions $B^*(w)$ and $B(w)$ have their angular boundary values equal 1 a.e. on $|w| = 1$, the angular boundary values exist also for $R(w)$ a.e. on $|z| = 1$.

Suppose then that $\overline{\mathbb{C}} \setminus D$ is a totally disconnected set and $z = \infty \in \overline{\partial D}$. With help of a fractional linear transformation $\varphi(z) = \zeta$ we can map D onto a domain Q such that $Q \ni \zeta = \infty$. As it is already proved, $R_{Q,\varphi(z_0)}(w)$ has its angular boundary values a.e. on $|w| = 1$ and they are different from $\varphi(\infty)$ (since $R_{Q,\varphi(z_0)} \neq \text{const}$). Hence, by the relation $R_{D,z_0}(w) = \varphi^{-1}(R_{Q,\varphi(z_0)}(w))$, we see that $R_{D,z_0}(w)$ has its angular boundary values a.e. on $|w|=1$. Summing up, the function $R(w)$ has its angular boundary values belonging to ∂D a.e. on $|w|=1$ in all the cases.

Now we are going to prove that, a.e. on $|w|=1$, these angular boundary values are regular boundary points of D. We start with applying the theorem of Evans (cf. [5], p. 76). By this theorem, if E is a compact set in $\mathbb{C}$ of logarithmic capacity zero, then there is a (positive) unit measure ν (Evans' measure), accumulated on E and such that the potential $U^{\nu}(x) = \int \log|\zeta - z|^{-1}\, d\nu(\zeta)$ is finite on $\mathbb{C}\setminus E$ and tends to $+\infty$ as z tends to an arbitrary point of E.

Firstly, suppose that $z=\infty \notin \overline{\partial D}$. Let E be a compact set on ∂D of logarithmic capacity zero and let ν denote the corresponding Evans' measure. Then the function U^{ν} is bounded from below on ∂D. If D is bounded, U^{ν} is bounded from below also in D; if D contains $z=\infty$, we have to replace U^{ν} by $-g(\infty,z)+U^{\nu}(z)$ which is harmonic and still bounded from below. Thus we can see that in D there is a harmonic function q, bounded from above, which tends to $-\infty$ as z tends from D to an arbitrary point of E, and such that for approaching from D an arbitrary point of $\partial D\setminus E$ all the limit values of q are finite and uniformly bounded from below on any piece of ∂D separable from E. Consider the function $q(R(w))$. It is bounded from above on $|w|<1$ and harmonic, so it has its angular boundary values finite a.e. on $|w| = 1$. Consequently, the set on $|w|=1$, where this function has its angular boundary values equal to $-\infty$, is linearly measurable and of linear measure zero. Since this set contains the set $\tilde{E}$ of all points $e^{i\theta}$ with the property that for each of them $R(w)$ has its angular boundary value that belongs to E, the set $\tilde{E}$ is linearly measurable and of linear measure zero.

Next, let A be a given F_σ-set on ∂D, of logarithmic capacity zero. Expressing, if necessary, A as a countable union of compact sets E_i, $i=1,2,\ldots$ we observe that the analogous union of $\tilde{E}_i$ is linearly measurable and of linear measure zero. Let $\tilde{A}$ be the set of all $e^{i\theta}$ with the property that for each of them $R(w)$ has its angular boundary value that belongs to A. Evidently $\tilde{A}$ coincides with the union of all $\tilde{E}_i$, so $\tilde{A}$ is linearly measurable and of linear measure zero. The set I of irregular boundary points of D is an F_σ-set (cf. [1], p. 244) and of logarithmic capacity zero. Thus the set of all $e^{i\theta}$, with the property that for each of them $R(w)$ has its angular boundary value which is itself an irregular boundary point of D, is a linearly measurable set and has linear measure zero.

The above statement has been established under the hypothesis that $z=\infty \notin \overline{\partial D}$. This assumption is, however, superfluous, as for $z=\infty \in \overline{\partial D}$ we can reduce the case to the previous one with help of a fractional linear transformation. Summing up all the conclusions established we arrive at the desired lemma.

5. Proof of Theorem 1. Without any loss of generality, we may assume that $f \neq \text{const}$. Let D be an arbitrary connected component of G and $z_0 \in D$, where $f'(z_0) \neq 0$. By Theorem 9 in [4] from our assumptions it follows that

$$(9) \qquad \omega_{\overline{G}}(f,\delta) \leq \beta\delta \quad \forall\ \delta > 0.$$

Let further E^0 denote the set of all those $\theta \in [0, 2\pi]$ for which there exist the angular boundary values $R(e^{i\theta})$ of the function $R(w) = R_{D,z_0}(w)$, which belong to ∂D and are regular boundary points of D. By the Lemma, E^0 is of measure 2π.

Let us fix a number $\tau \in (0,\pi)$ and, given functions f and $R(w)$, let

$$(10) \qquad F(w,\tau) = [f(R(w)) - f(R(we^{i\tau}))]/[R(w) - R(we^{i\tau})] \quad \text{for} \quad |w| < 1$$

be considered as a function of w. Denote by E^{τ} the set of all those $\theta \in E^0$ for which there are finite angular boundary values $R(e^{i(\theta+\tau)})$ of $R(we^{i\tau})$ satisfying the inequality $R(e^{i\theta}) \neq R(e^{i(\theta+\tau)})$. Here (cf. [3]) $\text{mes}\, E^{\tau} = 2\pi$, since otherwise (cf. [3]) $R(w) \equiv R(we^{i\tau})$ and $R'(0) = R'(0)e^{i\tau}$, what contradicts (7) and $\tau \in (0,\pi)$.

The function in the nominator of (10) has its angular boundary values finite on the boundary set of the total measure. By (9), the function (10) is bounded and holomorphic. By Lemma, on the boundary set of the total measure it has finite angular boundary values

$$(11) \qquad F(e^{i\theta}, \tau) = [f(R(e^{i\theta})) - f(R(e^{i(\theta+\tau)}))]/[R(e^{i\theta}) - R(e^{i(\theta+\tau)})] \quad \text{for} \quad \theta \in E^{\tau},$$

and hence it can be expressed as the Poisson integral

$$(12) \qquad F(w,\tau) = (1/2\pi)\int_0^{2\pi} F(e^{i\theta}, \tau)\, P(w,\theta)\,d\theta \quad \text{for} \quad |w| < 1$$

with the Poisson kernel $P(w,\theta) = \text{Re}[(e^{i\theta} + w)/(e^{i\theta} - w)]$. As shown in [3], the functions $R_{\tau}(e^{i\theta}) \underset{\text{def}}{=} R(e^{i(\theta+\tau)})$ of θ converge in measure to $R(e^{i\theta})$. This means that there is a sequence of $\tau_n \downarrow 0$ such that the functions $R_{\tau_n}(e^{i\theta})$ converge a.e. to $R(e^{i\theta})$, so, by the Lemma and relation (11),

(13) $\overline{\lim}_{\tau\to 0} |F(e^{i\theta}, \tau)| \le \overline{\lim}_{x\to 0} \omega_{\partial D}(f,x)/x$ for almost every $\theta \in [0, 2\pi]$.

From relations (12) and (13), and from the hypothesis of Theorem 1, which assures the uniform boundedness of the functions (11) with respect to $\theta \in E^{\tau}$ and all τ, we deduce the estimate $|f'(R(w))| = \lim_{\tau\to 0} |F(w,\tau)| \le \overline{\lim}_{x\to 0} \omega_{\partial D}(f,x)/x$ everywhere in the disc $|w| < 1$, that is,

$$|f'(z)| \le \overline{\lim}_{x\to 0} \omega_{\partial D}(f,x)/x \text{ everywhere in } D.$$

Consequently we arrive at (3) and this easily yields (2), thus concluding the proof.

Proof of the Corollary. It is sufficient to remark that (4) implies (1) with a suitable β. Hence we can apply Theorem 1 and, in the situation considered, the right-hand side of the estimate (3) vanishes and the relation (5) holds true.

6. Proof of Theorem 2. If $z = \infty$ is a boundary point for no connected component of G, from Theorem 1 (or from Theorem 9 in [4]) it follows that f satisfies the Lipschitz condition on $\overline{G}$.

If $z = \infty$ forms an one-point boundary component of G, then for any $r > 0$ we can construct, within the set $|z| > r$, a Jordan curve $\gamma_r \subset G$, separating the points $z = \infty$ and $z = 0$. Let G_r be that part of G which lies inside γ_r. The function f satisfies on ∂G_r the Lipschitz condition with some constant (not necessarily equal to β). As above, we can arrange that f satisfies the Lipschitz condition also on $\overline{G}_r$. The further proof of Theorem 2 has to be performed in this case, firstly, for G_r and then, owing to the arbitrariness of r, for G. Therefore, without any loss of generality, we may exclude from our considerations the case where $z = \infty$ forms an one-point boundary component of G, as it can be reduced to the previous case.

If $z = \infty$ belongs to a boundary component of one of the connected components of G, then from the additional condition given in Theorem 2 it follows that f satisfies the Lipschitz condition on $\overline{G}$. Thus in all the cases f satisfies the Lipschitz condition on $\overline{G}$. Without any loss of generality we may also assume that $(\partial G)_1 = \partial G$, since the isolated finite points are removable for f.

Let us fix a connected component D of G. From the Lemma and the proof of Theorem 1 we see that there is a sequence $\tau_n \downarrow 0$ such that the angular boundary values of the functions (6) and (10) satisfy the condition $\lim_{n\to\infty} F(e^{i\theta}, \tau_n) = f'_{\partial G}(R(e^{i\theta}))$ a.e. on $[0, 2\pi]$. Hence, by the properties of the function $F(e^{i\theta}, \tau)$ mentioned in the proof of Theorem 1, we get

$$\lim_{n\to\infty} \int_{E^{\tau_n}} F(e^{i\theta}, \tau_n)\, P(w,\theta)\,d\theta = \int_0^{2\pi} f'_{\partial G}(R(e^{i\theta}))\, P(w,\theta)\,d\theta.$$

The above formula, the equality (12), and the relations $\lim_{\tau\to 0+} F(w,\tau) = f'(R(w))$ for $|w| < 1$, resulting from (10), yield

$$(14) \qquad f'(R(w)) = (1/2\pi) \int_0^{2\pi} f'_{\partial G}(R(e^{i\theta}))\, P(w,\theta)\,d\theta \quad \text{for} \quad |w| < 1.$$

Consider now the complex-valued harmonic function $H(\zeta)$, defined as the solution of the generalized Dirichlet problem for G with the complex-valued boundary function $f'_{\partial G}(z)$. It is clear that $H(\zeta)$ is continuous on the union of G and the set of all regular boundary points of G, and at these points it attains the values $f'_{\partial G}(z)$. The function $H(R(w))$ is complex harmonic in $|w| < 1$ and bounded, so it has its angular boundary values $h(e^{i\theta})$ a.e. on $|w| = 1$ and

$$(15) \qquad H(R(w)) = (1/2\pi) \int_0^{2\pi} h(e^{i\theta})\, P(w,\theta)\,d\theta \quad \text{for} \quad |w| < 1.$$

a.e. on $[0, 2\pi]$

Since, by the Lemma, $\lim_{\varrho\uparrow 1} H(R(\varrho e^{i\theta})) = H(R(e^{i\theta})) = f'_{\partial G}(R(e^{i\theta}))$, then $h(e^{i\theta}) = f'_{\partial G}(R(e^{i\theta}))$. Hence, by (15), we have

$$H(R(w)) = (1/2\pi) \int_0^{2\pi} f'_{\partial G}(R(e^{i\theta}))\, P(w,\theta)\,d\theta \quad \text{for} \quad |w| < 1.$$

This relation, together with (14), gives $f'(R(w)) \equiv H(R(w))$ for $|w| < 1$, that is, $f'(\zeta) \equiv H(\zeta)$ in D.

The result obtained is valid for each connected component D of G, so $f'(\zeta) \equiv H(\zeta)$ in G. Consequently, the function f' can be continuously extended to all regular boundary points of ∂G by the values $H(z) = f'_{\partial G}(z)$. Finally, we observe that for every irregular boundary point z_0 of G, belonging to $(\partial G)_1 \setminus (\partial G)_*$, there is a portion of ∂G centred at z_0, removable for f, since this function is bounded on every bounded part of G. But $z_0 \in (\partial G)_1$, so the derivative $f'_{\partial G}(z_0)$ is well defined. Consequently, f' is continuously extendable to z_0 by the value $f'_{\partial G}(z_0)$. The proof is completed.

R e f e r e n c e s

[1] HAYMAN, W.K. and P.B. KENNEDY, Subharmonic functions I, Academic Press, London-New York-San Francisco 1976; Russian transl.: У. ХЕЙМАН, П. КЕННЕДИ, Субгармонические функции, Изд. "Мир", Москва 1980. (P. 244 in Engl. ed. corresponds to p. 262 in Russ. ed.)

[2] NEVANLINNA, R., Uniformisierung (Grundlehren der math. Wissenschaften 64), Springer-Verlag, Berlin-Göttingen-Heidelberg 1953; 2. Aufl., Berlin-Heidelberg-New York 1967; Russian transl.: Р. НЕВАНЛИННА, Униформизация, Изд. иностр. лит., Москва 1955. (Pp. 336-338 in Germ. eds. correspond to pp. 372-374 in Russ. ed.)

[3] TAMRAZOV, P.M. [ТАМРАЗОВ, П.М.], Контурные и телесные структурные свойства голоморфных функций комплексного переменного, Успехи мат. наук 28, 1 (1973), 131-161.

[4] ——, Контурно-телесные свойства голоморфных функций и отображений, Препринт 83.33, Институт математики АН Украинской ССР, Киев 1983, 17 pp.

[5] TSUJI, M., Potential theory in modern function theory, Maruzen Co., Tokyo 1959.

Институт математики
Академии наук УССР
ул. Репина 3
SU-252 601 Киев, СССР

TWO OPEN PROBLEMS CONNECTED WITH CAPACITIES

Promarz Melikovič Tamrazov (Kiev)

PROBLEM 1. Assume that G is an open set in the complex plane $\mathbb{C}$ whose boundary ∂G contains at least two points including a fixed point z_0. Let $f\colon \overline{G} \to \mathbb{C}$ be a bounded continuous function defined on $\overline{G} = G \cup \partial G$ and holomorphic in G. Suppose that μ is a positive nondecreasing function defined for $x > 0$ and satisfying the condition

$$(1) \qquad \sup_{x>0}[\mu(2x)/\mu(x)] < +\infty.$$

Prove that if

$$|f(z)| \leq \mu(|z - z_0|) \quad \forall\, z \in (\partial G) \setminus \{z_0\},$$

then

$$|f(z)| \leq c\mu(|z - z_0|) \quad \forall\, z \in \overline{G} \setminus \{z_0\}$$

with a finite c independent of z.

Remark 1. An analogous statement is proved in the following two cases: either if

1) The lower capacity density of ∂G in z_0 is positive [1]. This result contains solution of the problem posed in 1942 in Sewell's book or if

2) (1) is replaced by the condition that the function $\log \mu(e^t)$ is concave with respect to t [3]. In this case the result is proved with $c = 1$.

PROBLEM 2. Let (E^+, E^-) be a condenser in the complex plane $\mathbb{C}$ whose plates E^+, E^- are closed disjoint subsets of the interval $[-1, 1]$ of the real axis. Let (E_*^+, E_*^-) be a condenser whose plates E_*^+, E_*^- are the intervals $[d^+, 1]$, $[-1, d^-]$ of the real axis with d^+, d^- such that $-1 \leq d^-$, $d^+ \leq 1$, respectively; the logarithmic capacities of the plates satisfying the conditions

$$(2) \qquad \operatorname{Cap} E^+ = \operatorname{Cap} E_*^+, \quad \operatorname{Cap} E^- = \operatorname{Cap} E_*^-.$$

Prove that the logarithmic capacities of these condensers satisfy the inequality

$$\mathrm{Cap}(E^+,E^-) \geq \mathrm{Cap}(E_*^+,E_*^-)$$

and give the full description of the equality case.

Remark 2. The problem is a strengthening of Gončar's problem in which, instead of (2), there are the following conditions in terms of preserving the linear Lebesgue measures of the plates:

$$\mathrm{mes}\, E^+ = 1 - d^+, \quad \mathrm{mes}\, E^- = d^- + 1.$$

Gončar's problem has been solved in [2].

References

[1] TAMRAZOV, P.M., [ТАМРАЗОВ, П.М.], Контурные и телесные структурные свойства голоморфных функций комплексного переменного, Успехи Мат. наук 28, 1 (1973), 131-161.

[2] ——, Емкости конденсаторов. Метод перемешивания зарядов, Мат. сборник Н.С. 115 (157) (1981), 40-73.

[3] ——, Контурно-телесные свойства голоморфных функций и отображений, Препринт 83.33, Институт математики АН Украинской ССР, Киев 1983, 17 pp.

Институт математики
Академии наук УССР
ул. Репина 3
SU-252 601 Киев, СССР

EXTENSION OF CR-FUNCTIONS

Giuseppe Tomassini (Pisa)

Summary. Let Γ be a real hypersurface in $\mathbb{C}^n$, $n \geq 2$, oriented of class C^1, compact, connected with boundary $\partial\Gamma$. Suppose that $\partial\Gamma$ belongs to a C^∞-hypersurface M and there exists a relatively open subset A of M such that $\partial A = \partial\Gamma$, and that $\Gamma \cap M = \partial\Gamma$. Under these hypotheses, in a previous paper [2] G. Lupacciolu and the author have proved that if M is the zero-set of a pluriharmonic function on a neighbourhood of $\bar{D}$, every Lipschitz continuous CR-function f on $\Gamma^o = \Gamma \setminus \partial\Gamma$ extends uniquely by a function F, holomorphic on D, where D is an open set of $\mathbb{C}^n$ having $\Gamma \cup A$ as its boundary. Moreover, F is continuous on $D \cup \Gamma^o$. The present paper aims at generalizing this result to the case when M is Levi-flat or Levi-pseudoconcave with respect to D. The positive answer is obtained in two particular cases (Theorems 1 and 3).

1. Let Γ be a real hypersurface in $\mathbb{C}^n$, $n \geq 2$, oriented of class C^1, compact, connected with boundary $\partial\Gamma$. Assume the following conditions are verified:

I) $\partial\Gamma$ belongs to a C^∞-hypersurface M and there exists a relatively open subset A of M such that $\partial A = \partial\Gamma$,

II) $\Gamma \cap M = \partial\Gamma$.

Let D be the open set of $\mathbb{C}^n$ having $\Gamma \cup A$ as its boundary. In [2] the following theorem has been proved:

THEOREM. *Let M be the zero-set of a pluriharmonic function on a neighbourhood of $\bar{D}$. Then every Lipschitz continuous CR-function f on $\Gamma^o = \Gamma \setminus \partial\Gamma$ extends, in a unique way, by a function F, holomorphic on D and continuous on $D \cup \Gamma^o$.*

In particular, every holomorphic function on D, regular on $\bar{D}$

is completely determined by its values on Γ. Then it is natural to conjecture that such a property is still valid when M is Levi-flat or Levi-pseudoconcave with respect to D.

The aim of this paper is to give a positive answer to this conjecture in two particular cases (Theorems 1 and 3).

2. In the first case Γ and M verify the following conditions:

a) Γ is contained in the boundary of an open subset $\Omega \subset\subset \mathbb{C}^n$ defined by a C^∞-function ϕ, plurisubharmonic on a neighbourhood U of $\bar{\Omega}$;

b) M is the zero-set of a function $\rho \in C^\infty(U)$ and $\Gamma \subset \{z \in U : \rho(z) \geq 0\}$; moreover, on $\{z \in U : \rho(z) > 0\}$, ρ is plurisubharmonic and strictly plurisubharmonic if $n = 2$.

THEOREM 1. *Every Lipschitz continuous CR-function f on Γ^o extends by a function F, holomorphic on D and continuous on $D \cup \Gamma^o$.*

We make some remarks before going into the proof.

First, we may assume that U is a Stein domain, ϕ is plurisubharmonic on U, and, moreover, that f is continuous on a neighbourhood of Γ. For $\delta > 0$ we set $\Omega_\delta = \{z \in U : \phi(z) < \delta\}$ and $\Omega^+_\delta = \{z \in \Omega_\delta : \rho(z) > 0\}$ in such a way that $\Omega^+_0 = D$. Let $\rho_\varepsilon = \rho + \varepsilon(\phi - \delta)$, $\varepsilon > 0$; ρ_ε is plurisubharmonic on Ω^+_δ; therefore $\Omega^+_{\delta,\varepsilon} = \{z \ \Omega^+_\delta : \rho_\varepsilon(z) > 0\}$ is a Stein domain and $\mathcal{O}(\Omega^+_\delta)$ is a dense subspace of $\mathcal{O}(\Omega^+_{\delta,\varepsilon})$; [3]. For fixed $\varepsilon_0 > 0$ let $A^* \equiv A_{\varepsilon_0}$ be the intersection of $\bar{\Omega}$ with the hypersurface $\rho^* \equiv \rho_{\varepsilon_0} = 0$ and let $\hat{A}^*$ be the $\mathcal{O}(\Omega^+_\delta)$-envelope of A^*. $\hat{A}^*$ is contained in $\{z \in \Omega^+_\delta : \rho^*(z) \leq 0\}$ as follows from the fact that $\mathcal{O}(\Omega^+_\delta)$ is dense in $\mathcal{O}(\Omega^+_{\delta,\varepsilon})$ for every $\varepsilon > 0$.

Let $D(\varepsilon_0) = \{z \in D : \rho^*(z) > 0\}$ and for $g \in \mathcal{O}(\Omega_\delta)$ let $W_g = \{z \in \Omega^+_\delta : |g(z)| > \|g\|_{A^*}\}$; D is contained in $\bigcup_g W_g$.

LEMMA. *Let $\zeta \in \Omega^+_\delta$ be such that $\rho^*(\zeta) > 0$. Then there exist $g \in \mathcal{O}(\Omega^+_\delta)$ and a connected subset C verifying the following properties:*

1) $\bar{C} \subset \{z \in \bar{\Omega}_\delta : \rho^*(z) > 0\}$ *and* $C \cap D(\varepsilon_0)$, $G \cap (\Omega_\delta \setminus \bar{\Omega})$ *are connected*;

2) $G \cap \partial\Omega_\delta \neq \emptyset$ *and* $G \subset W'_g$, *the connected component of* W_g containing ζ;

3) $G \cap \{z \in \Omega^+_\delta : |g(z)| > \|g\|_{D(\varepsilon_0)}\} \neq \emptyset$.

Proof. As $\zeta \notin A^*$, there is a smooth, connected complex hypersurface Z of a neighbourhood of $\bar{\Omega}_\delta$ containing ζ and such that

$Z \cap \hat{A}^* = \emptyset$. Let $\psi = 0$ be its equation and let $Z_1 = Z \cap \Omega_\delta$. The closure of an irreducible component of Z_1 meets $\partial\Omega_\delta$; furthermore, because of the pseudoconvexity of $\partial\Omega_\delta$, the closure of an irreducible component of $Z_1 \cap (\Omega_\delta \setminus \bar{\Omega})$ also meets $\partial\Omega_\delta$. It follows that there exists a connected subset G of Z_1 containing ζ and such that $\bar{G} \cap \partial\Omega_\delta \neq \emptyset$ and $G \cap D(\varepsilon_0)$, $G \cap (\Omega_\delta \setminus \bar{\Omega})$ are connected.

Let h be a holomorphic function on Z_1 such that $\inf_{z\in C}|h(z)| > a > 0$ and $\sup_{z\in C}|h(z)| = +\infty$. Let H be a holomorphic extension of h on Ω_δ and set $g = H + \lambda\psi$ where $\lambda \in \mathcal{O}(\Omega_\delta^+)$. On a neighbourhood of $\hat{A}^*$ one has $g = \psi(H/\psi + \lambda)$ and H/ψ is holomorphic on $\hat{A}^*$. As $\hat{A}^*$ is $\mathcal{O}(\Omega_\delta^+)$-convex it is possible to choose λ in such a way that $\|g\|_{A^*} < a$. Then $\inf_{z\in C}|g(z)| > \|g\|_{A^*}$ and $\sup_{z\in C}|g(z)| = +\infty$ so that g and C satisfy the properties 1), 2), and 3) of the above lemma.

Now we are in position to prove Theorem 1.

3. <u>Proof of Theorem 1</u>. We use the same idea as in [2]. It is enough to prove that for every $\varepsilon_0 > 0$, f extends holomorphically on $D(\varepsilon_0) = \{z \in \Omega_0^+ : \rho_{\varepsilon_0}(z) > 0\}$. Let $\zeta_0 \in D(\varepsilon_0)$ and consider a function $g \in \mathcal{O}(\Omega_\delta^+)$ and a connected subset C as in the previous lemma. For $(z,\zeta) \in \Omega_\delta^+ \times \Omega_\delta^+$ we have $g(z) - g(\zeta) = \Sigma_{k=1}^n g_k(z,\zeta)(z_k - \zeta_k)$ where $g_1,\ldots,g_n \in \mathcal{O}(\Omega_\delta^+ \times \Omega_\delta^+)$. Let

$$\Phi_g(z,\zeta) = \frac{1}{g(z) - g(\zeta)} \sum_{k=1}^{n} g_k(z,\zeta)(z_k - \zeta_k)\,\Omega_k,$$

where

$$\Omega_k(z,\zeta) = \frac{(-1)^{n+k}}{(n-1)}\, C_n\, \frac{1}{(z_k-\zeta_k)|z-\zeta|}\, dz_1 \wedge \ldots \wedge dz_n$$

$$\wedge \left\{ \sum_{j=1}^{k-1} (-1)^j (z_j - \bar{\zeta}_j) d\bar{z}_1 \wedge \ldots \hat{j} \ldots \hat{k} \ldots \wedge d\bar{z}_n \right.$$

$$\left. + \sum_{j=k+1}^{n} (-1)^{j-1} (\bar{z}_j - \bar{\zeta}_j) d\bar{z}_1 \wedge \ldots \hat{k} \ldots \hat{j} \ldots \wedge d\bar{z}_n \right\},$$

$$C_n = (-1)^{n(n-1)/2}(n-1)!_{/(2\pi i)^n}, \quad k = 1,\ldots,n.$$

If $g(z) \neq g(\zeta)$ one has $\bar{\partial}\Phi_g(\cdot,\zeta) = \omega(\cdot,\zeta)$ where $\omega(\cdot,\zeta)$ is the Martinelli and Bochner Kernel.

Let $\Gamma_{\varepsilon_0} = \Gamma \cap \{z \in \Omega_\delta^+ : \rho^*(z) \geq 0\}$ and consider the function F_g

defined by the following formula:

$$(*)\qquad F_g(\zeta) = \int_{\Gamma_{\varepsilon_0}} f\omega(\cdot,\zeta) - \int_{\partial\Gamma_{\varepsilon_0}} f\Phi_g(\cdot,\zeta).$$

F_g is real analytic on $W_g'\setminus\Gamma$. Moreover, if $\zeta\in C$ and $|g(\zeta)|>\|g\|_{D(\varepsilon_0)}$, the kernels $\Phi_g(\cdot,\zeta)$ and $\omega(\cdot,\zeta)$ are both regular on a neighbourhood of Γ; thus, in view of Stokes' theorem, we have $F_g(\zeta)=0$. It follows that F_g vanishes on the connected component W_g'' of $W_g'\cap\overline{D(\varepsilon_0)}$ which contains $C\setminus C\cap\overline{D(\varepsilon_0)}$. Then, arguing as in [2] we can prove that F_g is holomorphic in a neighbourhood of $C\cap D(\varepsilon_0)$ and coincides with f on Γ.

Thus we have proved that there exist an open covering $\{U_i\}$ of $D(\varepsilon_0)$ and a family $\{F_i\}$ of functions, $F_i = F_{g_i}$, with the following properties:

(1) U_i is connected, $V_i = U_i\cap\Gamma\neq\emptyset$ and $\bigcup_i V_i = \Gamma\cap D(\varepsilon_0)$;

(2) F_i is holomorphic on U_i, continuous on $U_i\cup V_i$ and $F_i|V_i = f|V_i$; so, in order to end the proof we only have to verify that $F_i = F_j$ on $U_i\cap U_j\neq\emptyset$.

First assume $n\geq 3$. In view of the definition (*) we have

$$F_i(\zeta)-F_j(\zeta) = \int_{\partial\Gamma_{\varepsilon_0}} f[\Phi_i(\cdot,\zeta)-\Phi_j(\cdot,\zeta)],\quad \text{where}\quad \Phi_i=\Phi_{g_i}.$$

On a neighbourhood of Γ, the kernels Φ_i and Φ_j are real-analytic $(n,n-2)$-forms and $\bar\partial[\Phi_i(\cdot,\zeta)-\Phi_j(\cdot,\zeta)]=0$.

It is possible to write explicitly a $\bar\partial$-primitive $\Theta(\cdot,\zeta)$ of $\Phi_i(\cdot,\zeta)-\Phi_j(\cdot,\zeta)$. For instance in the case $n=3$ we have

$$\Omega_1-\Omega_2 = \frac{c_3}{2}\,\bar\partial\,\frac{\bar z_3-\bar\zeta_3}{(z_1-\zeta_1)(z_2-\zeta_2)|z-\zeta|^2}\,dz_1\wedge dz_2 = \bar\partial\Theta_{12}(z,\zeta),$$

$$\Omega_1-\Omega_3 = \frac{c_3}{2}\,\bar\partial\,\frac{(\bar z_2-\bar\zeta_2)}{(z_1-\zeta_1)(z_3-\zeta_3)|z-\zeta|^2}\,dz_1\wedge dz_2 = \bar\partial\Theta_{13}(z,\zeta),$$

and hence

$$\Phi_i(z,\zeta)-\Phi_j(z,\zeta) = \bar\partial\sum_{s<t}\frac{\lambda_s(z,\zeta)\mu_t(z,\zeta)(z_s-\zeta_s)(z_t-\zeta_t)\Theta_{st}(z,\zeta)}{(g_i(z)-g_i(\zeta))(g_j(z)-g_j(\zeta))},$$

where the λ_s' and μ_t' are the coefficients of the decompositions of $g_i(z)-g_j(\zeta)$ and $g_j(z)-g_j(\zeta)$, respectively. In particular, $\Theta(\cdot,\zeta)$ is real analytic where $g_i(z)\neq g_i(\zeta)$ and $g_j(z)\neq g_j(\zeta)$; in view of the fact that f is a CR-function we have $d(f\Theta(\cdot,\zeta)) = f[\Phi_i(\cdot,\zeta) - \Phi_j(\cdot,\zeta)]$ and, consequently,

$$F_i(\zeta) - F_j(\zeta) = \int_{\partial\Gamma_{\varepsilon_0}} f[\Phi_i(\cdot,\zeta) - \Phi_j(\cdot,\zeta)] = \int_{\partial\Gamma_{\varepsilon_0}} d(f\Phi(\cdot,\zeta)) = 0.$$

For $n = 2$, $\Phi_i(\cdot,\zeta) - \Phi_j(\cdot,\zeta)$ is a holomorphic (2,0)-form and the above argument does not work in this case. We shall proceed in another way.

From the first part of the proof it follows that there exist a covering $\{U_i\}$ of Γ by open sets of $\mathbb{C}^2$ and a family $\{F_i\}$ of holomorphic functions such that:

(3) $U_i \cap U_j$ is connected and if $U_i \cap U_j \neq \emptyset$ then $U_i \cap U_j \cap \Gamma \neq \emptyset$;

(4) F_i is holomorphic on $U_i \cap D$, continuous up to Γ and its boundary value is f.

Then if $U_i \cap U_j \neq \emptyset$ we have $F_i = F_j$; consequently f has a holomorphic extension on a neighbourhood of Γ in $\bar{D}$.

Let I be the set of the positive numbers ε such that f is extendable on $D(\varepsilon)$. Then $I \neq \emptyset$ and in view of the property that hypersurfaces $\rho_\varepsilon = 0$ are strictly pseudoconvex, it follows that $\inf I = 0$ ([1]). This completes the proof of Theorem 1.

4. The second case we wish to consider is that when M is a Levi-flat real analytic hypersurface. First we establish a result concerning the existence of a pluriharmonic equation for a bounded domain of M. We recall that M is foliated by complex hypersurfaces and furthermore if L is a leaf there exists a family of holomorphic functions $f_\alpha = u_\alpha + iv_\alpha$, $f_\alpha : B_\alpha \to C$ such that (cf. [4]):

(1) v_α is a local equation for M;

(2) the leaves of M are locally defined by $f_\alpha = c$, $c \in \mathbb{R}$, and $f_\alpha = 0$ is a local equation for L.

In particular the functions $f_{\alpha\beta} = f_\alpha f_\beta^{-1}$ are invertible on $B_\alpha \cap B_\beta$ and constant on the leaves. We may also assume that $u_\alpha u_\beta^{-1}$ is positive if $B_\alpha \cap B_\beta \cap L \neq \emptyset$.

Now consider the following sheaves of the groups:

$$\mathcal{T} = \{f \in \mathcal{O}_{|M} : f_{|M} \text{ is real}\}, \quad \mathcal{T}^* = \{f \in \mathcal{T} : f_{|M} > 0\}.$$

It is easy to check that the exponential map $f \mapsto \exp f$ gives an isomorphism $\mathcal{T} \overset{\sim}{\to} \mathcal{T}^*$.

<u>THEOREM 2</u>. <u>Let</u> A <u>be</u> <u>a</u> <u>bounded</u> <u>domain</u> <u>of</u> M. <u>Assume</u>:

(i) A <u>is simply connected</u>;

(ii) $\bar{A}$ <u>has a fundamental system of Stein neighbourhoods</u>.

<u>Then there exists a function</u> v, <u>pluriharmonic on a neighbourhood of</u> $\bar{A}$, <u>such that</u> $v = 0$, $dv \neq 0$ <u>on</u> $\bar{A}$.

5. <u>Proof of Theorem 2</u>. (1) Let $\{f_\alpha\}$ be a set of local holomorphic functions satisfying (1) and (2). Then $\xi = \{\xi_{\alpha\beta}\} = \{f_\alpha f_\beta^{-1}\}$ is a 1-cocycle with values in T^*. We shall prove that ξ is actually a coboundary. For this, in view of the isomorphism $T \overset{\sim}{\to} T^*$, it suffices to prove that $H^1(\bar{A}, T) = 0$.

Let $\{\xi_{\alpha\beta}\}$ be a 1-cocycle with values in T defined on an open covering $\{B'_\alpha\}$ of a neighbourhood of $\bar{A}$. Then $\xi_{\alpha\beta} = u_{\alpha\beta} + iv_{\alpha\beta}$ is holomorphic and $v_{\alpha\beta} = 0$ on M. Moreover, in view of (ii), we have $\xi_{\alpha\beta} = \psi_\beta - \psi_\alpha$ where ψ_α is holomorphic on B'. Let $\psi_\alpha = p_\alpha + iq_\alpha$; then if $B'_{\alpha\beta} = B'_\alpha \cap B'_\beta \cap M = \emptyset$ we have $q_\alpha = q_\beta$ on $B'_{\alpha\beta} \cap M$ and hence $\{q_{\alpha|M}\}$ defines a real analytic function q which is locally the restriction of a pluriharmonic function.

We claim that q admits a CR-conjugate p on a neighbourhood of $\bar{A}$ on M. Such a p exists locally and in order to find it globally we fix a point $z_0 \in \bar{A}$, a neighbourhood N_0 of z_0 in M, and a CR-conjugate p_0 of q on N_0.

Let $z \in \bar{A}$ and let $\gamma \subset \bar{A}$ be a continuous arc joining z_0 to z. Then p_0 can be extended as CR-conjugate of q along γ [we consider an open covering $N_0, \ldots, N_s$ of γ such that $N_j \cap N_{j+1}$ is connected and $N_j \cap N_{j+1} \cap \gamma \neq \emptyset$, $0 \leq j \leq s-1$; let p_1 be a CR-conjugate of q on N_1, then $p_0 - p_1$ is constant on $N_0 \cap N_1$ etc. ...].

Let p_γ be the CR-conjugate of q along γ. p_γ does not depend on "small variations" of γ and consequently it does not depend on γ, $\bar{A}$ being simply connected. Thus there is a real analytic function p such that $p + iq$ is a CR-function.

Let us denote by $\tilde{p} + i\tilde{q}$ the holomorphic extension of $p + iq$ and put $\psi' = \psi_\alpha - (\tilde{p} + i\tilde{q})$. Then $\psi'_\beta - \psi'_\alpha = \xi_{\alpha\beta}$ and ψ'_α is real on M so that $\{\xi_{\alpha\beta}\}$ is a coboundary with values in T. This proves that $H^1(A, T^*) = 0$.

(2) Let $f_\alpha f_\beta^{-1} = (u_\alpha + iv_\alpha)(u_\beta + iv_\beta)^{-1}$ and let $\{h_\alpha\}$, $h_\alpha = a_\alpha + ib_\alpha$ be a family of local sections of T^* such that $f_\alpha f_\beta^{-1} = (a_\beta + ib_\beta) \cdot (a_\alpha + ib_\alpha)^{-1}$. In particular $a_\alpha > 0$ and $b_\alpha = 0$ on M. It follows that $\{f_\alpha h_\alpha\}$ defines a function $u + iv$, holomorphic on a neighbourhood of

$\bar{A}$ and locally $u = a_\alpha u_\alpha - b_\alpha v_\alpha$, $v = a_\alpha v_\alpha + b_\alpha v_\alpha$. As $b_\alpha = 0$ on M and $v_\alpha = 0$ is a local equation for M, one has that v/v_α is a real function and $v/v_\alpha \neq 0$ near $L \cap \bar{A}$. In particular v is pluriharmonic on a neighbourhood of $\bar{A}$ and it defines M along $L \cap \bar{A}$. Since $\bar{A}$ is compact we can find finitely many pluriharmonic functions on a neighbourhood of $\bar{A}$, $v_1, \dots, v_k$ such that $v_1 = \dots = v_k = 0$ on M and $\{z \in \bar{A} : dv_1(z) = \dots = dv_k(z) = 0\} = \emptyset$. In view of Sard's lemma we can choose k real numbers $\lambda_1, \dots, \lambda_k$ in such a way that $\Sigma_{j=1}^{k} \lambda_j dv_j \neq 0$ on $\bar{A}$. Then the function $v = \Sigma_{j=1}^{k} \lambda_j v_j$ has the required properties.

Remark. In the general case for $\bar{A}$, a global pluriharmonic defining function could not exist.

6. With the same notations as in the beginning assume now M is real analytic and Levi-flat and that $\bar{A}$ verifies the conditions (i) and (ii) of Theorem 2. Then we get the following

THEOREM 3. *Let f be a Lipschitz continuous CR-function on Γ^{o}. Then f extends by a function F, holomorphic on D and continuous up to Γ^{o}.*

Proof. We may assume that f is continuous on Γ. Let v be a pluriharmonic function on a simply connected neighbourhood W of $\bar{A}$, which defines $W \cap M$ and such that $v > 0$ on $D \cap W$. Let u be a conjugate of v and let $g = u + iv$. For almost every $\varepsilon > 0$ the hypersurfaces $v = \varepsilon$ are smooth and transversal to Γ in such a way that $\gamma_\varepsilon = \Gamma \cap \{z \in W : v(z) = \varepsilon\}$ is smooth.

Let $\Gamma_\varepsilon = \{z \in \Gamma : v(z) \leq \varepsilon\}$, $W_\varepsilon = \{z \in W : 0 < v(z) < \varepsilon\}$; consider for $\zeta \in W_\varepsilon$ the function

$$F(\zeta) = \int_{\Gamma_\varepsilon} f\omega(\cdot,\zeta) - \int_{\partial\Gamma} f\Phi_g(\cdot,\zeta) - \int_{\gamma_\varepsilon} f\Phi_g(\cdot,\zeta)$$

$$= \int_{\Gamma_\varepsilon} f\omega(\cdot,\zeta) - \int_{\partial\Gamma_\varepsilon} f\Phi_g(\cdot,\zeta).$$

It is easy to prove that $F(\zeta) = 0$ for $\zeta \in W_\varepsilon \setminus D$ so that F is holomorphic on $W_\varepsilon \cap D$, continuous up to $\Gamma_\varepsilon \setminus \partial\Gamma$ and $F = f$ on $\Gamma_\varepsilon \setminus \partial\Gamma$.

Let $0 < \varepsilon' < \varepsilon$ be such that $v = \varepsilon'$ is smooth and transversal to Γ and let $F_{\varepsilon'}$ be the restriction of F to the hypersurface $v = \varepsilon'$. Put $f_{\varepsilon'} = f$ on $\Gamma \setminus \Gamma_\varepsilon$ and $f_{\varepsilon'} = F_{\varepsilon'}$ on $v = \varepsilon'$. $f_{\varepsilon'}$ is holomorphically extendable on $D \setminus \bar{W}_{\varepsilon'}$ and its extension does not depend on ε'. The theorem follows.

References

[1] HORMANDER, L., An introduction to complex analysis in several variables, Van Nostrand-Reinhold, Princeton 1966.

[2] LUPACCIOLU, G. and G. TOMASSINI, Un teorema di estensione per le CR-funzioni, Ann. Mat. Pura e Appl., to appear.

[3] NARASIMHAN, R., The Levi problem for complex spaces, Math. Ann. 142 (1961), 355-365.

[4] REA, C., Levi-flat submanifolds and holomorphic extension of foliations, Ann. Sci. Norm Sup. 26 (1972), 665-682.

Scuola Normale Superiore
Piazza dei Cavalieri, 7
I-56100 Pisa, Italy

ONE PARAMETER FAMILY OF OPERATORS ON A RIEMANNIAN MANIFOLD

Grigorios Tsagas and Apostolos Kobotis (Thessaloniki)

1. Introduction. Let (M,g) be a compact,orientable, Riemannian manifold of dimension n. We denote by $\overset{q}{\Lambda}(M)$ the vector space of exterior q-forms on M, where q=0,1,..,n. There are different differential operators on $\overset{q}{\Lambda}(M)$. To each of them corresponds a spectrum. We can also consider one parameter family of differential operators on $\overset{q}{\Lambda}(M)$ from which we obtain its spectrum.

The aim of the present paper is to study the influence of the spectrum of a special one parameter family of differential operators on $\overset{1}{\Lambda}(M)$ on the geometry of (M,g).

The whole paper contains five paragraphs.

In the second paragraph we study a second differential operator with leading symbol by the metric tensor g which acts on the set of cross sections $C^{\infty}(V)$, where V a vector bundle over the Riemannian manifold (M,g).

The special one parameter family of differential operators are studied in the third paragraph.

The fourth paragraph contains the influence of the spectrum of the differential operators, which have been studied in §3, on the geometry of special Riemannian manifolds.

In the last paragraph we study the relations between the spectrum of the differential operators, defined in §3, and the geometry of Kähler manifolds with constant holomorphic sectional curvature.

2. Let (M,g) be a compact, orientable, Riemannian manifold of dimension n, where g is the Riemannian metric on M. This Riemannian metric g in local coordinate system $(x^1,\dots,x^n)$ for a chart U,φ of M can be written

$$ds^2 = g_{ij}dx^i dx^j \tag{2.1}$$

We denote by $g^{-1} = (g^{ij})$ the metric on T^*M and dM the volume element of M.

Let V be a smooth vector bundle over M. We denote by

$$D : C^\infty(V) \longrightarrow C^\infty(V) \tag{2.2}$$

a second order differential operator with leading symbol by the metric tensor. If we use the local coordinate system $(x^1,\dots,x^n)$ for the chart (U,φ) and a local frame for V, then we can express D in the following form

$$D = -g^{ij}\partial^2/\partial x^i \partial x^j + P_k \partial/\partial x^k + Q \tag{2.3}$$

where P_k and Q are square matrices which are not invariantly defined but depend upon the choice of frame and local coordinates.

Let V_x be the fibre of V over x. For $t>0$ $\exp(-tD)$ is a well-defined infinitely smoothing operator which is of trace class in $L^2(V)$.

We denote by

$$K(t,D,x,y) : V_y \longrightarrow V_x \tag{2.4}$$

the Kernel function of $\exp(-tD)$, then we obtain

$$\exp(-tD)(u(x)) = \int_M K(t,D,x,y)(y(y))dM(y) \tag{2.5}$$

$$K(t,D,x,y) \text{ is smooth in } (t,x,y) \tag{2.6}$$

Now, we define

$$f(t,D,x) = \text{Trace}_{V_x}(K(t,D,x,y)) \tag{2.7}$$

$$f(t,D) = \text{Trace}_{L^2}(\exp(-tD)) = \int_M f(t,D,x)\,dM(x), \tag{2.8}$$

It is well known that as $t\to 0+$, then $f(t,D,x)$ has as an asymptotic expansion of the form

$$f(t,D,x) \sim (4\pi t)^{-n/2} \sum_{m=0}^{\infty} A_m(D,x)t^m \tag{2.9}$$

The coefficients $A_m(x,D)$ are smooth functions of x which can be computed functorially in terms of the derivatives of the total symbol of the differential operator D.

It can be easily proved that $A_m(D,x)$ is a local invariant of D. If we put

$$A_m(D) = \int_M A_m(D,x)\,dM(x) \tag{2.10}$$

then we have

$$A(t,D) \sim (4\pi t)^{-n/2} \sum_{m=0}^{\infty} A_m(D)t^m \tag{2.11}$$

If we assume that V has a smooth inner product on each fibre and if D is self-adjoint with respect to the fibre metric, then there is complete spectral decomposition of D into an orthonormal base of eigensections Θ_ν with corresponding eigenvalues λ_ν, $\nu=1,2,\ldots,\infty$

For such a D, we can express

$$f(t,D,x) = \sum_{\nu=1}^{\infty} \exp(-t\lambda_\nu)(\Theta_\nu,\Theta_\nu)(x) \sim$$

$$\sim (4\pi t)^{-n/2} \sum_{m=0}^{\infty} A_m(D,x)t^m \tag{2.12}$$

The set of all eigenvalues including their multiplicity is called spectrum of D and denoted by Sp(M,D).

Therefore we have

$$Sp(D,M) = \{0 \le \lambda_1 \le \lambda_2 \le \ldots < \infty\}$$

This spectrum is discrete and the multiplicity of each eigenvalue is finite since D is an elliptic operator.

$$f(t,D) = \sum_{\nu=1}^{\infty} \exp(-t\lambda_\nu) \sim (4\pi t)^{-n/2} \sum_{m=0}^{\infty} A_m(D)t^m$$

Therefore we conclude that the integrated invariants $A_m(D)$ depend only on the asymptotic behavior of the series $\sum_{\nu=1}^{\infty} \exp(-\lambda_\nu t)$ and hence are spectral invariant.

Let ∇_g be the Levi-Civita connection on TM. We extend ∇_g to tensors of all types. We identify TM with T*M using the metric g.

3. On the Riemannian manifold (M,g) we use the vector bundle $V=TM=T^*M$. For this vector bundle we consider two second order differential operators.

One of them is the Laplace operator

$$\Delta : C^\infty(TM) = \overset{1}{\Lambda}(M) \longrightarrow \overset{1}{\Lambda}(M) \qquad (3.1)$$

$$\Delta = d\delta + \delta d \; : \; \alpha \longrightarrow \Delta(\alpha) = (d\delta + \delta d)(\alpha) \qquad (3.2)$$

The other is the Bochner-Laplace operator D which for a chart (U,φ) with normal coordinate system $(x^1,\dots,x^n)$ takes the form

$$D = g^{ij}\nabla_i\nabla_j \qquad (3.3)$$

for which we also have

$$D : \overset{1}{\Lambda}(M) \longrightarrow \overset{1}{\Lambda}(M) \qquad (3.4)$$

$$D : \alpha \longrightarrow D(\alpha) \qquad (3.5)$$

From these two operators we construct one parameter family of differential operators which is defined by

$$N(\varepsilon) = \varepsilon\Delta + (1-\varepsilon)D \qquad (3.6)$$

which acts on $\overset{1}{\Lambda}(M)$ as follows

$$N(\varepsilon) = \varepsilon\Delta + (1-\varepsilon)D : \overset{1}{\Lambda}(M) \longrightarrow \overset{1}{\Lambda}(M) \qquad (3.7)$$

$$N(\varepsilon) = \varepsilon\Delta + (1-\varepsilon)D : \alpha \longrightarrow N(\varepsilon)(\alpha) = \varepsilon\Delta(\alpha) + (1-\varepsilon)D(\alpha) \qquad (3.8)$$

for every $\alpha \in \overset{1}{\Lambda}(M)$

After some calculations the first three coefficients, which are given by (2.10), for the differential operator, which is defined by (3.6), take the form ([7])

$$A_o(N(\varepsilon)) = n\mathrm{Vol}M \qquad (3.9)$$

$$A_1(N(\varepsilon)) = \frac{1}{6}\int_M (6\varepsilon-1)\tau dM \qquad (3.10)$$

$$A_2(N(\varepsilon)) = \frac{1}{360}\int_M \left[(60\varepsilon+5n)\tau^2 + (180\varepsilon^2-2n)|p|^2+(-30+2n)|R|^2\right]dM \quad (3.11)$$

where dM is the volume element on (M,g), R the curvature tensor field, p the Ricci tensor field, τ the scalar curvature, $|R|$ and $|p|$ are the norms of R and p respectively with respect to the metric g.

From the above we can put the following problem

PROBLEM 3.1. <u>What is the influence of Sp(M,N(e)) on the geometry of (M,g)?</u>

Answer.From the above we can conclude immediately the following

$$Sp(N(\varepsilon),M) \Longrightarrow \dim M \quad (3.12)$$

$$Sp(N(\varepsilon),M) \Longrightarrow \mathrm{Vol}M \quad (3.13)$$

The relations (3.12) and (3.13) can be stated as follows

If $Sp(N(\varepsilon),M) = Sp(N'(\varepsilon),M')$, then $\dim M = \dim M'$ (3.14)

If $Sp(N(\varepsilon),M) = Sp(N'(\varepsilon),M')$, then $\mathrm{Vol}M = \mathrm{Vol}M'$ (3.15)

4. Let (M,g),(M',g') be two compact orientable Riemannian manifolds for which we assume

$$Sp(N(\varepsilon),M) = Sp(N(\varepsilon),M') \quad (4.1)$$

Now we shall prove the following theorem

THEOREM 4.1. <u>We consider two compact, orientable Riemannian manifolds (M,g) and (M',g') with $Sp(N(\varepsilon),M) = Sp(N(\varepsilon),M')$, (which implies $\dim M= =\dim M'=n$). For every $n\geq 16$, there exists</u>

$$\varepsilon \in \mathbb{R} - \left[-\sqrt{\frac{-3+\sqrt{452}}{135}}, \sqrt{\frac{-3+\sqrt{462}}{135}}\right] \quad (4.2)$$

with

$$\varepsilon < -\sqrt{\frac{n^2-6n+60}{90(n-2)}} \quad \text{or} \quad \varepsilon > \frac{5n(n-1)-\sqrt{10(n-1)(n^2-3n+30)}}{30(n-1)} \quad (4.3)$$

or

$$\sqrt{\frac{n^2-6n+60}{90(n-2)}} < \varepsilon < \frac{5n(n-1) - \sqrt{10(n-1)(n^2-3n+30}}{30(n-1)} \qquad (4.4)$$

then (M,g) has constant sectional curvature k if and only if (M',g') has constant sectional curvature k' and k=k'.

It is known that the following formulas hold ([19])

$$|C|^2 = |R|^2 - \frac{4}{n-2}|P|^2 + \frac{2}{(n-1)(n-2)}\tau^2 \qquad (4.5)$$

$$|C|^2 = |P|^2 - \frac{1}{n}\tau^2 \qquad (4.6)$$

where C and G are the conformal curvature tensor field and the Einstein tensor field on M, respectively.

The formula (3.11), by means of (4.5) and (4.6), takes the form

$$A_2(N(\varepsilon)) = \frac{1}{360}\int_M \left[\Lambda_1|C|^2 + \Lambda_2|G|^2 + \Lambda_3\tau^2\right] dM \qquad (4.7)$$

where

$$\Lambda_1 = 2n - 30 \qquad (4.8)$$

$$\Lambda_2 = 180\varepsilon^2 - 2n + \frac{4}{n-2}(2n-30) \qquad (4.9)$$

$$\Lambda_3 = -60\varepsilon + 5n + \frac{1}{4}(180\varepsilon^2 - 2n) + \frac{2}{n(n-1)}(2n-30) \qquad (4.10)$$

We assume that the Riemannian manifold (M',g') has constant sectional curvature k'. Therefore we have

$$C' = 0, \qquad G' = 0 \qquad (4.11)$$

The formula (4.7) by means of (4.11) becomes

$$A'_2(N(\varepsilon)) = \frac{1}{360} \quad {}_{M'} \ \Lambda_3\tau'^2\, dM' \qquad (4.12)$$

From the relations (4.1), (4.7) and (4.12) we conclude that

$$\int_M \left[\Lambda_1|C|^2 + \Lambda_2|G|^2 + \Lambda_3\tau^2\right] dM = \int_{M'} \Lambda_3\tau'^2\, dM' \qquad (4.13)$$

In the relations (4.9) and (4.10) if we consider ε as a variable and n fixed, then for every $n \geq 16$, there exists

$$\varepsilon \in \mathbb{R} - \left[-\sqrt{\frac{-3+\sqrt{462}}{135}} \, , \sqrt{\frac{-3+\sqrt{462}}{135}} \right] \tag{4.14}$$

which satisfies one of the below relations

$$\varepsilon < -\sqrt{\frac{n^2-6n+60}{90(n-2)}} \tag{4.15}$$

or

$$\varepsilon > \frac{5n(n-1) + \sqrt{10(n-1)(n^2-3n+30}}{30(n-1)} \tag{4.16}$$

or

$$\sqrt{\frac{n^2-6n+60}{90(n-2)}} < \varepsilon < \frac{5n(n-1) - \sqrt{10(n-1)(n^2-3n+30}}{30(n-1)} \tag{4.17}$$

such that

$$\Lambda_1 > 0, \quad \Lambda_2 > 0, \quad \Lambda_3 > 0 \tag{4.18}$$

From the relation (4.1) we obtain

$$A_1(N(\varepsilon)) = A_1'(N(\varepsilon)) \tag{4.19}$$

which by virtue of (3.10) gives

$$\int_M \tau \, dM = \int_{M'} \tau' \, dM' \tag{4.20}$$

The relation (4.20) since $\tau' = \text{const.}$ implies

$$\int_M \tau^2 dM \geq \int_{M'} \tau'^2 \, dM' \tag{4.21}$$

From (4.13), (4.14), (4.15), (4.16), (4.17), (4.18) and (4.21) we conclude, when, $n \geq 16$, the following equalities

$$|C|^2 = 0, \quad |G|^2 = 0 \tag{4.22}$$

which imply

$$C = G = 0 \tag{4.23}$$

that is the Riemannian manifold (M,g) has constant sectional curvature k. From (4.20) we obtain k=k'.

From the above theorem we have the corollaries

COROLLARY 4.2. For each n≥16 there exist

$$\varepsilon \in \mathbb{R} - \left[-\sqrt{\frac{-3+\sqrt{462}}{135}}, \sqrt{\frac{-3+\sqrt{462}}{135}} \right] \tag{4.24}$$

with

$$\varepsilon < -\frac{n^2-6n+60}{90(n-2)} \tag{4.25}$$

or

$$\varepsilon > \frac{5n(n-1)+\sqrt{10(n-1)(n^2-3n+30}}{30(n-1)} \tag{4.26}$$

or

$$\sqrt{\frac{n^2-6n+60}{90(n-2)}} < \varepsilon < \frac{5n(n-1)-\sqrt{10(n-1)(n^2-3n+30)}}{30(n-1)} \tag{4.27}$$

such that the Euclidean sphere (S^n, g_o) is completely characterized from the spectrum of the differential operator $N(\varepsilon)$.

COROLLARY 4.3. The Euclidean sphere (S^{16}, g_o) is completely characterized from the spectrum of the operator $N(\varepsilon)$.

$$\varepsilon < -\frac{11}{63}, \text{ or } \varepsilon > \frac{120+\sqrt{357}}{63} \quad \text{or } \frac{11}{63} < \varepsilon < \frac{120-\sqrt{357}}{45} \tag{4.28}$$

THEOREM 4.4. Let (M,g), (M',g') be two compact, orientable Riemannian manifolds with the property $Sp(N(\varepsilon),M) = Sp(N(\varepsilon),M')$ (which implies $\dim M = \dim M' = n$). If

$$\varepsilon \in \mathbb{R} - \left[-\sqrt{\frac{-3+\sqrt{462}}{135}}, \sqrt{\frac{-3+\sqrt{462}}{135}} \right] \tag{4.29}$$

there exists n≥16 with the property

$$\frac{90\varepsilon^2+6-\sqrt{12(675\varepsilon^4+30\varepsilon^2-17}}{2} < n < \frac{90\varepsilon^2+6+\sqrt{12(675\varepsilon^4+30\varepsilon^2-17)}}{2} \tag{4.30}$$

if

$$\varepsilon \in \left[-\frac{11}{63}, -\frac{-3+\sqrt{462}}{135} \right] \quad \text{or} \quad \varepsilon \in \left[\frac{-3+\sqrt{462}}{135}, \frac{11}{63} \right] \tag{4.31}$$

then we have

$$16 \leq n < \frac{90\varepsilon^2 + \sqrt{12(675\varepsilon^4 + 30\varepsilon^2 - 17)}}{2} \tag{4.32}$$

<u>if</u>

$$\varepsilon < -\frac{11}{63} \quad \text{or} \quad \frac{11}{63} < \varepsilon < \frac{120 - \sqrt{35}}{45} \tag{4.33}$$

<u>then we obtain</u>

$$\sqrt[3]{-\frac{9}{2} + \sqrt{\Delta}} + \sqrt[3]{-\frac{9}{2} - \sqrt{\Delta}} + \frac{60\varepsilon+7}{15} < n < \frac{90\varepsilon^2 + \sqrt{12(675\varepsilon^4+30\varepsilon^2-17)}}{2} \tag{4.34}$$

<u>if</u>

$$\frac{120 - \sqrt{357}}{45} < \varepsilon < \frac{120 + \sqrt{357}}{45} \tag{4.35}$$

<u>then this implies</u>

$$16 \leq n < w\sqrt{-\frac{9}{2} + \sqrt{\Delta}} + w^2\sqrt{-\frac{9}{2} - \sqrt{\Delta}} + \frac{60\varepsilon + 7}{15} \tag{4.36}$$

if

$$\varepsilon > \frac{120 + \sqrt{35}}{45} \tag{4.37}$$

<u>then we have</u>

$$\sqrt[3]{-\frac{9}{2} + \sqrt{\Delta}} + \sqrt[3]{-\frac{9}{2} - \sqrt{\Delta}} + \frac{60\varepsilon + 7}{15} < n < \frac{90\varepsilon^2+6+\sqrt{12(675\varepsilon^4+30\varepsilon^2-17)}}{2} \tag{4.38}$$

<u>where</u>

$$w = -\frac{1}{2} - i\ \frac{3}{2} \tag{4.39}$$

$$q = \frac{54000\varepsilon^3 - 54000\varepsilon^2 + 17460\varepsilon - 39296}{3375} \tag{4.40}$$

$$\Delta = \frac{-648000\varepsilon^5 + 637000\varepsilon^4 - 7632000\varepsilon^3 + 559560\varepsilon^2 - 16260\varepsilon + 190673}{5625} \tag{4.41}$$

<u>such that (M,g) has constant sectional curvature k if and only if (M′,g′) has constant sectional curvature k′ and k=k′.</u>

Proof. This can be proved with the same technique as the theorem 4.1.

From the above theorem we obtain the corollary

COROLLARY 4.5. We assume that the conditions (4.29)-(4.41) are satisfied, then the Euclidean n-sphere (S^n, g_o) is completely characterized by the $Sp(N(\varepsilon), S^n)$.

Now we shall prove the following theorem

THEOREM 4.6. Let (M,g), (M',g') be two compact, orientable, Einstein manifolds with $Sp(N(\varepsilon),M) = Sp(N(\varepsilon),M')$, (which implies that $\dim M = \dim M' = n$). Then for each $n \geq 16$ there exists $\varepsilon \in \mathbb{R}$ which satisfies the inequalities

$$\varepsilon < \frac{5n(n-1) - \sqrt{10(n-1)(-^2-3n+30)}}{30(n-1)} \tag{4.42}$$

or

$$\varepsilon > \frac{5n(n-1) + \sqrt{10(n-1)(n^2-3n+30)}}{30(n-1)} \tag{4.43}$$

such that (M',g') has constant sectional curvature k' if and only if (M,g) has constant sectional curvature k' and $k=k'$.

Proof. From the assumption that (M,g) is an Einstein manifold we obtain $G=0$ and therefore formula (4.7) takes the form

$$A_2(N(\varepsilon)) = \frac{1}{360}\int_M (\Lambda_1 |C|^2 + \Lambda_3 \tau^2)\, dM \tag{4.44}$$

After some estimates we can prove that for every $n \geq 16$ there exists $\varepsilon \in \mathbb{R}$ which satisfies the inequalities (4.42) and (4.43) such that

$$\Lambda_1 > 0, \quad \Lambda_3 > 0 \tag{4.45}$$

We assume that the Einstein manifold (M',g') has constant sectional curvature $k' \neq 0$, then the formula (4.44) becomes

$$A_2'(N(\varepsilon)) = \frac{1}{360}\int_{M'} \Lambda_3 \tau'^2\, dM' \tag{4.46}$$

From the assumption of the theorem we obtain

$$A_1(N(\varepsilon)) = A_1'(N(\varepsilon)), \quad A_2(N(\varepsilon)) = A_2'(N(\varepsilon)) \tag{4.47}$$

The relations (4.47) by means of (3.10) and (4.44) and (4.45) become

$$\int_M \tau dM = \int_{M'} \tau' dM' \tag{4.48}$$

$$\int_M (\Lambda_1 |C|^2 + \Lambda_3 \tau^2) dM = \int_{M'} \Lambda_3 \tau'^2 dM' \tag{4.49}$$

Since τ' is constant from (4.48) we conclude that

$$\int_M \tau^2 dM \geq \int_{M'} \tau'^2 dM' \tag{4.50}$$

From (4.45),(4.49) and (4.50), when (4.42) and (4.43) are valid, we obtain

$$|C|^2=0 \Longrightarrow C=0 \tag{4.51}$$

that means the Einstein manifold (M,g) has constant sectional curvature k'. The relation (4.48) implies k=k'.

From this theorem we obtain the corollary

COROLLARY 4.7. Let (M,g) be a compact, orientable, Einstein manifold of dimension n. If $Sp(N(\varepsilon),M) = Sp(N(s),S^n)$, where (S^n,g_o) the standard Euclidean sphere. For every $n \geq 16$, there exists $\varepsilon \in \mathbb{R}$ with the properties

$$\varepsilon < \frac{5n(n-1) - \sqrt{10n(n-1)(n^2-3n+30)}}{30(n-1)}, \tag{4.52}$$

or

$$\varepsilon > \frac{5n(n-1) + \sqrt{10(n-1)(n^2-3n+30)}}{30(n-1)}. \tag{4.53}$$

such that (M,g) is isometric to the (S^n,g_o).

COROLLARY 4.8. We consider a compact, orientable, Einstein manifold (M,g) of dimension 16. We assume $Sp(N(\varepsilon),M) = Sp(N(\varepsilon),S^{16})$, where (S^{16},g_o) the standard Euclidean sphere of dimension 16. If the real number ε satisfies the inequalities

$$\varepsilon < \frac{120-\sqrt{357}}{45} \quad \text{or} \quad \varepsilon > \frac{120+\sqrt{357}}{45} \tag{4.54}$$

then (M,g) is isometric to (S^{16},g_o).

THEOREM 4.9. Let (M,g) and (M',g') be two compact, orientable Einstein manifolds with the property $Sp(N(\varepsilon),M) = Sp(N(\varepsilon),M')$, (which implies $\dim M=\dim M'=n$). If

(i) $$\varepsilon < \frac{120 - \sqrt{357}}{45} \tag{4.55}$$

then $$n \geq 16 \tag{4.56}$$

(ii) If

$$\frac{120 - \sqrt{357}}{45} < \varepsilon < \frac{120 + \sqrt{357}}{45} \tag{4.57}$$

then we obtain

$$n > \sqrt[3]{-\frac{9}{2} + \sqrt{\Delta}} + \sqrt[3]{-\frac{9}{2} - \sqrt{\Delta}} + \frac{60\varepsilon + 7}{45} \tag{4.58}$$

(iii) If

$$\varepsilon > \frac{120 + \sqrt{357}}{45} \tag{4.59}$$

then we have

$$16 \leq n \geq w\sqrt[3]{-\frac{9}{2} + \sqrt{\Delta}} + w^2\sqrt[3]{-\frac{9}{2} - \sqrt{\Delta}} + \frac{60\varepsilon + 7}{45} \tag{4.60}$$

$$w = -\frac{1}{2} + i\sqrt{\frac{3}{2}} \tag{4.61}$$

$$q = \frac{54000\varepsilon^3 - 54000\varepsilon^2 + 17460\varepsilon - 3926}{3375} \tag{4.62}$$

$$\Delta = \frac{-648000\varepsilon^5 + 637200\varepsilon^4 - 673200\varepsilon^3 + 559560\varepsilon^2 - 169260\varepsilon + 190673}{5265} \tag{4.63}$$

such that (M',g') has constant sectional curvature k'. If and only if (M,g) has constant sectional curvature k and $k=k'$.

Proof. This can be proved with the same technique as the theorem 4.6.

A consequence of the above theorem is the corollary

COROLLARY 4.10. Let (M,g) be a compact, orientable, Einstein manifold with the property $Sp(N(\varepsilon),M) = Sp(N(c),S^n)$, where (S^n,g_o) is the standard Euclidean sphere. If the conditions, which are given by (4.55)-(4.63), are satisfied, then (M,g) is isometric to (S^n,g_o).

5. Let (M,J,g), (M',J',g') be two compact Kähler manifolds with the property

$$Sp(N(\varepsilon),M) = Sp(N(\varepsilon),M'). \tag{5.1}$$

We study special Kähler manifolds whose geometry is determined by (5.1).

THEOREM 5.1. We consider two compact, Kähler manifolds (M,J,g) and (M',J',g') with the property $Sp(N(\varepsilon),M) = Sp(N(\varepsilon),M')$ (which implies $\dim M=\dim M'=n$). For every $2m=n>16$ there exists

$$\varepsilon \in \mathbb{R} - \left[-\sqrt{\frac{-10+\sqrt{304}}{45}} \, , \; \sqrt{\frac{-10+\sqrt{304}}{45}} \right] \tag{5.2}$$

with

$$\varepsilon < -\sqrt{\frac{n^2-12n+240}{90(n+4)}} \tag{5.3}$$

or

$$\varepsilon > \frac{5n(n+2)+\sqrt{10(n+2)(n^2-6n+120)}}{30(n+2)} \tag{5.4}$$

or

$$\sqrt{\frac{n^2-12n+240}{90(n+4)}} < \varepsilon < \frac{5n(n+2)-\sqrt{10(n+2)(n^2-6n+120)}}{30(n+2)} \tag{5.5}$$

such that (M',J',g') has constant holomorphic sectional curvature h' if and only if (M,J,g) has constant holomorphic sectional curvature h and $h=h'$.

Proof. Let B be the Bochner curvature tensor field on (M,J,g). It is known that the following relation holds ([16])

$$|B|^2 = |R|^2 - \frac{16}{n+4}|G|^2 + \frac{8}{(n+2)(n+4)}\tau^2 \tag{5.6}$$

The relation (3.12) by means of (4.6) and (5.6) takes the form

$$A_2(N(\varepsilon)) = \frac{1}{360}\int (\Sigma_1|B|^2 + \Sigma_2|G|^2 + \Sigma_3\tau^2)\,dM \tag{5.7}$$

where

$$\Sigma_1 = 2n-30 \tag{5.8}$$

$$\Sigma_2 = 180\varepsilon^2 - 2n + \frac{16}{n+4}(2n-30) \tag{5.9}$$

$$\Sigma_3 = -60\varepsilon+5n + \frac{1}{n}(180\varepsilon^2-2n) + \frac{8}{n(n+2)}(2n-30) \qquad (5.10)$$

In the relations (5.9) and (5.10) we consider ε as a variable and n constant.

Then we have for $n=2m\geq 16$, there exists

$$\varepsilon \in \mathbb{R} - \left[-\sqrt{\frac{-10+\sqrt{304}}{45}}, \sqrt{\frac{-10+\sqrt{304}}{45}}\right] \qquad (5.11)$$

which satisfies the inequalities

$$\varepsilon < -\sqrt{\frac{n^2-12n+240}{90(n+4)}} \qquad (5.12)$$

or

$$\varepsilon > \frac{5n(n+2) + \sqrt{10(n-2)\ n^2-6n+120)}}{30(n+2)} \qquad (5.13)$$

or

$$\sqrt{\frac{n^2-12n+240}{90(n+4)}} < \varepsilon < \frac{5n(n+2) - \sqrt{10(n+2)(n^2-6n+120)}}{30(n+2)} \qquad (5.14)$$

such that the inequalities valid

$$\Sigma_1 > 0, \quad \Sigma_2 > 0, \quad \Sigma_3 > 0 \qquad (5.15)$$

We assume that the Kähler manifold (M',J',g') has constant holomorphic sectional curvature h', which implies

$$|B'|^2 = 0, \quad |G'|^2 = 0 \qquad (5.16)$$

Therefore the formula (5.7) by means of (5.16) becomes

$$A_2'(N(\varepsilon)) = \frac{1}{360}\int_{M'} \Sigma_3 \tau'^2 dM' \qquad (5.17)$$

From (5.1) we obtain

$$A_1(N(\varepsilon)) = A_1'(N(\varepsilon)) \qquad (5.18)$$

which by means of (3.10) takes the form

$$\int_M \tau dM = \int_{M'} \tau' dM' \qquad (5.19)$$

From (5.19), since τ=constant, we obtain

$$\int_M \tau^2 dM \geq \int_{M'} \tau'^2 dM' \tag{5.20}$$

This relation (5.1) implies also the relation

$$A_2(N(\varepsilon)) = A_2'(N(\varepsilon)) \tag{5.21}$$

which by virtue of (5.7) and (5.17) yields

$$\int_M \left[\Sigma_1 |B|^2 + \Sigma_2 |G|^2 + \Sigma_3 \tau^2\right] dM = \int_{M'} \Sigma_3 dM' \tag{5.22}$$

The relation (5.22) by means of (5.11),(5.12),(5.13),(5.14), (5.15) and (5.20) implies

$$|B|^2 = 0, \quad |G|^2 = 0 \longrightarrow B = 0, \quad G = 0 \tag{5.23}$$

which give that the Kähler manifold (M,J,g) has constant holomorphic sectional curvature h. Finally the relation (5.19) implies h=h'.

From this theorem we have the corollaries

COROLLARY 5.2. For every $2m=n\geq 16$, there exists

$$\varepsilon \in \left[\sqrt{\frac{-10+\sqrt{304}}{45}} \; , \; \sqrt{\frac{-10+\sqrt{304}}{45}} \right] \tag{5.24}$$

which satisfies the inequalities (5.12),(5.13) and (5.14) such that the complex projective space $(\mathbb{P}^m(\mathbb{C}), J_o, g_o)$ with the Fubini-Study metric g_o is completely characterized by the $Sp(N(\varepsilon), \mathbb{P}^m(\mathbb{C}))$.

COROLLARY 5.3. The complex projective space $(\mathbb{P}^8(\mathbb{C}), J_o, g_o)$ is completely characterized by $Sp(N(\varepsilon), \mathbb{P}^8(\mathbb{C}))$, when

$$\varepsilon < -\frac{\sqrt{38}}{15} \quad \text{or} \quad \varepsilon > \frac{24+\sqrt{14}}{9} \tag{5.25}$$

or

$$\sqrt{\frac{38}{15}} < \varepsilon < \frac{24-\sqrt{14}}{9} \tag{5.26}$$

THEOREM 5.4. Let (M,J,g),(M',J',g') be two compact, Kähler Einstein manifolds with $Sp(N(\varepsilon),M) = Sp(N(\varepsilon),M')$ (this implies dimM=dimM'=n=2m).

For every

$$\varepsilon \in \mathbb{R} - \left[-\sqrt{\frac{-10+\sqrt{304}}{45}}\ ,\ \sqrt{\frac{-10+\sqrt{304}}{45}} \right] \tag{5.27}$$

there exists 2m=n with the properties

(i) If

$$\varepsilon < \left[-\sqrt{\frac{38}{15}} - \sqrt{\frac{-10+\sqrt{304}}{45}} \right] \text{ or } \varepsilon \left[-\sqrt{\frac{-10+\sqrt{304}}{45}}\ ,\sqrt{\frac{38}{15}} \right] \tag{5.28}$$

then we have

$$\frac{90\varepsilon^2+12-\sqrt{12(675\varepsilon^2+300\varepsilon^2-68)}}{45} < n < \frac{90\varepsilon^2+12+\sqrt{12(675\varepsilon^4+30\varepsilon^2-68)}}{45} \tag{5.29}$$

(ii) If

$$\varepsilon < -\sqrt{\frac{38}{15}} \text{ or } \sqrt{\frac{38}{15}} < \varepsilon < \frac{24-\sqrt{14}}{9} \tag{5.30}$$

then we obtain

$$16 \leq n < \frac{90\varepsilon^2+12+\sqrt{12(675\varepsilon^4+300\varepsilon^2-68)}}{2} \tag{5.31}$$

(iii) If

$$\frac{24-\sqrt{14}}{9} < \varepsilon < \frac{24+\sqrt{14}}{9} \tag{5.32}$$

then we have

$$\sqrt[3]{-\frac{9}{2}+\sqrt{\Delta}} + \sqrt[3]{-\frac{9}{2}+\sqrt{\Delta}} + \frac{60\varepsilon-8}{15} < n < \frac{90\varepsilon^2+12+\sqrt{675\varepsilon^4+300\varepsilon^2-68}}{2} \tag{5.33}$$

(iv) If

$$\varepsilon > \frac{24+\sqrt{14}}{9} \tag{5.34}$$

then we obtain

$$16 \leq n < w\sqrt[3]{-\frac{9}{2}+\sqrt{\Delta}} + w^2\sqrt[3]{-\frac{9}{2}-\sqrt{\Delta}} + \frac{60\varepsilon-8}{15} \tag{5.35}$$

or

$$\sqrt[3]{-\frac{9}{2}+\sqrt{\Delta}} + \sqrt[3]{-\frac{9}{2}-\sqrt{\Delta}} + \frac{60\varepsilon-8}{15} < n < \frac{90\varepsilon^2+12+\sqrt{12(675\varepsilon^4+300\varepsilon^2-68)}}{2} \tag{5.36}$$

where

$$w = -\frac{1}{2} - i\frac{\sqrt{3}}{2},\quad q = \frac{54000\varepsilon^3+27000\varepsilon^2+52560\varepsilon-165296}{3375} \tag{5.37}$$

$$\Delta = \frac{-648000\varepsilon^5 - 108000\varepsilon^4 + 45210\varepsilon^3 - 657600\varepsilon^2 - 2161920\varepsilon + 3373956}{5626} \tag{5.38}$$

such that (M,J,g) has constant holomorphic sectional curvature h if and only if (M',J',g') has constant holomorphic sectional curvature h' and h=h'.

Proof. This can be proved with the same technique as the theorem 5.1.

From this theorem we obtain the corollary

COROLLARY 5.5. We assume that the conditions (5.27)-(5.38) are satisfied. Then the complex projective space $(\mathbb{P}^m(\mathbb{C}), J_o, g_o)$ is completely characterized by the $Sp(N(\varepsilon), \mathbb{P}^m(\mathbb{C}))$.

THEOREM 5.6. We consider two compact, Kähler Einstein manifolds (M,J,g) and (M',J',g') with the property $Sp(N(\varepsilon),M) = Sp(N(\varepsilon),M')$ (which implies $\dim M = \dim M' = n = 2m$). For every $2m = n \geq 16$, there is $\varepsilon \in \mathbb{R}$ with the properties

$$\varepsilon < \frac{5n(n+2) - \sqrt{10(n+2)(n^2-6n+120)}}{30(n+2)} \tag{5.39}$$

or

$$\varepsilon > \frac{5n(n+2) + \sqrt{10(n+2)(n^2-6n+120)}}{30(n+2)} \tag{5.40}$$

such that (M,J,g) has constant holomorphic sectional curvature if and only if (M',J',g') has constant holomorphic sectional curvature h' and h=h'.

Proof. From the fact the Kähler manifold (M,J,g) is an Einstein we conclude that

$$G = 0 \tag{5.41}$$

The formula (5.7) by means of (5.41) becomes

$$A_2(N(\varepsilon)) = \frac{1}{360} \int_M (\Sigma_1 |B|^2 + \Sigma_3 \tau^2)\, dM \tag{5.42}$$

If $2m = n \geq 16$ is given, then there exists $\varepsilon \in \mathbb{R}$ with the properties

$$\varepsilon < \frac{5n(n+2) - \sqrt{10(n+2)(n^2-6n+30)}}{30(n+2)} \tag{5.43}$$

or

$$\varepsilon > \frac{5n(n+2) + \sqrt{10(n+2)(n^2-6n+30)}}{30(n+2)} \tag{5.44}$$

such that

$$\Sigma_1 > 0 \ , \quad \Sigma_3 > 0 \tag{5.45}$$

We assume that the Einstein Kähler manifold (M′,J′,g′) has constant holomorphic sectional curvature h′. This implies B′=0 and hence formula (5.42) takes the form

$$A_2'(N(\varepsilon)) = \frac{1}{360}\int_{M'} \Sigma_3 \tau'^2 \, dM' \tag{5.46}$$

From (5.1) we obtain

$$A_2(N(\varepsilon)) = A_2'(N(\varepsilon)) \tag{5.47}$$

which by means of (5.42) and (5.46) becomes

$$\int_M \left[\Sigma_1 |B|^2 + \Sigma_3 \tau^2\right] dM = \int_{M'} \Sigma_3 \tau'^2 \, dM' \tag{5.48}$$

The relation (5.48) by virtue of (5.20),(5.43),(5.44) and (5.45) yields

$$|B|^2 = 0 \Longrightarrow B = 0 \tag{5.49}$$

which implies that the Einstein Kähler manifold (M,J,g) has constant holomorphic sectional curvature h. With the same technique as in the theorem 5.1. we obtain h=h′.

From this theorem we obtain the corollaries

COROLLARY 5.7. Let (M,J,g) be a compact, Kähler Einstein manifold of dimension n=2m. We consider the complex projective space ($\mathbb{P}^m(\mathbb{C}), J_o, g_o$) with the Fubini-Study metric g_o. We assume that we have the condition $Sp(N(\varepsilon),M) = Sp(N(\varepsilon),\mathbb{P}^m(\mathbb{C}))$. For every 2m=n, there exists $\varepsilon \in \mathbb{R}$ with the properties

$$\varepsilon < \frac{5n(n+2) - \sqrt{10(n+2)(n^2-6n+30)}}{30(n+2)} \tag{5.50}$$

or

$$\varepsilon > \frac{5n(n+2) + \sqrt{10(n+2)(n^2-6n+30)}}{30(n+2)} \tag{5.51}$$

such that (M,J,g) is holomorphically isometric to $(\mathbb{P}^m(\mathbb{C}),J_o,g_o)$.

COROLLARY 5.8. We consider the Kähler Einstein manifold (M,J,g) and the complex projective space $(\mathbb{P}^8(\mathbb{C}),J_o,g_o)$ with the property $Sp(N(\varepsilon),M) = Sp(N(\varepsilon),\mathbb{P}^8(\mathbb{C}))$. Then there exists $\varepsilon \in \mathbb{R}$ which satisfies the inequalities

$$\varepsilon < \frac{24-\sqrt{14}}{9} \quad \text{or} \quad \varepsilon > \frac{24+\sqrt{14}}{9} \tag{5.52}$$

such that (M,J,g) is holomorphically isometric to $(\mathbb{P}^8(C),J_o,g_o)$.

THEOREM 5.9. Let (M,J,g), (M',J',g') be two compact, Kähler Einstein manifolds with the property $Sp(N(\varepsilon),M) = Sp(N(\varepsilon),M')$ (which implies $\dim M=\dim M'=n=2m$). For every $\varepsilon \in \mathbb{R}$, there exists $2m=n\geq 16$ which is defined as follows

(i) If
$$\varepsilon < \frac{24-\sqrt{14}}{9} \tag{5.53}$$

then we obtain
$$n \geq 16 \tag{5.54}$$

(ii) If
$$\frac{24-\sqrt{14}}{9} < \varepsilon < \frac{24+\sqrt{14}}{9} \tag{5.55}$$

then we have
$$n > \sqrt[3]{-\frac{9}{2}+\sqrt{\Delta}} + \sqrt[3]{-\frac{9}{2}+\sqrt{\Delta}} + \frac{60\varepsilon-8}{15} \tag{5.56}$$

(iii) If
$$\varepsilon > \frac{24+\sqrt{14}}{9} \tag{5.57}$$

then we obtain
$$16 \leq n < w\sqrt[3]{-\frac{9}{2}+\sqrt{\Delta}} + w^2\sqrt{-\frac{9}{2}-\sqrt{\Delta}} + \frac{60\varepsilon-8}{15} \tag{5.58}$$

or
$$n > \sqrt[3]{-\frac{9}{2}+\sqrt{\Delta}} + \sqrt[3]{-\frac{9}{2}-\sqrt{\Delta}} + \frac{60\varepsilon-8}{15} \tag{5.59}$$

where

$$w = -\frac{1}{2} - i\sqrt{\frac{3}{2}} \qquad q = \frac{54000\varepsilon^3+27000\varepsilon^2+52560\varepsilon-165296}{3375} \tag{5.60}$$

$$\Delta = \frac{-647000\varepsilon^5-108000\varepsilon^4+45210\varepsilon^3-657600\varepsilon^2-2161920\varepsilon-3373952}{5625} \tag{5.61}$$

such that (M,J,g) has constant holomorphic sectional curvature h if and only if (M',J',g') has constant holomorphic sectional curvature h' and h=h'.

From the theorem we have the corollary

COROLLARY 5.10. Let (M,J,g) be a compact, Kähler Einstein manifold with the property $Sp(N(\varepsilon),M) = Sp(N(\varepsilon),\mathbb{P}^m(\mathbb{C}))$, where $(\mathbb{P}^m(\mathbb{C}),J_o,g_o)$ is the complex projective space with Fubini-Study metric g_o. If the conditions (5.53)-(5.60) are satisfied, then (M,J,g) is holomorphically isometric to $(\mathbb{P}^m(\mathbb{C}),J_o,g_o)$.

References

[1] ATIYAH, M., BOTT, R. and PATODI, V.K., On the heat equation and the index theorem, Invent. Math. 19 (1973), 279-330.

[2] BERGER, M., GAUDACHON, P. and MAZET, E., Le spectre d'une variété riemannienne, Lecture Notes in Math., No. 194, Springer-Verlag, Berlin-Heidelberg-New York, 1971.

[3] DONNELLY, H., Symmetric Einstein spaces and spectral geometry, Indiana Univ. Math. J. 24 (1974/75), 603-606.

[4] DONNELLY, H., The differential form spectrum of hyperbolic spaces, Manusripta Mathematica 33 (1981), 365-385.

[5] EJIRI, N., A construction of non-flat, compact irreducible Riemannian manifolds which are isospectral but not isometric, Math. Z. Vol. 212, (1979), 207-212.

[6] FEGAN, H.D., The spectrum of the Laplacian for forms over a Lie group, Pacific J. Math. 90 (1980), 373-387.

[7] GILKEY, P., The spectral geometry of symmetric spaces, Trans. of the A.M.S. 255 (1977), 341-353.

[8] GILKEY, P., Curvature and the eigen-values of the Laplacian for elliptic complexes, Advances in Math. 10 (1973), 344-382.

[9] GILKEY, P., Spectral geometry and the Kähler condition for complex manifolds, Inventiones math. 26, (1974), 231-258.

[10] GILKEY, P., The spectral geometry of a Riemannian manifold, J. Diff. Geom. Vol. 10, No. 4, (1975), 601-618.

[11] LEVY-BRUHL, A., Spectre du Laplacien de Hodge-de Rham sur $\mathbb{C}P^n$, Bull. Sc. Math. 140, (1980), 135-143.

[12] PATODI, V.K., Curvature and the eigenforms of the Laplace operator, J. Differential Geometry 5 (1971), 233-249.

[13] PATODI, V.K., Curvature and the fundamental solution of the heat equation, J. Indian Math. Soc. 34 (1970), 269-285.

[14] SEELEY, R.T., Complex powers of an elliptic operator, Proc. Symp. Pure Math. Vol. 10, Amer. Math. Soc., (1967), 288-307.

[15] TSAGAS, Gr., On the spectrum of the Bochner-Laplace operator on the 1-forms on a compact Riemannian manifold, Math. Z. 164, (1978), 153-157.

[16] TSAGAS, Gr., The spectrum of the Laplace operator for a special complex manifold, Lecture Notes in Mathematics ≠ 838, Global Differential Geometry and Global Analysis, Proceedings, Berlin 1979, 233-238.

[17] TSAGAS, Gr. and KOCKONOS, K., The geometry and the Laplace operator on the exterior 2-forms on a compact Riemannian manifold, Proc. Amer. Math. Soc. 73 (1979), 109-116.

[18] TSAGAS, Gr., The geometry and the Bochner-Laplace operator on the exterior 2-forms on a compact Riemannian manifold, Tensor, Vol. 36, (1982), 73-78.

[19] TSAGAS, Gr., The spectrum of the Laplace operator for a special Riemannian manifold, Kodai Math. J., Vol. 4, No. 3 (1981), 377,382.

Department of Mathematics
Faculty of Technology
Aristotelian University
Thessaloniki, Greece

HOLOMORPHIC EXTENSIONS OF FUNCTIONS ON SUBMANIFOLDS: A GENERALIZATION OF H. LEWY'S EXAMPLE

Wolfgang Tutschke (Halle)

Contents   page

Summary. In the present paper the differential equation

$$\frac{\partial u}{\partial t} = \sigma(x,y,t)\frac{\partial u}{\partial x} + \tau(x,y,t)\frac{\partial u}{\partial y}$$

is investigated from a new point of view. Instead of the holomorphy of the coefficients, another sufficient condition is obtained which ensures the existence of at least two linearly independent solutions such that any solution may be extended to a holomorphic function.

Introduction. The general solution of many partial differential equations in the plane depends on an arbitrary holomorphic function. If u and v are two arbitrary complex-valued solutions of the differential equation

$$(1) \qquad \sum_{i=1}^{3} A_i \frac{\partial u}{\partial \alpha_i} = 0$$

and if Φ denotes an arbitrary holomorphic function in u and v, then (u,v) is also a solution of the same differential equation. Conversely, Lewy proved in his paper [3] that an arbitrary solution may be represented as superposition of two special solutions and a suitably chosen holomorphic function in two complex variables. In order to construct this holomorphic function H. Lewy defined a really three-dimensional manifold S in $\mathbb{C}^2$ by giving two different from each other complex-valued solutions, and extended an arbitrary solution to a holomorphic function in $\mathbb{C}^2$ existing on one side of S. This construction needs the existence of at least two linearly independent complex-valued solutions. In the case of holomorphic coefficients this existence follows immediately from the Cauchy-Kovalevska theorem, otherwise it has to be postulated.

On the other hand, there are first order differential equations with infinitely differentiable coefficients without any solution. This was demonstrated also by Lewy in his paper [4] containing a corresponding example. Recently, (necessary and sufficient) conditions for the solvability of differential equations have been deduced (cf. [1]). Further, the Cauchy-Kovalevska theorem was generalized to the case of generalized analytic functions as intial functions (cf. [6] and also [2]).

In the present paper the differential equation (1) will be regarded in a new way. Instead of the holomorphy of the coefficients there will be derived another sufficient condition ensuring the existence of at least two linearly independent solutions, such that any solution may be extended to a holomorphic function *).

1. Statement of the result

Regard the space $\mathbb{R}^3$, in which the independent variables will be denoted by x, y, and t. Give a differential equation of type (1), which can be written in the form

$$\frac{\partial u}{\partial t} = \sigma(x,y,t)\frac{\partial u}{\partial x} + \tau(x,y,t)\frac{\partial u}{\partial y}, \tag{2}$$

where σ and τ are given complex-valued coefficients, and u is the unknown complex-valued solution. Denote by G a given bounded domain of the (x,y)-plane. Further, let T be any positive number. Suppose

*) During the 8th Conference on Analytic Functions (Błażejewko 1982) I gave a lecture on the solution of initial value problems in classes of generalized analytic functions. Professor P. Dolbeault asked me whether the theory of generalized analytic functions might be also applied to the embedding of solutions of partial differential equations into higher-dimensional complex euclidean spaces. In this way my attention was drawn to the problem being studied in this paper.

that the coefficients σ and τ of the differential equation (2) fulfil the following conditions:

a) σ and τ are continuously differentiable in $G \times [0,T]$,

b) the inequalities

(3) $\tau \neq 0$,

(4) $\sigma \neq i\tau$,

(5) $\mathrm{Im}(\bar{\sigma}\tau) > 0$

are fulfilled in $G \times [0,T]$,

c) the quotient $(\sigma + i\tau)/(\sigma - i\tau)$ has to be independent on t,

d) everywhere in $G \times (0,T)$ the given coefficients σ and τ fulfil the inequality

$$\mathrm{Re}\,\sigma \cdot \left(\sigma \frac{\partial \bar{\tau}}{\partial x} + \tau \frac{\partial \bar{\tau}}{\partial y} - \frac{\partial \bar{\tau}}{\partial t} - \bar{\sigma} \frac{\partial \tau}{\partial x} - \bar{\tau} \frac{\partial \tau}{\partial y}\right) \neq \mathrm{Re}\,\tau \cdot \left(\sigma \frac{\partial \bar{\sigma}}{\partial x} + \tau \frac{\partial \bar{\sigma}}{\partial y} - \frac{\partial \bar{\sigma}}{\partial t} - \bar{\sigma} \frac{\partial \sigma}{\partial x} - \bar{\tau} \frac{\partial \sigma}{\partial y}\right). \tag{6}$$

Now define

$$q(z) = -\frac{\sigma + i\tau}{\sigma - i\tau}$$

in G and $q(z) = 0$ outside G, $z = x + iy$, and regard the Beltrami equation

$$\frac{\partial \zeta}{\partial \bar{z}} = q(z) \frac{\partial \zeta}{\partial z}, \tag{7}$$

which in view of (5) is proved to be uniformly elliptic. By virtue of assumption c) the solutions $\zeta = \zeta(z)$ of the Beltrami equation do not depend on the variable t. In the following $\zeta = \zeta(z)$ denotes a fixed homeomorphic solution, for instance the basic homeomorphism (cf. [7]). Then the following theorem holds:

THEOREM. *Suppose that the expression*

$$\frac{\mathrm{Im}(\sigma\bar{\tau})}{\bar{\tau}} \frac{\partial \bar{\zeta}}{\partial x} \tag{8}$$

defines an anti-holomorphic function in G *for every* t, $0 < t < T$. *Then there are two complex-valued solutions* u *and* $\tilde{u}$ *of the differential equation* (2), *for which the rank of the matrix*

$$M = \begin{bmatrix} \frac{\partial u}{\partial x} & \frac{\partial u}{\partial y} & \frac{\partial u}{\partial t} \\ & & \\ \frac{\partial \tilde{u}}{\partial x} & \frac{\partial \tilde{u}}{\partial y} & \frac{\partial \tilde{u}}{\partial t} \end{bmatrix}$$

is 2. Thus every solution of (2) can be extended to a holomorphic function defined on one side of S, where the manifold S defined by u and $\tilde{u}$ is embedded in $\mathbb{C}^2$.

2. Sketch of the proof

First write the differential equation (2) as real system

$$(9) \qquad \begin{aligned} -\frac{\partial u_2}{\partial y} + a_{11}\frac{\partial u_1}{\partial x} + a_{12}\frac{\partial u_1}{\partial y} &= f_1, \\ \frac{\partial u_2}{\partial x} + a_{21}\frac{\partial u_1}{\partial x} + a_{22}\frac{\partial u_2}{\partial y} &= f_2, \end{aligned}$$

where $u = u_1 + iu_2$. Then we have

$$a_{11} = |\sigma|^2/\mathrm{Im}(\bar{\sigma}\tau),$$

$$a_{12} = a_{21} = \mathrm{Re}(\sigma\bar{\tau})/\mathrm{Im}(\bar{\sigma}\tau),$$

$$a_{22} = |\tau|^2/\mathrm{Im}(\bar{\sigma}\tau),$$

$$f_1 = \mathrm{Re}(\bar{\sigma}\frac{\partial u}{\partial t})/\mathrm{Im}(\bar{\sigma}\tau),$$

$$f_2 = \mathrm{Re}(\bar{\tau}\frac{\partial u}{\partial t})/\mathrm{Im}(\bar{\sigma}\tau).$$

Introducing ζ instead of z as a new variable, the system (9) may be rewritten in canonical form. Since

$$\Delta = a_{11}a_{22} - \tfrac{1}{4}(a_{12} + a_{21})^2$$

is identically equal to $\underline{1}$, the reduction to the canonical form does not demand the replacement of u_1, u_2 by new functions

$$u_1, \quad u_2 - \frac{a_{12} - a_{21}}{2} u_2$$

(cf. [7]). Consequently, we get a canonical differential equation for the original function u, namely

$$(10) \qquad \frac{\partial u}{\partial \bar{\zeta}} = \frac{1}{2}\frac{1}{J}\left(f_1\frac{\partial \zeta}{\partial x} + f_2\frac{\partial \zeta}{\partial y}\right),$$

where J is the Jacobian. On the other hand, the Beltrami equation (7) is equivalent to

$$\sigma \frac{\partial \zeta}{\partial x} + \tau \frac{\partial \zeta}{\partial y} = 0, \tag{11}$$

such that

$$J = \frac{\operatorname{Im}(\bar{\sigma}\tau)}{|\tau|^2} \left|\frac{\partial \zeta}{\partial x}\right|^2$$

and (10) becomes

$$\frac{\partial u}{\partial t} = 2i \frac{\operatorname{Im}(\sigma\bar{\tau})}{\bar{\tau}} \frac{\partial \bar{\zeta}}{\partial x} \frac{\partial u}{\partial \bar{\zeta}}.$$

If (8) is anti-holomorphic, then the operator on the right-hand side of the last equation maps the space of all anti-holomorphic functions into itself. Regarding the scale of Banach spaces of anti-holomorphic functions, the initial value problem for the last differential equation can be solved. Thus, we may choose the initial values of two solutions u and $\tilde{u}$ in such a way that the rank of M equals 2. In view of (6) the manifold S defined by u and $\tilde{u}$ is (maybe after a quadratic transformation) strongly pseudoconvex, such that Lewy's method (see [3]) is applicable. This method leads to the wanted extension.

3. Concluding remarks

a) The expression (8) depends on σ and τ, but it contains also the function $\zeta = \zeta(z)$ that cannot be expressed explicitely by σ and τ. In order to make the anti-holomorphy of (8) better applicable we regard a given diffeomorphism $\zeta = \zeta(z)$. This diffeomorphism can be used as a solution of the regarded Beltrami equation if σ and τ are connected by the equation (11). Define

$$\lambda(z) = \frac{\partial \bar{\zeta}}{\partial y} \operatorname{Im}\left(\frac{\partial \bar{\zeta}}{\partial x} / \frac{\partial \bar{\zeta}}{\partial y}\right),$$

then (8) is anti-holomorphic if and only if σ fulfils the differential equation

$$\frac{\partial \lambda}{\partial z} \sigma + \lambda \frac{\partial \sigma}{\partial z} = 0. \tag{12}$$

It means that an extension of an arbitrary solution of (2) can be constructed if σ fulfils the last differential equation and, further, if τ is given by (11).

In the special case $\zeta(z) = z$ for every z the relation (11) says that $\tau = i\sigma$. In this case we have, moreover,

$$\lambda(z) = -i$$

and σ is proved to be anti-holomorphic. In the case $\tau = i\sigma$ the two conditions (3) and (4) are identical.

b) Since the initial value problem is also solvable in more general scales of Banach spaces, the method described above is also applicable to differential equations being more general than the equation (2).

References

[1] EGOROV, Ju.V., Conditions for the solvability of differential equations with simple characteristics, Soviet Math. Dokl. 17 (1976), 1194-1197 and DAN 229 (1976), 1310-1312.

[2] LANCKAU, E. and W. TUTSCHKE, Complex analysis. Methods, trends, and applications, Berlin 1983.

[3] LEWY, H., On the local character of the solutions of an atypical linear differential equation in 3 variables and a related theorem for regular functions of 2 complex variables, Math. Ann. 64 (1956), 514-522.

[4] ——, An example of a smooth linear partial differential equation without solution, Math. Ann. 66 (1957), 155-158.

[5] TRÈVES, F., Basic linear partial differential equations, New York-San Francisco-London 1975.

[6] TUTSCHKE, W., A problem with initial values for generalized analytic functions depending on time (generalizations of the Cauchy-Kovalevska and Holmgren theorems), Soviet Math. Dokl. 25 (1983), 201-205 and DAN 262 (1982), 1081-1085.

[7] VEKUA, I.N., Generalized analytic functions, Reading 1962 and Moscow 1959.

Martin-Luther-Universität
Halle-Wittenberg
Sektion Mathematik
Universitätsplatz 6
DDR-4020 Halle, GDR

A REMARK ON THE CONVERGENCE OF SERIES OCCURING IN THE CONSTRUCTION OF θ-FUNCTIONS ON A TOROIDAL GROUP

Georges G. Weill (Brooklyn, N.Y.)

Summary. To find whether on a toroidal group $\mathbb{C}^n/_\Gamma$ any meromorphic function can be written as the quotient of two θ-functions associated with the lattice Γ one has to study the existence of $\mathbb{Z}^n$-periodic entire solutions of a system of difference equations. The formal series solutions are shown to converge "generically". This gives a partial answer to a question of A. Andreotti as to find "good conditions" for the convergence of the series that occur in the construction of the θ-functions associated with Γ.

1. Any abelian complex Lie group of dimension n is known to be isomorphic to $\mathbb{C}^n/_\Gamma$, where Γ is a discrete subgroup of $\mathbb{C}^n$ (and conversely any $\mathbb{C}^n/_\Gamma$ is such a Lie group).

An abelian complex Lie group G is said to be toroidal if all holomorphic functions on G are constant. In particular, complex tori are toroidal.

Let $\vec{p}_1,\dots,\vec{p}_r$ be a basis for Γ; those vectors are linearly independent over $\mathbb{R}$ and generate Γ as a $\mathbb{Z}$-module. The $(n\times r)$ matrix P whose columns are $\vec{p}_1,\dots,\vec{p}_r$ is called a period basis for the lattice generated by its columns in $\mathbb{C}^n$.

It is known [3] that if $X = \mathbb{C}^n/_\Gamma$ is a toroidal group then $r = n + s$ with $1 \leq s \leq n$, and one can choose a period basis of the form

$$P = (I_n\, Q),$$

where Q is an $n\times s$ matrix. $\mathbb{C}^n/_{P\mathbb{Z}^{n+s}}$ is toroidal if and only if ${}^t\sigma Q \notin \mathbb{Z}^s$ for each $\sigma \in \mathbb{Z}^n \sim \{o\}$.

2. The problem of representing the meromorphic functions on G as quotients of θ-functions (cf. [1]) leads to solving the following system of difference equations: we are given $Q = (\vec{p}_1,\dots,\vec{p}_s)$ a complex $n\times s$ matrix such that ${}^t\sigma Q \notin \mathbb{Z}^s$ for all $\sigma \in \mathbb{Z}^n \sim \{o\}$ and $\mathbb{Z}^n$-periodic holomorphic functions $a_j : \mathbb{C}^n \to \mathbb{C}$, $j = 1,\dots,s$, and we want to find a $\mathbb{Z}^n$-periodic holomorphic function $g : \mathbb{C}^n \to \mathbb{C}$ such that

$$g(\mathfrak{z} + \vec{p}_j) - g(\mathfrak{z}) = a_j(\mathfrak{z}), \quad j = 1,\dots,s.$$

It is known [3] that if

$$a_i(\mathfrak{z} + \vec{p}_j) + a_j(\mathfrak{z}) = a_j(\mathfrak{z} + \vec{p}_i) + a_i(\mathfrak{z}), \quad i,j = 1,\dots,s$$

and if the constant terms of the Fourier expansions of all a_j vanish, then a necessary and sufficient condition for the solvability of the problem for all such a_j is the existence of constants $C > 0$, $a \geq 0$ such that

$$\| {}^t\sigma Q + {}^t\tau \| \geq C \exp(-a\|\sigma\|)$$

for all $\sigma \in \mathbb{Z}^n \sim \{o\}$ and all $\tau \in \mathbb{Z}^n$. The Fourier series defining the θ-functions will then converge.

3. In Number Theory [2] an n-tuple of real numbers $(\alpha_1,\dots,\alpha_n)$ is said to be very well approximable if there exists $\delta > 0$ such that the inequality

$$|\alpha_1\rho_1 + \dots + \alpha_n\rho_n + p| < \frac{1}{\|\rho\|^{n+\delta}}, \quad \|\rho\| > 0$$

has infinitely many solutions $(\rho_1,\dots,\rho_n, p) \in \mathbb{Z}^{n+1}$. It is known [2] that the very well approximable n-tuples form a set of Lebesgue measure zero in $\mathbb{R}^n$.

Definition. A matrix of periods for a toriodal group is said to be "generic" if at least one of its columns p_k satisfy the following condition: $\operatorname{Re}\vec{p}_k$ or $\operatorname{Im}\vec{p}_k$ is a not very well approximable n-tuple of real numbers.

Generic matrices form a subset of $\mathbb{R}^{2ns}$, whose complement has Lebesgue measure zero.

THEOREM. Let P be a generic matrix of periods for a toroidal

<u>group. Then the series defining the θ-functions which are solutions of the above system of difference equations converge.</u>

P r o o f. Assume $\operatorname{Re} p_k = (\alpha_1, \ldots, \alpha_n)$ is not very well approximable. Then

$$|\alpha_1 \rho_1 + \ldots + \alpha_n \rho_n + p| < \frac{1}{\|\rho\|^{n+\delta}}$$

for only finitely many $(\rho_1, \ldots, \rho_n, p) \in \mathbb{Z}^{n+1}$.

Then one can choose $C > 0$ and $a \geq 0$ such that

$$|\alpha_1 \rho_1 + \ldots + \alpha_n \rho_n + p| \geq C \exp(-a \|\rho\|)$$

for all $(\rho_1, \ldots, \rho_n, p) \in \mathbb{Z}^{n+1}$ and the necessary and sufficient condition for convergence is satisfied. If $\operatorname{Im} \vec{p}_k = (\beta_1, \ldots, \beta_n)$ is not very well approximable we consider

$$|\beta_1 \rho_1 + \ldots + \beta_n \rho_n + p| < \frac{1}{\|\rho\|^{n+\delta}}, \quad p = 0,$$

and obtain the same result.

<u>COROLLARY</u>. <u>If</u> P <u>is generic, then the meromorphic functions on</u> $\mathbb{C}^n/\Gamma$ <u>can be written as quotients of two</u> θ-<u>functions</u>.

R e f e r e n c e s

[1] CONFORTO, F., Abelsche Funktionen und algebraische Geometrie, Springer-Verlag, Berlin 1956.

[2] SCHMIDT, W.M., Approximation to algebraic numbers, L'Enseignement Mathématique XVII (3-4) (1971), 187-253.

[3] VOGT, C., Geradenbündel auf toroiden Gruppen, Dissertation, Dusseldorf 1981.

Polytechnic Institute of New York
333 Jay Street
Brooklyn, N.Y. 11201, U.S.A.

Vol. 1008: Algebraic Geometry. Proceedings, 1981. Edited by J. Dolgachev. V, 138 pages. 1983.

Vol. 1009: T.A. Chapman, Controlled Simple Homotopy Theory and Applications. III, 94 pages. 1983.

Vol. 1010: J.-E. Dies, Chaînes de Markov sur les permutations. IX, 226 pages. 1983.

Vol. 1011: J.M. Sigal. Scattering Theory for Many-Body Quantum Mechanical Systems. IV, 132 pages. 1983.

Vol. 1012: S. Kantorovitz, Spectral Theory of Banach Space Operators. V, 179 pages. 1983.

Vol. 1013: Complex Analysis – Fifth Romanian-Finnish Seminar. Part 1. Proceedings, 1981. Edited by C. Andreian Cazacu, N. Boboc, M. Jurchescu and I. Suciu. XX, 393 pages. 1983.

Vol. 1014: Complex Analysis – Fifth Romanian-Finnish Seminar. Part 2. Proceedings, 1981. Edited by C. Andreian Cazacu, N. Boboc, M. Jurchescu and I. Suciu. XX, 334 pages. 1983.

Vol. 1015: Equations différentielles et systèmes de Pfaff dans le champ complexe – II. Seminar. Edited by R. Gérard et J.P. Ramis. V, 411 pages. 1983.

Vol. 1016: Algebraic Geometry. Proceedings, 1982. Edited by M. Raynaud and T. Shioda. VIII, 528 pages. 1983.

Vol. 1017: Equadiff 82. Proceedings, 1982. Edited by H. W. Knobloch and K. Schmitt. XXIII, 666 pages. 1983.

Vol. 1018: Graph Theory, Łagów 1981. Proceedings, 1981. Edited by M. Borowiecki, J.W. Kennedy and M.M. Sysło. X, 289 pages. 1983.

Vol. 1019: Cabal Seminar 79–81. Proceedings, 1979–81. Edited by A. S. Kechris, D. A. Martin and Y. N. Moschovakis. V, 284 pages. 1983.

Vol. 1020: Non Commutative Harmonic Analysis and Lie Groups. Proceedings, 1982. Edited by J. Carmona and M. Vergne. V, 187 pages. 1983.

Vol. 1021: Probability Theory and Mathematical Statistics. Proceedings, 1982. Edited by K. Itô and J.V. Prokhorov. VIII, 747 pages. 1983.

Vol. 1022: G. Gentili, S. Salamon and J.-P. Vigué. Geometry Seminar "Luigi Bianchi", 1982. Edited by E. Vesentini. VI, 177 pages. 1983.

Vol. 1023: S. McAdam, Asymptotic Prime Divisors. IX, 118 pages. 1983.

Vol. 1024: Lie Group Representations I. Proceedings, 1982–1983. Edited by R. Herb, R. Lipsman and J. Rosenberg. IX, 369 pages. 1983.

Vol. 1025: D. Tanré, Homotopie Rationnelle: Modèles de Chen, Quillen, Sullivan. X, 211 pages. 1983.

Vol. 1026: W. Plesken, Group Rings of Finite Groups Over p-adic Integers. V, 151 pages. 1983.

Vol. 1027: M. Hasumi, Hardy Classes on Infinitely Connected Riemann Surfaces. XII, 280 pages. 1983.

Vol. 1028: Séminaire d'Analyse P. Lelong – P. Dolbeault – H. Skoda. Années 1981/1983. Edité par P. Lelong, P. Dolbeault et H. Skoda. VIII, 328 pages. 1983.

Vol. 1029: Séminaire d'Algèbre Paul Dubreil et Marie-Paule Malliavin. Proceedings, 1982. Edité par M.-P. Malliavin. V, 339 pages. 1983.

Vol. 1030: U. Christian, Selberg's Zeta-, L-, and Eisensteinseries. XII, 196 pages. 1983.

Vol. 1031: Dynamics and Processes. Proceedings, 1981. Edited by Ph. Blanchard and L. Streit. IX, 213 pages. 1983.

Vol. 1032: Ordinary Differential Equations and Operators. Proceedings, 1982. Edited by W. N. Everitt and R. T. Lewis. XV, 521 pages. 1983.

Vol. 1033: Measure Theory and its Applications. Proceedings, 1982. Edited by J. M. Belley, J. Dubois and P. Morales. XV, 317 pages. 1983.

Vol. 1034: J. Musielak, Orlicz Spaces and Modular Spaces. V, 222 pages. 1983.

Vol. 1035: The Mathematics and Physics of Disordered Media. Proceedings, 1983. Edited by B.D. Hughes and B.W. Ninham. VII, 432 pages. 1983.

Vol. 1036: Combinatorial Mathematics X. Proceedings, 1982. Edited by L. R. A. Casse. XI, 419 pages. 1983.

Vol. 1037: Non-linear Partial Differential Operators and Quantization Procedures. Proceedings, 1981. Edited by S.I. Andersson and H.-D. Doebner. VII, 334 pages. 1983.

Vol. 1038: F. Borceux, G. Van den Bossche, Algebra in a Localic Topos with Applications to Ring Theory. IX, 240 pages. 1983.

Vol. 1039: Analytic Functions, Błażejewko 1982. Proceedings. Edited by J. Ławrynowicz. X, 494 pages. 1983

Vol. 1040: A. Good, Local Analysis of Selberg's Trace Formula. III, 128 pages. 1983.

Vol. 1041: Lie Group Representations II. Proceedings 1982–1983. Edited by R. Herb, S. Kudla, R. Lipsman and J. Rosenberg. IX, 340 pages. 1984.

Vol. 1042: A. Gut, K. D. Schmidt, Amarts and Set Function Processes. III, 258 pages. 1983.

Vol. 1043: Linear and Complex Analysis Problem Book. Edited by V.P. Havin, S.V. Hruščëv and N.K. Nikol'skii. XVIII, 721 pages. 1984.

Vol. 1044: E. Gekeler, Discretization Methods for Stable Initial Value Problems. VIII, 201 pages. 1984.

Vol. 1045: Differential Geometry. Proceedings, 1982. Edited by A.M. Naveira. VIII, 194 pages. 1984.

Vol. 1046: Algebraic K-Theory, Number Theory, Geometry and Analysis. Proceedings, 1982. Edited by A. Bak. IX, 464 pages. 1984.

Vol. 1047: Fluid Dynamics. Seminar, 1982. Edited by H. Beirão da Veiga. VII, 193 pages. 1984.

Vol. 1048: Kinetic Theories and the Boltzmann Equation. Seminar, 1981. Edited by C. Cercignani. VII, 248 pages. 1984.

Vol. 1049: B. Iochum, Cônes autopolaires et algèbres de Jordan. VI, 247 pages. 1984.

Vol. 1050: A. Prestel, P. Roquette, Formally p-adic Fields. V, 167 pages. 1984.

Vol. 1051: Algebraic Topology, Aarhus 1982. Proceedings. Edited by I. Madsen and B. Oliver. X, 665 pages. 1984.

Vol. 1052: Number Theory. Seminar, 1982. Edited by D.V. Chudnovsky, G.V. Chudnovsky, H. Cohn and M.B. Nathanson. V, 309 pages. 1984.

Vol. 1053: P. Hilton, Nilpotente Gruppen und nilpotente Räume. V, 221 pages. 1984.

Vol. 1054: V. Thomée, Galerkin Finite Element Methods for Parabolic Problems. VII, 237 pages. 1984.

Vol. 1055: Quantum Probability and Applications to the Quantum Theory of Irreversible Processes. Proceedings, 1982. Edited by L. Accardi, A. Frigerio and V. Gorini. VI, 411 pages. 1984.

Vol. 1056: Algebraic Geometry. Bucharest 1982. Proceedings, 1982. Edited by L. Bădescu and D. Popescu. VII, 380 pages. 1984.

Vol. 1057: Bifurcation Theory and Applications. Seminar, 1983. Edited by L. Salvadori. VII, 233 pages. 1984.

Vol. 1058: B. Aulbach, Continuous and Discrete Dynamics near Manifolds of Equilibria. IX, 142 pages. 1984.

Vol. 1059: Séminaire de Probabilités XVIII, 1982/83. Proceedings. Edité par J. Azéma et M. Yor. IV, 518 pages. 1984.

Vol. 1060: Topology. Proceedings, 1982. Edited by L. D. Faddeev and A.A. Mal'cev. VI, 389 pages. 1984.

Vol. 1061: Séminaire de Théorie du Potentiel. Paris, No. 7. Proceedings. Directeurs: M. Brelot, G. Choquet et J. Deny. Rédacteurs: F. Hirsch et G. Mokobodzki. IV, 281 pages. 1984.

Vol. 1062: J. Jost, Harmonic Maps Between Surfaces. X, 133 pages. 1984.

Vol. 1063: Orienting Polymers. Proceedings, 1983. Edited by J.L. Ericksen. VII, 166 pages. 1984.

Vol. 1064: Probability Measures on Groups VII. Proceedings, 1983. Edited by H. Heyer. X, 588 pages. 1984.

Vol. 1065: A. Cuyt, Padé Approximants for Operators: Theory and Applications. IX, 138 pages. 1984.

Vol. 1066: Numerical Analysis. Proceedings, 1983. Edited by D.F. Griffiths. XI, 275 pages. 1984.

Vol. 1067: Yasuo Okuyama, Absolute Summability of Fourier Series and Orthogonal Series. VI, 118 pages. 1984.

Vol. 1068: Number Theory, Noordwijkerhout 1983. Proceedings. Edited by H. Jager. V, 296 pages. 1984.

Vol. 1069: M. Kreck, Bordism of Diffeomorphisms and Related Topics. III, 144 pages. 1984.

Vol. 1070: Interpolation Spaces and Allied Topics in Analysis. Proceedings, 1983. Edited by M. Cwikel and J. Peetre. III, 239 pages. 1984.

Vol. 1071: Padé Approximation and its Applications, Bad Honnef 1983. Prodeedings. Edited by H. Werner and H.J. Bünger. VI, 264 pages. 1984.

Vol. 1072: F. Rothe, Global Solutions of Reaction-Diffusion Systems. V, 216 pages. 1984.

Vol. 1073: Graph Theory, Singapore 1983. Proceedings. Edited by K.M. Koh and H.P. Yap. XIII, 335 pages. 1984.

Vol. 1074: E.W. Stredulinsky, Weighted Inequalities and Degenerate Elliptic Partial Differential Equations. III, 143 pages. 1984.

Vol. 1075: H. Majima, Asymptotic Analysis for Integrable Connections with Irregular Singular Points. IX, 159 pages. 1984.

Vol. 1076: Infinite-Dimensional Systems. Proceedings, 1983. Edited by F. Kappel and W. Schappacher. VII, 278 pages. 1984.

Vol. 1077: Lie Group Representations III. Proceedings, 1982–1983. Edited by R. Herb, R. Johnson, R. Lipsman, J. Rosenberg. XI, 454 pages. 1984.

Vol. 1078: A.J.E.M. Janssen, P. van der Steen, Integration Theory. V, 224 pages. 1984.

Vol. 1079: W. Ruppert. Compact Semitopological Semigroups: An Intrinsic Theory. V, 260 pages. 1984

Vol. 1080: Probability Theory on Vector Spaces III. Proceedings, 1983. Edited by D. Szynal and A. Weron. V, 373 pages. 1984.

Vol. 1081: D. Benson, Modular Representation Theory: New Trends and Methods. XI, 231 pages. 1984.

Vol. 1082: C.-G. Schmidt, Arithmetik Abelscher Varietäten mit komplexer Multiplikation. X, 96 Seiten. 1984.

Vol. 1083: D. Bump, Automorphic Forms on GL (3,IR). XI, 184 pages. 1984.

Vol. 1084: D. Kletzing, Structure and Representations of Q-Groups. VI, 290 pages. 1984.

Vol. 1085: G.K. Immink, Asymptotics of Analytic Difference Equations. V, 134 pages. 1984.

Vol. 1086: Sensitivity of Functionals with Applications to Engineering Sciences. Proceedings, 1983. Edited by V. Komkov. V, 130 pages. 1984

Vol. 1087: W. Narkiewicz, Uniform Distribution of Sequences of Integers in Residue Classes. VIII, 125 pages. 1984.

Vol. 1088: A.V. Kakosyan, L.B. Klebanov, J.A. Melamed, Characterization of Distributions by the Method of Intensively Monotone Operators. X, 175 pages. 1984.

Vol. 1089: Measure Theory, Oberwolfach 1983. Proceedings. Edited by D. Kölzow and D. Maharam-Stone. XIII, 327 pages. 1984.

Vol. 1090: Differential Geometry of Submanifolds. F
1984. Edited by K. Kenmotsu. VI, 132 pages. 1984.

Vol. 1091: Multifunctions and Integrands. Proceedings,
by G. Salinetti. V, 234 pages. 1984.

Vol. 1092: Complete Intersections. Seminar, 1983. Edited
and R. Strano. VII, 299 pages. 1984.

Vol. 1093: A. Prestel, Lectures on Formally Real Fields. XI
1984.

Vol. 1094: Analyse Complexe. Proceedings, 1983. Edité
R. Gay et Nguyen Thanh Van. IX, 184 pages. 1984.

Vol. 1095: Stochastic Analysis and Applications. Procee
Edited by A. Truman and D. Williams. V, 199 pages. 198

Vol. 1096: Théorie du Potentiel. Proceedings, 1983. E
Mokobodzki et D. Pinchon. IX, 601 pages. 1984.

Vol. 1097: R.M. Dudley, H. Kunita, F. Ledrappier, Éco
Probabilités de Saint-Flour XII – 1982. Edité par P.L. He
396 pages. 1984.

Vol. 1098: Groups – Korea 1983. Proceedings. Edited b
and B.H. Neumann. VII, 183 pages. 1984.

Vol. 1099: C.M. Ringel, Tame Algebras and Integral Quadr
XIII, 376 pages. 1984.

Vol. 1100: V. Ivrii, Precise Spectral Asymptotics for Elliptic
Acting in Fiberings over Manifolds with Boundary. V, 2
1984.

Vol. 1101: V. Cossart, J. Giraud, U. Orbanz, Resolution
Singularities. VII, 132 pages. 1984.

Vol. 1102: A. Verona, Stratified Mappings – Structure an
lability. IX, 160 pages. 1984.

Vol. 1103: Models and Sets. Proceedings, Logic Colloqu
Part I. Edited by G.H. Müller and M.M. Richter. VIII, 484 pa

Vol. 1104: Computation and Proof Theory. Proceedin
Colloquium, 1983, Part II. Edited by M.M. Richter, E. Börge
schelp, B. Schinzel and W. Thomas. VIII, 475 pages. 198

Vol. 1105: Rational Approximation and Interpolation. Pro
1983. Edited by P.R. Graves-Morris, E.B. Saff and R.
XII, 528 pages. 1984.

Vol. 1106: C.T. Chong, Techniques of Admissible Recursi
IX, 214 pages. 1984.

Vol. 1107: Nonlinear Analysis and Optimization. Proceedir
Edited by C. Vinti. V, 224 pages. 1984.

Vol. 1108: Global Analysis – Studies and Applications I.
Yu.G. Borisovich and Yu.E. Gliklikh. V, 301 pages. 1984.

Vol. 1109: Stochastic Aspects of Classical and Quantum
Proceedings, 1983. Edited by S. Albeverio, P. Combe and M
Collin. IX, 227 pages. 1985.

Vol. 1110: R. Jajte, Strong Limit Theorems in Non-Co
Probability. VI, 152 pages. 1985.

Vol. 1111: Arbeitstagung Bonn 1984. Proceedings. Edited b
bruch, J. Schwermer and S. Suter. V, 481 pages. 1985.

Vol. 1112: Products of Conjugacy Classes in Groups.
Z. Arad and M. Herzog. V, 244 pages. 1985.

Vol. 1113: P. Antosik, C. Swartz, Matrix Methods in An
114 pages. 1985.

Vol. 1114: Zahlentheoretische Analysis. Seminar. Herau
von E. Hlawka. V, 157 Seiten. 1985.

Vol. 1115: J. Moulin Ollagnier, Ergodic Theory and Statistica
ics. VI, 147 pages. 1985.

Vol. 1116: S. Stolz, Hochzusammenhängende Mannigfa
und ihre Ränder. XXIII, 134 Seiten. 1985.